T0133444

Geographic Information Systems to Spatial Data Infrastructure

Geographic Information Systems to Spatial Data Infrastructure

A Global Perspective

Edited by
Ian Masser

CRC Press
Taylor & Francis Group
Boca Raton London New York

CRC Press is an imprint of the
Taylor & Francis Group, an **informa** business

CRC Press
Taylor & Francis Group
6000 Broken Sound Parkway NW, Suite 300
Boca Raton, FL 33487-2742

First issued in paperback 2022

ISBN 13: 978-1-03-247521-9 (pbk)
ISBN 13: 978-1-138-58462-4 (hbk)

DOI: 10.1201/9780429505904

Library of Congress Cataloging-in-Publication Data

Names: Masser, Ian editor.
Title: From Geographic information systems to Spatial data infrastructures : a global perspective / Ian Masser, editor.
Description: Boca Raton : CRC Press | Taylor & Francis Group , 2019. | "A CRC title, part of the Taylor & Francis imprint, a member of the Taylor & Francis Group, the academic division of T&F Informa plc." | Includes bibliographical references and index.
Identifiers: LCCN 2019014247| ISBN 9781138584624 (Hardback : acid-free paper) | ISBN 9780429505904 (eBook)
Subjects: LCSH: Geographic information systems. | Spatial data infrastructures.
Classification: LCC G70.212 .F75 2019 | DDC 910.285–dc23
LC record available at https://lccn.loc.gov/2019014247

Visit the Taylor & Francis Web site at
http://www.taylorandfrancis.com

and the CRC Press Web site at
http://www.crcpress.com

Contents

GLOBAL CITIZEN AWARD
Professor IAN MASSER

The GSDI Global Citizen Award recognizes individuals who have provided exemplary thought leadership and substantive contributions in promoting informed and responsible use of geospatial information and technologies for the benefit of humanity and for fostering Spatial Data Infrastructure developments that support sustainable social, economic, and environmental systems integrated from local to global scales.

Professor Emeritus Ian Masser has served the national, regional and global geospatial communities for more than 30 years, as both a distinguished academic in UK and Dutch universities and as a global thought leader during development of Spatial Data Infrastructure concepts globally.

- At national level, Ian was appointed to the UK Economic and Social Research Council early in his career, to coordinate the Council's Regional Research Laboratory Initiative, where he played a major role in establishing GIS-based research across the United Kingdom.

- At the European level, Ian was Co-Director of the European Science Foundation's GISDATA scientific programme, which helped to establish a European GIS research community, and was Founder President of the Association of GI Laboratories in Europe (AGILE). He was also a key contributor to early development work that led to the pan-European SDI known as INSPIRE (Infrastructure for Spatial Information in the European Community), an initiative with which he was to remain involved for more than two decades. Ian was also an early President of the European Organisation for Geographic Information (EUROGI), which represented national GI Associations across Europe, especially in relations with the European Commission and other European institutions regarding geo information policy developments in Europe.

- At global level, in 2002 Prof Masser became President of the Global Spatial Data Infrastructure initiative which had started in 1996, two years prior to formation of the GSDI Association in 2004, and provided significant support and guidance in development of what was to become one of the premier global organisations focusing on SDI.

For exemplary service to the geospatial community,
the GSDI Association recognizes Professor Ian Masser as an exemplary Global Citizen.

Dr David Coleman, President, GSDI Association

November 2016

Preface

The starting point for this collection of my papers on Geographical Information Systems and Spatial Data Infrastructures came soon after I learnt in October 2016 from the president of the Global Spatial Data Infrastructure Association, Professor David Coleman, that I had been awarded the Association's Global Citizen Award. It was the greatest moment in my academic life.

The GSDI Association occasionally recognizes globally an individual who has provided exemplary thought leadership and substantive worldwide contributions in promoting informed and responsible use of geographic information and geospatial technologies for the benefit of society and fostering spatial data infrastructure developments that support sustainable social, economic, and environmental systems integrated from local to global scales. The recipient, if any, is selected by past recipients of the award and the GSDI Association executive committee which consists of its past president, president, and president-elect. I am only the fourth person to be nominated for the award.

I felt very honoured to be chosen to receive the award. Figure 0.1 contains the citation which highlights the extent to which the adjudicators thought that I had served the national, regional and global geospatial communities for more than 30 years, as both a distinguished academic in UK and Dutch universities and as a global thought leader during development of Spatial Data Infrastructure concepts globally.

This citation made me think about the papers that I had written during the last 30 years. These fell broadly into two categories; those concerned with the emergence and diffusion of geographic information systems technologies after the publication of the Chorley Report on Handling Geographic Information in 1987 up to the millennium, and those dealing with what has become known as the spatial data infrastructure phenomenon after that time. The papers can also be divided into essentially more theoretical and more practical contributions to the debates that were taking place all over the world during the two periods. In the case of all the published papers, I have added a final section updating some of the main events since the time of their original publication. Apart from this, I have not changed anything in the original papers, other than to delete some references to obsolete websites.

Part I Geographic Information Systems

The first chapter introduces some of the basic concepts underlying my own research and activities relating to GIS in practice. It is a publication co-authored with Henk Ottens on urban planning and GIS, which sets out my credentials from my starting point as an urban planner. From this it can be seen that, from the outset my research has been driven by policy considerations and specific applications. Because both geographical information systems and spatial data infrastructures tend to be technology driven, an important component of my work has been to consider appropriate organisational structures to facilitate the exploitation of these technologies.

Under the heading of theory, I have grouped together four papers. The first two of these relate to large-scale research networks. The first of these describes the UK Economic and Social Research Council's Regional Research Laboratory initiative that I coordinated together with Michael Blakemore from 1986 to 1991. This catapulted me into a key national position in a rapidly developing research field. In due course, my experience with the RRL initiative in the UK led to my appointment as co-director with Francois Salgé of the European Science Foundation's GISDATA Scientific Programme in 1993 that is the subject of the second paper. Between 1993 and 1997, this programme ran alongside the US National Science Foundation funded National Centre for Geographic Information and Analysis as well as playing an important part in establishing a truly European GIS research community. Consequently, my contribution to this section contains my own assessment of the Data Integration component of the scientific research that was carried out during the GISDATA programme.

Over the next few years, I also began to explore some of the broader issues arising out of the introduction of GIS technologies with various colleagues. One of these lines of investigation was to explore some possible scenarios of GIS with Michael Wegener using the scenario approach that we had already utilised in our book *The Geography of Europe's Futures* (Masser and Wegener 1992). The third paper of this group describes four contrasting scenarios of GIS diffusion: a *Trend* scenario based on the incremental diffusion of information systems, a *Market* scenario based on the total commodification of information which restricts access to the more powerful, a *Big Brother* scenario in which surveillance and control pervades all aspects of life, and a *Beyond GIS* scenario in which information in the public domain contributes to greater democratisation and grassroots empowerment. Another project with Bob Barr considered some contrasting concepts of information, in general, with particular reference to geographic information. Consequently, the final chapter in this group, co-authored with Bob Barr, starts off with a discussion of classical views regarding information as a resource and then moves on to examine three other alternative perspectives which see it as a commodity, an asset and an infrastructure, respectively.

The four papers that fall into the geographic information systems practice category reflect some of my concerns relating to the development and implementation of GIS. As noted above, the starting point for much of the discussion was the publication in 1988 of the Chorley Report to the British Government on Handling Geographic Information. The Committee's most important recommendations were those relating to digital topographic mapping, the linking and availability of data, and the role of government. The basic arguments underlying the Report are reviewed in the chapter in relation to the Government's response to the Report's recommendations. Although the Government largely failed to implement its recommendations, this document proved extremely influential in kick-starting GIS research in Britain and elsewhere.

This chapter is followed by a summary of the findings of the larger study of British local government practice that I carried out with Heather Campbell. The results of our telephone survey of all 514 local authorities indicate that one in six local authorities in Great Britain had acquired GIS facilities by April 1991. While this work was under way, Heather and I were asked to participate in a workshop that was organised by the US National Center for Geographic Information and Analysis in San Diego on data sharing. Our presentation there was later extended for a subsequent presentation at the first GISRUK conference at Keele in 1993, which is the third chapter in this group. This draws upon the findings of the 12 case studies that formed part of the research described in the previous paper. The case for more corporate approaches which maximize the benefits to be obtained from information sharing is evaluated and an alternative case which highlights the operational advantages of departmental approaches is considered; the findings of this survey show that there is a 50:50 split between more corporate and departmental approaches in current British practice.

The final contribution to this group with Max Craglia discusses some of the findings of comparative European research on the diffusion of GIS in local government in different European countries that were presented at one of the GISDATA workshops. This includes nine national case studies of GIS adoption in Denmark, France, Germany, the UK, Greece, Italy, the Netherlands, Poland and Portugal. The analysis shows that two key factors account for most of the differences between the nine countries in GIS adoption. The first of these measures the links between digital-data availability and GIS diffusion, while the second reflects the professional cultures surrounding GIS applications in local government.

Part II Spatial Data Infrastructures

The first chapter in the spatial data infrastructure section of the book papers contains only one paper which complements the paper on urban planning and GIS in the introduction to the first volume. In September 1999, I was asked by the Rector of the International Institute for Geo-Information Science and Earth Observation in Enschede to make a public presentation to mark the opening of the 1999–2000 academic year. My presentation highlights the challenges for urban planners in less-developed countries that are presented by urban growth over the next few decades, and describes some of the ways they can respond to this challenge with the help of remote sensing and geographic information systems. In the process it gives an overview of the theoretical and policy issues that underlie the development of spatial data infrastructures which are the subject of this section of my papers.

As a result of my participation in the discussions about the development of the US national spatial data infrastructure prior to the publication of President Clinton's Executive Order in 1994 and my involvement in the Global Spatial Data Infrastructure Conference organisation, I found myself in a unique position with access to SDI documentation from all over the world. This provided the database for the first chapter in this section which examines the main features of the 11 national SDIs that were being developed in different parts of the world during the 1990s in Australia, Canada, Indonesia, Japan, Korea, Malaysia, Netherlands, Portugal, Qatar, United Kingdom and United States. The findings of this evaluation demonstrate that there is no single recipe for a successful SDI, as each SDI is the product of a unique set of national contextual and operational circumstances.

These questions are also the subject of the next chapter. This is the keynote paper on the future of SDIs that I was invited to present at a workshop organised by the International Society for Photogrammetry and Remote Sensing (ISPRS) in Hangzhou. It reviews the main milestones of SDI development over the last two decades and SDI experiences in different parts of the world, as well as explores some of the main strengths and weaknesses of current SDI practices in relation to the perceived opportunities and threats that are likely to emerge in the foreseeable future.

The next two chapters deal with some general questions about the nature of spatial data infrastructures and spatial data infrastructure research. The first of these was written for the first issue of the first scientific journal dealing specifically with spatial data infrastructures research. Under the heading of 'What's special abut SDI-related research?' it explores some issues relating to SDI diffusion, SDI evolution, data sharing and multi-level stakeholder participation. The other paper in this section considers some of the impacts of innovations in communications and information technology during this period on the nature of SDIs and explores the changes that have taken place in the notion of an SDI over the last 15 years. It also discusses the changing nature of SDI implementation in the context of the concepts of multi-level governance that have been developed by political scientists. Underlying this discussion is the realisation that SDI development and implementation is very much a social process of learning by doing. The concluding section of the paper considers the challenges facing SDI implementation and identifies a number of dilemmas that have yet to be resolved (see also Masser, 2005).

The final group of chapters in this book consider various developments of spatial data infrastructures in practice. The first of these, co-authored with Max Craglia, discusses the batch of initiatives at the European level that emerged in the nineties with respect to the establishment of a policy framework for geographic information. It particularly focuses on the GI2000 debate which surrounded the creation of a European geographic information infrastructure. From this it can be seen that significant progress had been made in many areas during the nineties and also that the European Union played an increasingly important role in this field. The last section of the chapter updates the discussion with reference to the period between 1998 and 2002. This includes the scandals that led to the resignation of all the EU Commissioners, the end of the GI2000 discussions and my personal involvement in the activities of EUROGI and the GINIE project, as well as the first steps towards the development of a new multinational European SDI framework which eventually became the INSPIRE initiative.

The other chapters in this section consider various aspects of SDI practice. The second chapter considers the Indian National Geospatial Data Infrastructure that was launched in 2001 at a workshop held in New Delhi. I participated in this workshop and noted the extent to which the debates at the workshop highlighted four issues that would need special consideration by those involved. In order of priority these were the nature of the machinery for coordination, the need to develop metadata services, the importance of capacity building initiatives and the need to promote data integration. Looking back after nearly 20 years, however, it is clear that not a lot has happened despite the enthusiasm shown at the 2001 workshop. A National Spatial Data Infrastructure has come into being in name, but it has lacked the authority to get the main government departments to cooperate fully in such a project. For these reasons, this paper is included in the book as a warning to those who argue that the development and implementation of SDIs is a straightforward task once Government support is obtained for such a project.

The third chapter, co-authored with Max Craglia, explores some of the geographic information (GI) policy issues associated with the unique set of circumstances created by the potential enlargement of the European Union (EU) from 15 to as many as 28 countries. It describes the main features of the procedures devised by the EU for evaluating progress towards EU membership and provides an overview of the countries which have applied for membership and discusses in detail the emergence of a national spatial data infrastructures in Bulgaria, Hungary, Lithuania, and Slovenia. The final section of the chapter updates the discussion with reference to the post accession situation and considers the current situation in the four case study countries with respect to the implementation of the INSPIRE Directive.

The fourth chapter in this section considers the sub-national dimension of SDI development and implementation. The overall discussion regarding SDIs has been dominated by national initiatives and the achievements of many sub-regional SDIs have often been neglected in the process. This deficiency was rectified to some extent by the eSDI-Net+ project which ran from 2007 to 2010. This project brought together a substantial number of SDI players in a Thematic Network and provided a platform for the communication and exchange of ideas and experiences between different stakeholders involved in the creation and use of SDIs throughout Europe. Co-authored with Joachim Rix and Swetlana Fast, this chapter discusses some of the main findings of this project and considers their implications for future research.

The final chapter in this collection rounds off the discussion of spatial data infrastructures in practice with reference to the development and ongoing implementation of Directive 2007/2/EC of the European Parliament and of the Council of 14 March 2007 establishing an Infrastructure for Spatial Information in the European Community (INSPIRE) to create an SDI for improving environmental data management in the European Community by 2021. This brings the 28 Member States together to build an SDI based on 34 related data themes. The origins and development of INSPIRE have been described in some detail elsewhere (Masser and Crompvoets, 2015), and this chapter, also written with Joep Crompvoets, identifies four main lessons that can be learnt from the overall framework that has been created for the INSPIRE implementation process. These are (1) the establishment of a legal framework for SDI development, (2) the procedures developed for creating the range of implementing rules that is required for the implementation process, (3) the mechanisms developed for monitoring progress in each of the member states and (4) the methods used to evaluate the overall performance of the initiative.

References

Masser, I. 2005. *GIS Worlds: Creating Spatial Data Infrastructures.* Redlands CA: Esri Press.

Masser I., and J. Crompvoets. 2015. *Building European Spatial Data Infrastructures.* Redlands CA: Esri Press.

Masser I., O. Sviden, and M. Wegener. 1992. *The Geography of Europe's Futures.* London: Belhaven Press.

Acknowledgements

I would like to acknowledge the part that my wife Suzy has played in making it possible for me to complete this book. I would also thank my daughter, Sophia Fox, for her invaluable help in resolving various computing problems that came up during the preparation of the manuscript. Without her assistance I doubted several times whether I would ever complete this task.

I would also like to express my appreciation for the efforts of the staff at CRC Press and Taylor and Francis, particularly Irma Shagla Britton for encouraging me to start this project and also for keeping the show on the road to completion with the help of Rebecca Pringle.

As this book draws heavily on materials that I have published over the last 30 years, consequently, it is important to acknowledge the original sources of each chapter.

- Sections 1.1 to 1.5 of Chapter 1 were originally published as I. Masser and H. Ottens, 1999. 'Urban Planning and GIS' in *Geographical Information and Planning*, J. Stilwell S. Geertman and S. Openshaw (Eds.), Springer, Berlin, 25–42.

- Sections 2.1 to 2.4 of Chapter 2 were originally published as I. Masser, 1990. The Regional Research Laboratory Initiative: An Overview. *International Journal of Information Resource Management 1*, 3, 1–12.

- Sections 3.1 to 3.3 of Chapter 3 formed part of a presentation by I. Masser entitled 'Data Integration Research: Overview and Future Prospects' presented at the Final GISDATA Conference, in Le Bischenberg, France in September 1997.

- Sections 4.1 to 4.4 of Chapter 4 were originally published as M. Wegener and I. Masser, 1996. 'Brave New GIS Worlds' in *GIS Diffusion: The Adoption and Use of Geographical Information Systems in Local Government in Europe*, I. Masser, H. Campbell and M. Craglia (Eds.), Taylor and Francis, London, 9–21.

- Sections 5.1 to 5.7 of Chapter 5 were originally published as R. Barr and I. Masser, 1997. 'Geographic Information: A Resource, a Commodity, an Asset or an Infrastructure?' in *Innovations in GIS, 4*, Z. Kemp (Ed.), Taylor and Francis, London, 234–248.

- Sections 6.1 to 6.5 of Chapter 6 were originally published as I. Masser, 1988. The Development of Geographic Information Systems in Britain: The Chorley Report in Perspective, *Environment and Planning B, 15*, 489–494.

- Sections 7.1 to 7.5 of this chapter were originally published as I. Masser and H. Campbell, 1993. GIS in Local Government: Some Findings from Great Britain, in *Geographic Information Handling: Research and Applications,* P.M. Mather (Ed.), Wiley, Chichester, 273–286.

- Sections 8.1 to 8.4 of Chapter 8 were originally published as I. Masser and H. Campbell, 1994. Information Sharing and Implementation of GIS, in *Innovations in GIS,* M. Worboys (Ed.), Taylor and Francis, London, 217–227.

- Sections 9.1 to 9.3 of Chapter 9 were originally published as M. Craglia and I. Masser, 1996. The Diffusion of GIS in Local Government in Europe, in *Geographic Information Research: Bridging the Atlantic,* M. Craglia and H. Couclelis (Eds.), Taylor and Francis, London, 92–107.

- Sections 10.1 to 10.5 of Chapter 10 were originally published as I. Masser, 2001. Managing Our Urban Future: The Role of Remote Sensing and Geographic Information Systems, *Habitat International, 25,* 503–512.

- Sections 11.1 to 11.4 of Chapter 11 were originally published as I. Masser, 1999. All Shapes and Sizes: The First Generation of National spatial Data Infrastructures, *International Journal of Geographic Information Systems, 13,* 67–84.

- Sections 12.1 to 12.4 of Chapter 12 were originally published as I. Masser, 2005. The Future of SDIs, *International Archives of Photogrammetry, Remote Sensing and Spatial Information Sciences, 36,* (4/W6), 7–15.

- Sections 13.1 to 13.5 of Chapter 13 were originally published as I. Masser, 2006 What's Special about SDI Related Research? *International Journal of Spatial Data Infrastructure Research, 1,* 14–23.

- Sections 14.1 to 14.6 of Chapter 14 were originally published as I. Masser, 2009. Changing Notions of Spatial Data Infrastructures, in *SDI Convergence: Research, Emerging Trends, and Critical Assessment,* B. van Loenen, J. W. J. Bessemer and J. A. Zevenbergen (Eds.), Netherlands Geodetic Commission, Delft, 219–228.

- Sections 15.1 to 15.4 of Chapter 15 were originally published as M. Craglia and I. Masser, 1997. A European Policy Framework for Geographic Information, *Computers Environment and Urban Systems, 21,* 393–406.

- Sections 16.1 to 16.4 of Chapter 16 were originally published as I. Masser, 2001. The Indian National Geospatial Data Infrastructure, *GIM International 15, 8,* 37–39.

- Sections 17.1 to 17.5 of Chapter 17 were originally published as M. Craglia and I. Masser, 2002. Geographic Information, and the Enlargement of the European Union, *URISA Journal, 14,* 2, 43–52.

- Sections 18.1 to 18.4 of Chapter 18 were originally published as J. Rix, S. Fast and I. Masser, 2010. Operational SDIs: The Subnational Dimension in the European Context, *Proceedings of the Fourth INSPIRE Conference*, Krakow.
- Chapter 19 was written specially for this volume by I. Masser and J. Crompvoets.

About the Editor

 **Emeritus Professor Ian Masser** was educated in geography and town planning at Liverpool University. He received his PhD in 1975 and was awarded a LittD in 1993 by the same University. From 1986 to 1991 he was national coordinator of ESRC's Regional Research Laboratory initiative and he co-directed the European Science Foundation's GISDATA scientific programme from 1992 to 1997. His publications include more than twenty books and over 300 contributions to conference proceedings, books and scientific journals.

He retired from the position of Professor of Urban Planning and Management at ITC in the Netherlands in 2002. Prior to that he was Professor of Town and Regional Planning at the University of Sheffield for nearly 20 years. He was the Founder Chairman of the Association of Geographic Information Laboratories in Europe (AGILE) from 1998 to 1999, President of the European Umbrella Organisation for Geographic Information (EUROGI) from 1999 to 2003 and the first President of the Global Spatial Data Infrastructure Association from September 2002 to February 2004.

Prof Masser has held visiting positions at universities in Australia, Germany, Greece, Italy, Iraq, Japan, Switzerland and the United States and at government research centres in Korea and Malaysia. He has also acted as advisor to a large number of national and international agencies including the European Commission, Eurostat, OECD, UNCRD and the World Bank.

In December 2016 he received the GSDI Association's Global Citizen award for his substantive worldwide contributions in promoting informed and responsible use of geographic information and geospatial technologies for the benefit of society and fostering spatial data infrastructure developments that support sustainable social, economic and environmental systems integrated from local to global scales.

List of Contributors

Dr. Robert Barr is Visiting Professor in the Geographic Data Science Lab at Liverpool University and Chairman of Manchester Geomatics, a company spun out of the Manchester Regional Research Laboratory which he co-founded while working at the University of Manchester. In 2008 he was awarded an OBE for services to geography.

Heather Campbell is Professor and Director of the School of Community and Regional Planning at the University of British Columbia. She is Senior Editor of *Planning Theory and Practice* and a Fellow of the Academy of Social Sciences. Her research interests focus on how public policy interventions concerned with cities and regions can produce better outcomes and how research can better support transformational change.

Dr. Massimo Craglia Craglia is a lead scientist at the European Commission's Joint Research Centre, Digital Economy Unit. His current research interests are in assessing the contribution of geospatial data and earth observation to the digital economy, and exploring the opportunities for new forms of governance created by the digital transformation of society.

Prof. Joep Crompvoets holds a chair in 'Information Management in the public sector' at the KU Leuven Public Governance Institute (Belgium). He is also secretary-general of EuroSDR, a European spatial data research network linking national mapping agencies with research institutes and universities for the purpose of applied research.

Swetlana Fast holds a Diploma in Economics and worked on industrial and R&D projects in the area of Spatial Data Infrastructure at Fraunhofer Institute for Computer Graphics Research, IGD, and the Technical University of Darmstadt, Germany for about 10 years. Currently, she is living in Norway, working as a Project Manager in Industrial Robotics at PPM Robotics AS in Trondheim.

Henk Ottens is Emeritus Professor of Human Geography, Utrecht University. He is also a Member of the ITC Foundation, University of Twente; Chair of the Accreditation Panels NVAO Accreditation Organisation of the Netherlands and Flanders; Past President KNAG Royal Dutch Geographical Society; and Past President EUGEO Association of Geographical Societies in Europe.

Dr. Joachim Rix is Deputy Head of Competence Center Spatial Information Management at Fraunhofer Institute for Computer Graphics Research IGD and the Technical University of Darmstadt. He has more than 25 years of experiences in applied research and development in national and international projects. His current activities are in the area of spatial information management with a focus on data harmonization and management, 3D visualization and interactive analysis of 3D city models and land management.

Michael Wegener Until 2003, Michael Wegener was Professor and Director of the Institute of Spatial Planning at the University of Dortmund. Since 2003 he is a partner in Spiekermann and Wegener Urban and Regional Research (S&W), Dortmund. His main research fields are modelling of urban and regional development and European urban systems.

Part I

Geographic Information Systems
Introduction

1

Urban Planning and Geographic Information Systems

Ian Masser and Henk Ottens

1.1 Introduction

The juxtaposition of words in the chapter title reflects the sequence of the argument in this chapter. Urban planning comes first because it has a long history as an activity which makes extensive use of geographic information. This extends from the sanitary maps that were made by the precursors of the modern planners in the 1830s and 1840s in Britain and the United States to the multi-purpose multi-user geographic information systems (GIS) that have been implemented in many of today's cities. In many ways, the needs of planning have actually anticipated the development of GIS. For example, Lewis Keeble (1952, 70) argues in his manual for the new generation of British planners created by the 1947 Town and Country Planning Act:

> 'There are two ways in which interrelated survey subjects can be compared: the first is by means of overlays, the second by means of combination or sieve maps' which suggest 'the metaphorical straining of all the land in the area under consideration through a series of sieves – standards of unsuitability – that which passes through all the sieves being prima facie the most suitable for the purpose in question and that which passes through the fewest the least suitable'.

GIS come second in the title of the chapter because they are a set of relatively new technologies dating from 1960s, which are currently transforming the handling of geographic information in fields such as urban planning. Even though the concepts underlying such systems have been in existence for over 30 years, however, it is only in last 10 years that technological developments in computing have made it possible for urban planners to exploit the opportunities opened up by these systems (Coppock and Rhind 1991). Consequently, virtually all the GIS systems that are currently in use in urban planning have been developed since 1990 (Masser and Craglia 1996, 228).

Given the recent nature of most GIS applications in urban planning and the potential that had been opened up by recent technological develop-ments – 'the biggest step forward in the handling of geographic informa-tion since the invention of the map' according to Britain's Chorley Committee (Department of the Environment 1987, 8) – this chapter con-siders the relationships between urban planning and GIS with particular reference to two related questions:

1. How will geographic information technologies change urban planning?
2. How will the needs of urban planning shape the future development of GIS, given that it is very much an applications-led set of technologies?

This chapter is divided into four main sections. In the first of these, the two predominant traditions that have emerged in urban planning over the last century and a half are identified and their distinguishing characteristics are discussed. These are the plan-making and the administrative traditions, respectively. Section 1.3 discusses the rela-tionships between urban planning and GIS in the context of each of these two planning traditions, while Section 1.4 examines the implica-tions of the previous discussion for the two questions posed earlier. Section 1.5 discusses developments in both GIS and planning with particular reference to Britain and the Netherlands over the last twenty five years. To avoid any confusion regarding the terminology used in this chapter, it should be noted that the term 'urban planning,' which is used throughout refers to spatially oriented urban planning in its broadest sense, and therefore also encompasses both regional- and local-level planning activities.

1.2 The Two Planning Traditions

Geographic information is fundamental to the practice of urban planning. It has been estimated that about 80% of all the information that is used by planners today is geographical, either in the sense that it directly makes use of topographic or other types of maps or in the sense that it contains a key geographic reference in the form of a coordinate reference or a street address or a reference to a particular administrative area. However, it is possible to distinguish between two different ways in which geographic information is collected, stored and utilised for plan-ning purposes in GIS. These reflect what can be termed the plan-making and the administrative traditions, respectively. The former was the domi-nant tradition of urban planning prior to the Second World War and continues to this day alongside the latter that has taken over as the dominant tradition wherever urban planning functions have become a formal statutory requirement of local government.

1.2.1 The Plan-Making Tradition

Prior to the Second World War, the concept of survey before plan was prevalent in most urban planning activities. Planning pioneers such as Patrick Geddes (1949) regarded a comprehensive survey of an area's social, cultural and environmental traditions as an essential pre-requisite for planning:

> We cannot too fully survey and interpret the city for which we are to plan ... Its civic character, its collective soul, thus in some measure discerned and entered into, its active daily life may be more fully touched, and its economic efficiency more vitally stimulated.

Similarly, Patrick Abercrombie (1933) argued in his influential text on Town and Country Planning that the planning survey was of critical importance in the planning process, and his surveys of cities such as Sheffield (1924) and regions such as the Doncaster (1922) provided practical demonstrations of his thinking. It is also worth noting that the chapter of his book which deals with the planning survey is the only one that comes near to offering direct guidance on planning methods in a practical sense, and that he had comparatively little to say at all about the other stages of the planning process.

As time went on, however, the notion of survey became increasingly linked with plan-making as an activity. The mysterious nature of the link between survey and plan is particularly well captured in another influential text on urban planning, Raymond Unwin's (1909, 193) *Town Planning in Practice*:

> There will come a stage when the main lines of the plan as determined on the site exist in a flexible condition in [the planner's] mind, when he feels the need of something more definite. This is the time for his designing genius to seize the ductile mass of requirements, conditions and necessities and, anchoring itself to the few absolutely fixed points, brushing aside minor obstacles or considerations where necessary, modifying or bowing to the major ones as each case seems to require, to mould the whole into some ordinary and beautiful design.

Nowadays, urban planning is conceived as an activity with three main components: (socio-economic) analysis, (technical) design and (political) decision-making. This is reflected in the teams that make the plans, where often several disciplines are represented, such as planners, geographers, economists, environmental scientists, (city) architects, civil engineers, and public administration and policy specialists. Within the team, the primarily analytical approach of the social scientists has to be interwoven with the design approach of the technical scientists:

> In the analytical approach, the purpose is to determine optimal adjustments to the current spatial structure in order to resolve an existing planning problem. Global planning objectives and restrictions are reduced to well-defined criteria, variables, and threshold values. Next, this formalized model is applied to empirical data. The focus of the design approach is on shaping new spatial structure. Through visualization of successive conceptual designs, the mental maps of the participating planners gradually evolve into a coherent perspective for the plan area.
>
> *(Schuur 1994, 97)*

Although planning systems and approaches in Europe differ from country to country, governments nearly always produce three types of plans:

- *Policy reports*: These plans are documents that set out spatial and spatially relevant policies for a large region or a country in a global, generalised and often strategic way. In general, they are indicative and comprehensive in nature, but sectoral reports are also produced (e.g. for transportation infrastructure, major business areas, nature conservation, etc.). The reports often contain only sketch maps. The documents serve the need to formulate policies that spatially integrate sectoral policies and to explore spatial opportunities and restrictions. This type of plans is found in countries with a well-developed planning system and tradition. The series of national physical planning reports in the Netherlands is well known in this respect (Hall 1977, pp. 103–117; van der Cammen et al. 1988). But there are also examples at the supranational level like the Europe 2000 study (CEC 1991).

- *Structure plans*: This is a very common type of spatial plan. Typically, structure plans are designed for medium-sized regions (counties, provinces, urban areas, major urban extensions, new towns, etc.). Structure or master plans deal with spatial development for 10- to 20-years periods, based on a comprehensive analysis of land requirements and spatial opportunities. Future development is visualised in maps, often with a level of detail well above the parcel level. The maps contain future land use patterns, urbanisation patterns, employment and service centre structures, ecological structures and main transportation networks. In most countries these regional plans are not legally binding and are used as guidelines for detailed spatial plans. They also play a role when local plans have to be approved at the regional level.

- *Zoning plans*: At the local level, plans that regulate the use of land and buildings at parcel level are the most important planning documents. The plans are physical in nature, they refer primarily to the built environment. Often these plans have some kind of legal status and are the basis for land development (building plans) and building control. Zoning or land allocation plans contain more or less detailed directives regarding permitted land use categories, building densities, building heights, building alignments, etc. In many countries, zoning plan regulations are to a certain extent negotiable in a development process. This reflects different roles assigned to local plans: instruments to provide certainty at the cost of inflexibility or flexible instruments to promote development with limited *a priori* certainty for the parties involved. The Dutch planning legislation very much takes the first approach, the British system the second (Thomas et al. 1983).

In most cases, planning departments are responsible for the process to produce the spatial plans just mentioned. Small departments often do not have the capacity and/or expertise to make these plans themselves. In those cases, external researchers, consultants or developers become partners in the plan-making process. External expertise is also called upon for complex and politically sensitive plan-making exercises. Planning legislation often requires or indicates a regular production and updating of plans and, especially for regional plans, a full coverage of the area of jurisdiction. Nevertheless, in many countries plan-making has the character of a - rather ad hoc activity. The production of local plans, however, is often directly related to building, renewal or conservation projects. In those cases, making local plans tends to evolve to a primarily technical and administrative type of activity with a routine character.

Apart from these, often statutory, plans, planning departments are regularly involved in planning studies. These can be background studies as part of a plan-making process or studies for specific problems, issues or projects. Even more than plans, planning studies are produced and organised on an ad hoc basis. A well-known recent example is the study report of the European Union on the development of central and capital cities and regions in North Western Europe (CEC 1996).

1.2.2 The Administrative Tradition

The Second World War marked a turning point in the evolution of urban planning throughout the world as governments increasingly came to recognise the need for more systematic land use controls in both urban and rural areas and took the necessary steps to make planning a statutory responsibility of local government. The impact of such measures is particularly dramatic in Britain where the 1947 Town and Country Planning Act gave far-reaching powers to local authorities to control development in their areas through the approval or rejection of applications submitted by potential developers, while at the same time imposing a statutory duty on all authorities to prepare development plans for their areas. This system differed from the pre-war system in two fundamental ways: first, in the powers given to local authorities to control development which were backed up by the effective nationalisation of the rights of private developers to compensation in the event of a refusal of planning permission, and second, in that every local authority was obliged to prepare a development plan for their areas and to submit this to government for approval, whereas under the pre-war system authorities could decide for themselves whether or not to prepare plans.

The implications of developments such as these and similar developments in other countries, like the Netherlands and Denmark, are that urban planning became a local government administrative activity with its own professional culture. The extent to which this occurred varies between countries according to the division of responsibilities between central and local government and the nature of local government itself. In the Netherlands, for example, there is a strong central government dimension to planning which is evident in the series of Notes setting out national strategies which have been periodically released by the National Physical Planning Agency since the early sixties (see, for example, Faludi and van der Valk 1994). In Britain, in contrast, central government has played a much less prominent role in strategic planning and, as a result, planning is almost exclusively a local government activity.

There are also major differences between countries with respect to the nature of local government itself. Whereas Britain is divided up into unitary and two-tier authorities with an average population of around 150,000, the average size of a French municipality is only 1,500. Under these circumstances, the potential of the former for the development of an in-house professional planning culture is considerably greater than the latter which is heavily dependent upon both centrally provided and private sector consultancy services (Masser and Craglia 1996).

Despite the differences, the characteristics of urban planning as an administrative activity from the standpoint of geographic information remain broadly similar although the outcomes are much less homogeneous in France than they are in Britain or the Netherlands. It should also be noted that the notion of planning itself as an administrative activity has changed substantially over the last 50 years and that the emphasis has shifted from a physical plan-led state interventionist mode of thinking towards a view of planning, and local government as a whole, as a facilitating or an enabling mechanism with respect to both public and private sector development (Thornley 1991). Initiating, influencing and participating in spatial development has become more important than rigidly guiding and controlling physical change (Ottens 1990). Public–private partnerships and the emergence of semi-public 'development corporations' to facilitate plan implementation are exponents of this reorientation.

1.2.3 Evaluation

The main features of the two planning traditions are summarised in Table 1.1. From this it can be seen that the plan-making tradition is essentially task oriented in nature. Its primary purpose is to produce tangible products in the form of sets of proposals or plans for approval by a higher-level authority or elected body. Typically, such proposals or plans are prepared by groups specifically set up for the purpose and once they have been approved the group is disbanded or substantially reduced in size. Under these circumstances plan-making is carried out to an agreed time frame with explicit terms of reference for a specific client.

As Table 1.1 shows, there are plenty of examples of the plan-making tradition in operation in practice today. Probably the most dramatic of these relates to the construction of new towns on green field sites where a considerable amount of preparatory work is needed prior to the implementation of the plan. Once the implementation process gets under way in such projects the demand for plan-making are essentially reduced to the task of monitoring the progress of the project. Other examples of the plan-making tradition in practice include the preparation of both strategic structure plans and more detailed local plans. As in the case of new towns, once the elements of such plans have been agreed the emphasis shifts away from plan-making towards the management of plan implementation. Many of the features of the plan-making tradition can also be seen in the thematic studies that are often commissioned by planning agencies and other bodies. These include transportation studies and environmental impact analyses which are also carried out to an agreed time frame with explicit terms of reference for specific clients.

In contrast to the plan-making tradition, the administration tradition has grown up in response to the statutory planning duties imposed on local or regional government. Although this may give rise to plan-making activities at different points in time, the most important feature of the administrative tradition is that these duties require some form of permanent organisational

TABLE 1.1

1 Key Features of the Two Planning Traditions

	Plan-Making Tradition	**Administrative Tradition**
Primary purpose	Development of blue prints and strategies for future spatial development	Control of ongoing spatial development and links to technical and administrative management
Organisational requirements	Carried out by ad hoc teams set up for the purpose	Continuous in nature: require some form of permanent organisational structure.
Examples	New town plans Structure plans Thematic studies	Planning agencies Zoning plans Approval systems

structure to discharge them. In this context, the tangible products that are produced as a result of plan-making form only one component in a much wider range of continuing activities. Often, the planning activities are directly related to administrative (e.g. cadastre) and technical (e.g. public works) management tasks.

The primary examples of the administrative tradition in practice are the planning agencies themselves. In the Netherlands, for example, these include the National Physical Planning Agency, the provincial planning departments and the municipal planning departments which have been set up to fulfil specific planning mandates. In addition to these general purpose agencies, there are also a number of special purpose agencies such as those associated with the national parks in most countries.

Table 1.2 sets out some of the GIS requirements associated with the two planning traditions. From this it can be seen that there are major differences between them with respect to their information requirements and also that they are likely to require different special features. Applications in the plan-making tradition are generally dedicated single user in nature. Their objectives are defined by those of the plan-making task and the GIS that is implemented is likely to be used exclusively by the group set up for this purpose. Consequently, although the number of users in practice may be greater than one person, their system requirements are likely to be similar in nature, hence the term single user in this instance. In cases such as these, the information requirements are likely to be determined by the nature of the application itself. A number of key variables associated with the application are likely to be identified and data for these variables assembled mainly from secondary and external sources. Where information for a particular variable is not available from these sources proxy variables will be identified to fill the gap. Special features of GIS applications in the plan-making tradition are likely to be the incorporation of applications

TABLE 1.2

GIS Requirements Associated with the Two Planning Traditions

	Plan-Making Tradition	Administrative Tradition
System requirements	Single purpose Single user analytical GIS	Multi-purpose Multi-user management GIS
Information requirements	Determined by specific application Both internal and external data	Planning transactions Management information at all levels Primarily internal information
Special features	Application specific analytical tools Visualisation and display capabilities	Quality assurance for database maintenance and updating Metadata. Technical mapping

specific analytical and modelling tools such as multi-criteria analysis or transportation models in the operational systems. Given that plan-making tasks are also very much product-oriented, special attention is also likely to be given to visualisation and display facilities for presentation purposes.

In contrast to the plan-making tradition, the system requirements of the administrative tradition are for multi-purpose multi-user GIS, given that the system must provide management information to a wide variety of users who will use it for many different purposes. The focal point of systems developed in the administrative tradition, unlike those implemented in the plan-making tradition, is likely to be the database itself. This is likely to contain a considerable amount of data from primary as well as data from secondary sources, including planning control transactions. Under these circumstances, the special features of GIS applications of this kind are likely to be closely linked to the maintenance and updating of the database itself. This can be regarded as a local spatial data infrastructure for all planning activities in urban areas. Given that much of this data will be derived from primary sources quality assurance mechanisms are likely to be of particular importance. Similarly, given the diversity of potential users involved and the wide range of possible applications involving data held in the database, the provision of metadata facilities to help end users find the data they need is likely to be a special feature of such systems. The cartographic capabilities of these systems are primarily geared towards an efficient production of technical and administrative maps for use by specialists.

It should also be noted that there are important differences between the two traditions with respect to their development over time. Most projects in the plan-making tradition are likely to be limited life projects and the GIS applications that are developed will be essentially static in character. In contrast the multi-purpose multi-user systems that are implemented in the administrative tradition are likely to be constantly evolving over time in response to the changing requirements of management.

Some of the implications of these differences in practice will be explored in greater detail in the next two sections of this chapter with respect to each of these urban planning traditions.

1.3 GIS and the Plan-Making Tradition

In order to discuss the role of GIS in plan-making, first information needs will be elaborated. Assuming a rational model of planning, the nature of urban plans requires the processing of a combination of substantive information on the area for which the plan is made and policy information about the political decision-making process that accompanies the successive steps in the plan-making process. This is illustrated in Table 1.3. Van

TABLE 1.3

Information Types for Spatial Plan-Making

Area information	Policy information
Retrospective information	Evaluative information
Current information	Diagnostic information
Prospective information	Conclusive information

Lammeren distinguishes in this respect between four types of knowledge that are required in the process (Eweg 1994, 16–18):

- 'object' knowledge, related to the nature, location and the functioning of the area which is subject to planning;
- 'normative' knowledge, necessary to develop intentions, goals, etc.;
- 'process' knowledge, dealing with the procedures and logistics of the plan-making process;
- 'method' knowledge, related to data and information processing for the generation of plans.

A first conclusion therefore is that a GIS, which is designed to process locationally referenced data, needs to be complemented by other information systems in order to process all necessary plan-making information. Document processing, project management and workgroup information systems are the most obvious other support tools that have to be available. But a spreadsheet programme to produce financial calculations and overviews might also be useful.

The information needed in plan-making relates to three temporal domains: retrospective, current and prospective (Batty 1993, 59–61; Webster 1993, 711–713). First, relevant recent developments have to be analysed. This includes an ex-post evaluation of operative spatial policies. Next, current issues and problems have to be assembled and assessed both in the area and the policy domain, a mainly descriptive and diagnostic exercise. Finally, possible future developments have to be explored and designed. This prospective information has to be evaluated and the decision-making process about desirable directions of development has to be documented. Especially policy, structure and local plans of a strategic character require this full temporal information range. The substantive area information to be processed can be further detailed. Perloff (1980, 163–177) provides a useful summary of substantive information required when producing a comprehensive urban plan:

- Contextual information (on the nation and region relevant for the urban area).
- Monitoring information (on processes of change within the urban area)

- Needs information (on needs for improvements and ways of achieving them)
- Problem information (on special problems and difficulties and ways of mitigating them)
- Future information (on opportunities and possibilities for the area)

These information categories have to be gathered and processed for at least four subject domains: population (cultural assets, households, income, welfare), economic activity (production, distribution, labour market), environment (land use, buildings, infrastructure, nature) and public finances (tax base, investments, service provision, welfare payments).

The common denominator for all this information is the locational dimension as all information has to contribute to a spatially articulated plan. Therefore, GIS is an obvious tool for handling this information.

Webster (1993, 1994) has made a thorough analysis of the usefulness of GIS in the urban planning process. His conclusions are that GIS is particularly useful for the generation of descriptive and prescriptive information: the analysis of the present state of the plan area and the evaluation of scenarios for future development. The visualisation, data management and geographical modelling capabilities are the main attraction of using a GIS for planning purposes. Efficient visualisation in the form of thematic maps is particularly useful when plans have to be discussed in political decision-making and public participation processes. GIS database management can greatly improve cumbersome manual data handling procedures in planning offices. Geographical modelling refers to the standard locational and topological object and layer manipulation functions found in off-the-shelf analytical GIS software (overlay, buffer, spatial query, etc.). These tools can be used for urban plan-making relevant activities like suitability mapping and evaluation of spatially specified goals. For generating predictive information (projections, forecasts, scenarios, etc.). GIS is much less suited, mainly because it lacks elaborate spatial modelling functionality.

In the same vein, Batty (1993) argues that when GIS is used as a planning support system (PSS), its database and visualisation capabilities need to be harnessed by traditional spatial models (see also Klosterman 1997). Examples of functions that are needed in many plan-making projects and planning studies are geostatistics, exploratory spatial analysis, interpolation and smoothing methods, econometric and regression analysis, and elaborate network analysis (Openshaw 1990). Various strategies can be followed to enhance the processing capabilities of GIS, from loose coupling of model software to full integration of modelling modules (Batty 1993). After hardware and software costs have come down considerably, and processing power of desktop computers continues to increase dramatically, data availability and software complexity remain two major obstacles to widespread use of GIS for plan-making. As plan-making is very much an ad hoc activity that often requires a considerable research effort, data needs are generally large and difficult to identify in advance. Moreover, a large part of the data needed cannot be derived from internal technical and administrative databases of local government agencies but have to be obtained from external registers and statistical and map data-bases or through surveys. As data files from different sources have to be integrated through a common geometric frame, pre-processing is always a complex and time-consuming activity. The extensive data manipulation and modelling functionality of analytical GIS software makes learning curves for users of these packages steep. The combination of complex data, complex software and complex planning problems and procedures requires genuine specialists. As only large agencies have sufficient resources to put dedicated GIS units in place, a lot of GIS-supported plan-making and planning research is still carried by consultants and universi-ties. On the other hand, highly standard personal computer software that is capable of handling geographical information has proven to be very useful as support tool in many planning projects where data volumes are modest and data modelling is rather simple and standard.

1.4 GIS and the Administrative Tradition

There are clear parallels between the GIS requirements that are set out in Table 1.2 and those of any management information system. This is particularly evident in the concept of the multi-purpose multi-user system and the accompanying desire to make information available to users of all kinds and at all levels of the organisation. Three main levels of information requirement have been identified in the management literature (see, for example, Anthony 1956), and similar levels of decision-making can be observed in most planning agencies. These are:

- *strategic planning*: primarily concerned with the identification of objectives and the allocation of resources by high-level managers in the organisation;
- *managerial control*: primarily concerned with the utilisation of these resources by middle-level managers;
- *operational control*: primarily concerned with the day-to-day opera-tions of low-level managers.

The extent to which these levels in the decision-making hierarchy are identifiable in practice is likely to depend on the size of the planning agency, but what is more important from the point of view of the devel-opment of a multi-purpose multi-user GIS is the different information characteristics that they represent. These are summarised in Table 1.4 with respect to the top and bottom levels of the organisational hierarchy on the assumption that the middle-level characteristics will fall somewhere in between these two extremes. From this it can be seen that strategic planners mainly require summary information on many different topics to carry out their duties while operational controllers at the other end of the spectrum required detailed information on only a few topics. These differences are also reflected in the accuracy and frequency of information required by the two levels of decision maker. Whereas strategic planners are likely to be satisfied with a tolerable level of accuracy and periodic updates, those involved in operational control are likely to need highly accurate information which is updated in real time to carry out their tasks.

From Table 1.4 it can also be seen that there are important differences between the time frames of the information used by strategic planners and operational controllers. The former is primarily concerned with anticipating future developments so that the necessary steps can be taken to respond to them while the latter are largely concerned with evaluating past experience to monitor progress towards implementation. These differences can also be seen in the nature of the sources required by each type of decision maker. Most of the information used by operational controllers is likely to come from internal sources where strategic planners are likely to make more use of external than internal sources for their purposes.

Given the conflicting nature of these requirements, it will be apparent that the development of multi-purpose multi-user GIS by urban planning agencies is likely to bring with it disadvantages as well as advantages in practice and that both need to be considered in more detail. Table 1.5 summarises some of the advantages and disadvantages associated with the development of multi-purpose multi-user GIS. From this it can be seen that one of the most important advantages of developing a multi-purpose multi-user GIS is that it enables the integration of formerly separate

TABLE 1.4

Information Characteristics Associated with the Strategic Planning and Operational Control Levels of Urban Planning

	Strategic Planning	**Operational Control**
Scope	Summary information on many topics	Detailed information on a few topics
Accuracy	Tolerable	Highly accurate
Frequency	Periodic	Real time
Timeframe	Anticipating future developments	Past experience
Sources	Largely external	Largely internal

Source: Adapted from Grimshaw (1994/95).

datasets. This is one of the most powerful properties of GIS technology according to Rhind (1992):

> All GIS experience thus far strongly suggest that the ultimate value is heavily dependent upon the association of one dataset with one or more others, thus in the EEC's CORINE (and perhaps in every [other]) project the bulk of the success and value came from linking datasets together.

Although no-one would dispute such statements in principle, there are likely to be limits in practice to the extent that disparate datasets need to be integrated, and it should also be noted that additional costs are likely to be associated with their integration. This does not rule out the need for greater integration but suggests that unrestricted efforts to integrate data may not necessarily be in the interests of the planning agency.

Similar views have been expressed with respect to the increased capacity for data sharing that is seen as another of the main advantages of developing multi-purpose multi-user GIS. In principle, again, as Onsrud and Rushton (1995, xiv) have pointed out:

> The value and social utility of geographic information is important because the more it is shared the more it is used and the greater become society's ability to evaluate and address the wide range of pressing problems to which such information may be applied.

Once again, the principle makes a lot of sense but the need for data sharing is likely to vary considerably in practice. As shown in Table 1.4, operational controllers have little to gain from sharing information on a large scale and they may regard the additional costs of cleaning up and documenting the information they collect so that it can be shared with others

TABLE 1.5

Advantages and Disadvantages of Multi-Purpose Multi-User GIS

Advantages	Disadvantages
Integration of formerly separate data sets	Differences in priorities among users
Increased capacity for data sharing	Disagreements over data standards, access, equipment training etc.
Improved access to information	
More informed decision-making	
Increased efficiency due to reduced duplication of effort	

Source: Adapted from Campbell and Masser (1995, Tables 3.3 and 3.5).

as outweighing the benefits to be obtained by gaining access to other datasets.

Similar qualifications can be raised regarding the other three advantages cited in Table 1.5. Improved access to information may also result in challenges to existing administrative structures. More informed decision-making does not automatically mean better decisions in itself and some measure of duplication of effort is not necessarily wasteful in itself under some circumstances. In practice, too greater emphasis on reducing duplication can result in decreased rather than increased organisational efficiency.

In all these areas, there are both costs and benefits for the different groups involved in multi-purpose multi-user GIS development and the advantages of this strategy may turn into disadvantages as soon as a large enough proportion of these groups perceives it to bring more costs rather than benefits. Table 1.5 also shows that there are some specific disadvantages associated with developing such systems. If a large number of users are involved there are likely to be differences in priorities in users as well as disagreements as to data standards, access, equipment and training. These may dramatically affect the prospects of successful implementation in a similar way to the circumstances surrounding the failure to implement the EDA programme in Oakland, California, that were graphically described in Pressman and Wildawsky's (1973) classic work on policy implementation. For this reason, Campbell and Masser (1995, 35–38) have argued that most of the GIS literature on this subject is based on preconceptions as to how organisations ought to operate rather than how they operate in practice. They claim that the technological determinist and managerial rationalist perspectives which have tended to dominate thinking about GIS implementation have three important weaknesses. First, they fail to take account of variations in the extent to which the potential of technological innovations such as GIS is realised in practice. The findings from other research in a wide range of organisational settings show that the outcomes are rarely only positive or completely negative. As noted above, the successful implementation of a multi-purpose multi-user GIS is likely to be much more a matter of the relative trade-offs between the costs and benefits to those involved (see, for example, Moore 1993). Second, there is plenty of evidence from other fields to show that the outcomes of implementing the same technology varies so markedly that it must be concluded that it is not the technology itself that determines the results of this process but the particular organisational and institutional circumstances surrounding its implementation (see, for example, Eason 1988). Finally, both the technological and managerial perspectives of technological innovation are narrowly focused on equipment and methods and take a similar mechanistic view of the context in which GIS is expected to operate. Once again there is plenty of evidence from historical evaluations of technological evaluation to indicate the multifaceted nature of this process and the degree to which the adoption and utilisation of technology is dependent not so much on their intrinsic qualities but rather on social attitudes and institutional circumstances (see, for example, Bijker et al. 1987).

To overcome these deficiencies, Campbell and Masser (1995, 37) argue that it must be recognised that the implementation of multi-purpose multi-user GIS in organisations is far more complex and problematic than is implied by the technological determinist or the managerial rationalist perspectives and that a social interactionist approach is required to understand socio political realities of organisational life:

> The fundamental assumption of the social interactionist approach is that technologies are not independent of the environments in which they are located but rather only gain meaning from their context. The adoption of effective implementation of an innovation [such as GIS] is therefore the result of interaction between the technology and potential users within a particular cultural and organisational arena.

Such an approach makes it necessary to develop a user-centred system design philosophy which starts from the assumption that organisational rather than technical issues are most likely to threaten effective system implementation.

With these considerations in mind, Campbell and Masser (1995) identify three key factors that must be taken into account by organisations to increase their chance of successful implementing multi-purpose multi-user GIS. First, an information management structure that identifies the needs of users and takes account of the resources and values of the organisation. It must be emphasised that an information management strategy is very different from an information technology strategy in that it is primarily concerned with the information needs of users rather than the technology that is needed to deliver this information to them. An information management strategy is not necessarily a written document that has received formal approval but can also take the form of a set of shared priorities and attitudes towards information which is deeply embedded in the culture of the organisation. Above all, it requires a thorough understanding of the needs of users and an awareness that are likely to be available for implementing such a strategy both now and in the future. Second, commitment to and participation in the implementation of the system by individuals at all levels of the organisation. This factor reflects the issues raised above in connection with user-centred philosophies towards the implementation of computer-based systems in general. In such philosophies, a critical issue is the need to increase awareness amongst potential users given that most users in planning agencies are likely to be passive users for whom the technology is only a means to an end rather than an end in itself. Finally, there is the requirement of an ability to cope with change. It must be recognised that change and instability are inherent in the life of any organisation. Consequently, an ability to cope with changing organisational goals and turnover of personnel is also a factor which is of vital importance for successful GIS implementation.

Huxhold (1991) provides a detailed account of the development of the multi-purpose, multi-user urban GIS for the city of Milwaukee which meets these criteria. He shows how the Milwaukee system has been used in a wide range of applications for service delivery, policy setting and urban management. These include map updating, zoning, reapportionment, building inspection, workload balancing, solid waste collection routing, housing management, liquor licensing, environmental health and library facilities planning. A subsequent book by Huxhold and Levinsohn (1995) provides a great deal of practical advice for those concerned with the management of GIS projects of this kind. This includes not only matters of strategic planning and implementation planning for GIS but also questions relating to staffing and training needs and budgeting for GIS operation and maintenance

1.5 Discussion

After elaborating the current role of GIS in the plan-making and administrative traditions of urban planning and indicating some problem areas, it is time to return to the two questions posed in the beginning of this chapter: How will GIS technologies change urban planning? How will the needs of urban planning shape the development of GIS?

There is little doubt that GIS will become an ever more essential part of the software suite that every researcher and planner engaged in urban plan-making will have available on his desktop computer. GIS will increasingly be used in combination and integrated with other popular office software. On the other hand, dedicated systems, with ample analytical and modelling capabilities, will continue to function as central information handling facilities in specialised research and development units in large agencies, research institutes and consultancy firms. But also, here, GIS will become integrated into a network-based information and communication environment. The development of data infrastructures for geographical information – greatly promoted by President Clinton's Executive Order of April 1994 to develop a National Spatial Data Infrastructure – may have a dramatic impact on the way GIS is going to be used in plan-making. If most of the technical, financial and legal restrictions on the use of internal and external databases is removed, a truly data rich environment will emerge (Masser 1998). A wealth of geometric, thematic (both real time transaction and statistical data) and procedural information will become available for analysis. Moreover, it can be expected that real time data from earth, landscape and event observations, including high-resolution satellite data, will further broaden the range of available data sources. Data-rich environments will undoubtedly stimulate the development and use of standardised monitoring systems for change detection, early warning and policy evaluation and of more elaborate modules for quick scanning, datamining and exploratory data analysis. Finally, the visualisation capabilities of GIS need to be and will be more fully exploited. Multimedia environments and Internet technologies for negotiating and bargaining, public participation and political decision-making can greatly benefit from GIS-based visualisation, especially when they support interactive and collaborative ways of working. Dynamic three-dimensional virtual reality type visualisations can be useful elements of those systems.

Administrative planning activities are very much affected by advancements in computer support that follow from so-called 're-engineering' programs. These programs aim at integrating systems and adapting organisation structures to make business processes more efficient and effective. Data warehousing, client-server and intranet technologies are often used for the newly developed information infrastructures. Further, more versatile database management systems and Enterprise Resource Planning software form the software core of the new information systems. These approaches are now well accepted in vertical markets (industrial production, logistic chains) but are gradually also introduced for administrative and service-oriented business processes. These, to some extent, recentralising development strategies bear risks for GIS applications. The lessons learned in the past and summarised above should not be forgotten. In this line of development, GIS-based planning applications will benefit from the development of data warehouses in local governments and regional spatial data infrastructures. These new facilities have the potential for opening up planning by enabling anyone who is interested to access key data. Internet access for citizens and businesses to the information contained in urban plans and real estate registers and GIS supported processing of development and building applications are obvious examples. Further, in a highly networked information handling environment, planning – and GIS – activities will become more tightly interwoven with activities in other sectors of local government.

Because the use of GIS has diffused to a broad range of application fields, the influence urban planning requirements can exercise on GIS development with the major vendors has been reduced. This is, *a fortiori*, the case for needs related to the use of GIS in urban planning research. On the other hand, because of demand pressure, systems have become more open, flexible and adaptable, enabling users and small software and service companies to specialise in niche markets by developing dedicated systems (e.g. for transportation and virtual reality applications) and add-on modules for standard general-purpose software (e.g. for vector-raster conversion). Also, the now widely recognised power of geographic data integration and cartographical visualisation, basic features of each GIS, continues to put commercial developers to include this kind of spatial functionality in both off-the-shelf and custom-made information systems, including Internet/WWW applications (The Enterprise 1997). As a consequence, GIS 'middleware' has become a fast growing new software market. Statistical packages and spreadsheets are well known examples. Database companies and computer-aided design (CAD) software vendors have extended the range of datatypes they support to include geometric data. Further, feature-based spatial data models will become more important for geographical databases. This has led to new architectures for GIS software, where geometric data is stored in standard relational database management systems with either a separate or an integrated layer with geographic intelligent software tools (ESRI 1996). It surely is not urban planning alone that commanded these developments, but as an established and prominent segment of the GIS user community, experiences with GIS in urban planning undoubtedly have contributed to the development lines we witness today.

1.6 Urban Planning and GIS – 25 Years Later

To begin with, a reminder of the order of events in this chapter which begins

> The juxtaposition of words in the title reflects the sequence of the argument of this chapter. Urban planning comes first because it has long history as an activity which makes extensive use of geographic information. ... In many ways the needs of planning have actually anticipated the development of geographic information systems.
>
> *(Masser and Ottens 1998)*

There are many similarities between these views and those of Yeh (1999). Yeh also considers the use of GIS in different functions and stages of urban planning and discusses factors influencing the use of GIS in urban planning. His conclusion is based upon a deep understanding of the issues involved:

> To use GIS more effectively in planning, planners must be trained not so much in the operation of the system, but in how to make use of the data and functions of GIS in different processes of planning and plan evaluation.
>
> *(p. 887)*

To some extent, however, the juxtaposition of words in the title of our chapter is wishful thinking because the reality is that developments in GIS itself have tended to dominate most of the discussion. This reflects the extent to which GIS is a technology driven field. This has been very much the case during the last 20 years as a result of the massive changes that have occurred in computer technology and network services over this period. This appears to be particularly the case in the United States. For example, it can be seen in a paper by Drummond and French (2008) entitled 'The future of GIS in planning', which claims that 'This article outlines important changes in geospatial technology to initiate a discussion of how the planning profession can best respond to these challenges and opportunities' (p. 161). Its conclusion is:

> The world of GIS is changing. With the advent of mass market GIS, the needs of planners are less central than they were previously to commercial GIS vendors. However, there are exciting new opportunities provided by web-based systems and open-source geospatial software.
>
> *(p. 161)*

The article is followed by two commentaries from Klosterman (2008) and Ferreira (2008), which are more optimistic about the impact of these developments than Drummond and French. In Klosterman's view,

> I am much less concerned than Drummond and French about the future role of GIS in planning. GIS, broadly defined, has evolved significantly in the past in ways that have been very beneficial for planning, and I expect this to be true in the future.
>
> *(p. 175)*

He highlights the extent to which planners from around the world are already developing GIS-based planning support systems (PSS) designed for planners' needs and points out that 'These developments parallel those in the broader GIS industry, where general-purpose GIS data and tools are increasingly being adapted to meet the particular needs of professionals in fields such as retail sales, real estate, and health care' (p. 175). Ferreira also takes a more optimistic view. His main concern is

> that the danger to planners from the growth of mass-market GIS is not so much that planners will be irrelevant to its future. Rather, we should be concerned that GIS may become embedded in metropolitan information infrastructures in ways that are poorly suited to planning. Without sufficient urban planning input, the core datasets, access tools, and protocols may not support the types of collaborative planning, urban modelling, urban design, and community empowerment that we want to encourage.
>
> *(p. 179)*

One matter that comes out from this discussion is the impression that GIS has become ubiquitous in urban planning in America. According to Drummond and French (2008, 161), 'GIS is now a standard item in planners' tool kits. Nevertheless, a web-based survey of practitioners in Wisconsin's public planning agencies by Gocmen and Ventura (2010), found that organisational and institutional issues are more pertinent than technological barriers to GIS utilisation although improved access to training may alleviate many of the barriers planners face in using GIS in general and enable them to incorporate more sophisticated GIS functions in their work.

On the European side of the Atlantic, some profound changes have taken place during the last 20 years regarding the statutory framework and the role of urban planning itself. In the United Kingdom, for example, Lord and Tewdwr-Jones 2012, p. 345) have chronicled 'the cumulative effects of the sustained programme of neo-liberalization to which urban and environmental planning has been subjected in England over a period spanning approximately the last 15 years.' Between the election of the Blair government in 1997 and the election of the Coalition government in 2010, they point out:

> England witnessed an intense 13 years in which successive governments introduced three separate legislative reforms of planning (in 2004, 2008 and 2009) and embarked on a series of high-profile reviews of the planning system to identify and ameliorate what was increasingly branded the 'failure' of planning.
>
> *(p. 346)*

The findings of their analysis highlight the extent to which 'planning has been such an intractable issue for all governments that have sought its reform irrespective of the particularities of their political agenda' (p. 345).

Urban planning has also been weakened and changed during this period in its traditional stronghold – the Netherlands. From 2002, centre-left cabinets ruled the country and promoted a leaner government. More importantly, the banking and public debt crisis (2007–2011) gave economic and financial recovery the highest priority. Among the decisions taken were a decentralisation of physical planning power and less public involvement in building and housing. A new Environment and Planning Act is currently being implemented, integrating dozens of existing laws and directives (Ministry of Infrastructure and Environment 2017). This law will simplify plan-making and planning administration and will give provinces, municipalities and the private sector more influence in spatial planning. As a consequence, physical planning has ended up as only a Directorate-General in the Ministry of Infrastructure and Environment and the National Planning Agency has become part of a much wider Netherlands Environmental Assessment Agency. A strategic vision document will provide the basis for flexible actions and a draft first vision document is already available on the web (Ministry of the Interior and Kingdom Relations 2018). However, while interest in planning decreased, the information infrastructure for the field has improved substantially. The availability and accessibility of the data for all types of users has greatly improved as web services became operational at all levels of government. The use of generic planning support systems remained limited, but specialised GIS applications are being developed and used more widely (Pelzer 2015).

However, it should be noted that there is nothing new about these concerns. For example, more than 30 years ago Klosterman (1985) published a paper entitled 'Arguments for and against planning'. In this chapter he examined four types of argument that have been made for and against planning in a modern democratic society: economic arguments, pluralist arguments, traditional arguments and neo-Marxist arguments. In the light of his analysis, he concludes:

> The preceding discussion has examined a variety of arguments for and against planning. Underlying this apparent diversity is an implicit consensus about the need for public sector planning to perform four vital social functions-promoting the common or collective interests of the community, considering the external effects of individual and group action, improving the information base for public and private decision making, and considering the distributional effects of public and private action.
>
> *(p. 15)*

Consequently, it will come as no surprise to find that, irrespective of the current debates surrounding such ideologies, urban planning practice in British and Dutch local government has a great deal in common with the processes described 20 years ago and the main features of the plan-making and the administrative traditions can still be observed. However, the overall scope of urban planning has changed during this period, and there is a greater emphasis on sustainability and environmental issues. The development of web-based decision support systems has also extended the range of GIS applications as has the emergence of participatory GIS which are context- and issue-driven rather than technology-led and seek to emphasise community involvement in the production and/or use of geographical information. In the eyes of Dunn (2007), for example, 'Participatory GIS celebrates the multiplicity of geographical realities rather than the disembodied, objective and technical "solutions" which have tended to characterize many conventional GIS applications.' With respect to the status of GIS in urban planning, it is also worth noting that the UK Government's current National Careers Service information sheet on the town planner (http//: nationalcareerservice.direct.gov.uk/job-profiles/townplanner) has IT skills in its job description. It also includes the following description under the heading of what planners do:

You'll use survey techniques, geographical information systems (GIS), and computer aided design (CAD) to draw up plans and make recommendations for land use.

Obviously, many things have changed since the original paper was written 20 years ago, but GIS remains an important component of the planner's tool kit in both Europe and the United States.

References

Abercrombie, P., 1933. *Town and country planning*, London: Thornton Butterworth.

Anthony, R.N., 1956. *Planning and control systems*, Boston: Harvard Business School Press.

Batty, M., 1993. Using geographic information systems in urban planning and policymaking, in Fischer, M.M. and Nijkamp, P. (eds.), *Geographic information systems, spatial modelling, and policy evaluation*, Berlin: Springer, 51–69.

Bijker, W.E., T.P. Hughes, and T.J. Pinch (eds.), 1987. *The social construction of technological systems: new directions in the sociology and history of technology*, Cambridge: MIT Press.

Campbell, H., and Masser, I., 1995. *GIS and organisations: how effective are GIS in practice?* London: Taylor and Francis.

Commission of the European Communities, 1991. *Europe 2000: outlook for the development of the community's territory*, DG XVI, Brussels.

Commission of the European Communities, 1996. Prospects for the development of central and capital cities and regions, *Regional Development Studies*, DG XVI, Office for Publications of the European Communities, Luxembourg.

Coppock, T., and D. Rhind, 1991. The history of GIS, in Maguire, D., Rhind, D., and Goodchild, M. (eds.), *Geographical information systems: principles and applications*, Volume 1, London: Longman, 21–43.

Department of the Environment, 1987. *Handling geographic information*, Report of the Committee of Enquiry chaired by Lord Chorley, London: Her Majesty's Stationery Office.

Drummond W.J., and S.P. French, 2008. The future of GIS in planning: converging technologies and diverging interests, *Journal of the American Planning Association*, 74, 2, 161–174.

Dunn, C., 2007. Participatory GIS, *Progress in Human Geography*, 31, 5, 616–637.

Eason, K.D., 1988. *Information technology and organisational change*, London: Taylor and Francis.

The Enterprise, the Next Generation, and the OCX Files, 1997. *Geo info systems showcase 4*, 10–13.

ESRI, 1996. *Spatial database engine*, Redlands: Environmental Systems Research Institute.

Eweg, R., 1994. *Computer supported reconnaissance planning*, Wageningen: Agricultural University.

Faludi, A., and A. van der Valk, 1994. *Rule and order: Dutch planning doctrine in the twentieth century*, Dordrecht: Kluwer.

Ferreira, J., 2008. Comment on Drummond and French: GIS evolution – are we messed up by mashups? *Journal of the American Planning Association*, 74, 2, 177–179.

Geddes, P., 1949. *Cities in evolution*, edition new and revised, London: Williams and Northgate.

Gocmen, Z.A., and S.J. Ventura, 2010. Barriers to GIS use in planning, *Journal of the American Planning Association*, 76, 2, 172–183.

Grimshaw, D.J., 1994. *Bringing geographical information systems into business*, Harlow: Longmans.

Hall, P., 1977. *The world cities*, London: Weidenfeld & Nicolson.

Huxhold, W.E., 1991. *An introduction to urban geographic information systems*, Oxford: Oxford University Press.

Huxhold, W.E., and A.G. Levinsohn, 1995. *Managing geographical information system projects*, Oxford: Oxford University Press.

Keeble, L., 1952. *Principles and practice of town and country planning*, London: Estates Gazette.

Klosterman, R.E., 1985. Arguments for and against planning, *Town Planning Review*, 56, 1, 5–20.

Klosterman, R.E., 1997. Planning support systems: a new perspective on computer aided planning, *Journal of Planning Education and Research*, 17, 45–54.

Klosterman, R.E., 2008. Comment on Drummond and French: another view of the future of GIS, *Journal of the American Planning Association*, 74, 2, 174–176.

Lord, A., and M. Tewdwr-Jones, 2012. Is planning 'under attack': chronicling the deregulation of urban and environmental planning in England, *European Planning Studies*, 22, 2, 345–361.

Masser, I., 1998. *Governments and geographic information*, London: Taylor and Francis.

Masser, I., and Craglia, M., 1996. A comparative evaluation of GIS diffusion, in local government in nine European countries, in Masser, I., Campbell, H., and Craglia, M. (eds.), *GIS diffusion: the adoption and use of geographic information systems in local government in Europe*, London: Taylor and Francis, 211–233.

Ministry of Infrastructure and Environment, 2017. The Environment and Planning Act: Explanatory memorandum, www.government.nl/topics/spatial-planning-and-infrastructure/revision-of-environment-planning-laws

Ministry of the Interior and Kingdom Relations, 2018. Scope and level of detail SEA report: National Environmental Planning Strategy, www.denationaleomgevings visie.nl/publicaties/default.aspx#folder=937064 (last accessed 16 February 2019)

Moore, G.C., 1993. Implications from MIS research for the study of GIS diffusion, in Masser, I. and Onsrud, H.J. (eds.), *Diffusion and use of geographic information technologies*, Dordrecht: Kluwer, 77–94.

National Careers Service, no date. *Job profile of a town planner: spatial planner, planner, urban designer, planning officer*, http://:nationalcareerservice.direct.gov.uk/job-profiles/townplanner (last accessed 16 February 2019)

Onsrud, H.J., and Rushton, G. (eds.), 1995. *Sharing geographic information*, New Brunswick: Centre for Urban Policy Research.

Openshaw, S., 1990. Spatial analysis and geographical information systems, in Scholten, H.J. and Stilwell, J.C.H. (eds.), *Geographical information systems for urban and regional planning*, Dordrecht: Kluwer, 153–163.

Ottens, H.F.L., 1990. The application of geographical information systems in urban and regional planning, in Scholten, H.J. and Stilwell, J.C.H. (eds.), *Geographical information systems for urban and regional planning*, Dordrecht: Kluwer, 15–22.

Pelzer, P., 2015. Usefulness of planning support systems: conceptual perspectives and practitioners' experiences, Ph.D. thesis, Department of Human Geography and Spatial Planning, Utrecht University.

Perloff, H.S, 1980. *Planning the post-industrial city*, Chicago/Washington: Planners Press/APA.

Pressman, J.L., and A.B. Wildawsky, 1973. *Implementation: how great expectations in Washington are dashed in Oakland*, Berkeley: University of California Press.

Rhind, D., 1992. Data access, charging and copyright and their implications for geographic information systems, *International Journal of GIS, 6,* 13–30.

Schuur, J., 1994. Analysis and design in computer-aided physical planning, *Environment and Planning B: Planning and Design, 21,* 97–108.

Thomas, D., J. Minnet, S. Hopkins, S. Hamnet, A. Faludi, and D. Barrel, 1983. *Flexibility and commitment in planning*, The Hague: Martinus Nijhoff.

Thornley, A., 1991. *Urban planning under Thatcherism: the challenge of the market*, London: Routledge.

Unwin, R., 1909. *Town planning in practice*, London: Fisher Unwin.

van der Cammen, H., R. Groeneweg, and G. van de Hoef, 1988. Randstad Holland, in H. van der Cammen (ed.), *Four metropolises in Western Europe*, Assen: Van Gorcum, 117–175.

Webster, C.J., 1993. GIS and the scientific inputs to urban planning: part 1: description, *Environment and Planning B: Planning and Design, 20,* 709–728.

Webster, C.J., 1994. GIS and the scientific inputs to urban planning: part 2: prediction and prescription, *Environment and Planning B: Planning and Design, 21,* 145–157.

Yeh, A.G.-O., 1999. Urban planning and GIS, in Longley, P.A., M.F. Goodchild, D. J. Maguire, and D.W Rhind (eds.), *Geographical information systems: principles, techniques, management and applications*, 2nd edition, abridged, Chichester: John Wiley and Sons.

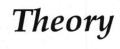

Theory

2

The Regional Research Laboratory Initiative

An Overview

Ian Masser

2.1 Introduction

The Regional Research Laboratory (RRL) Initiative was launched in 1987 by the British Economic and Social Research Council to establish regional centres of excellence in fields of data handling, database management, geographical analysis, software development, training and advice. It is one of the key components of the Council's corporate strategy for data resource management, a strategy that was first laid out in a joint ESRC/US National Science Foundation report, Large-scale Data Resources for the Social Sciences (1986).

The main objectives of the RRL Initiative are embodied in its title:

- the laboratory dimension indicates the intention to establish a resource base for research and policy analysis;
- the research dimension highlights the need for work on methodological issues relating to the management of large-scale data resources;
- the regional dimension recognises the need to develop local centres of expertise which are capable of meeting the distinctive requirements of the different regional research communities within the United Kingdom.

This chapter describes the main features of the trial and main phases of the Initiative and evaluates the experience that has been gained so far in relation to the three objectives reflected in its title. The last sections consider some of the possible long-term impacts of the RRL Initiative in relation to these objectives and compares the research strategy developed in the Initiative with that being implemented in the US by the National Center for Geographic Information and Analysis.

2.2 The Trial Phase

2.2.1 Overview

The trial phase of the RRL Initiative lasted from February 1987 to October 1988. It had an initial budget of £150,000. During this phase, four centres came into operation covering Scotland, Wales and South-west England, Northern England and South-east England. These involved multidisciplinary teams of researchers from Birkbeck College, London School of Economics, the Universities of Edinburgh, Lancaster and Newcastle, the University of Wales College of Cardiff, and Bath University Computer Services. All but the Scottish RRL involved split site operations.

The designated trial-phase RRLs were asked to develop and consolidate work in progress within the framework of the Initiative with a view to producing tangible products that could be evaluated by the wider research community at an early stage. The overall strategy underlying this phase was to build up operational experience and identify issues which would have to be taken into account in planning main phase operations (Masser 1988).

An important task for all concerned was to establish the corporate identity of the Initiative with reference to other organisations in the data resources management field. Of particular importance in this respect was the position of the RRLs relative to the ESRC Data Archive at the University of Essex. At an early stage in the trial phase agreement was reached whereby the RRLs would act as regional 'shop windows' for datasets acquired by the Archive regarding the procedures to be followed with respect to data acquisition, documentation and charging.

To promote collective thinking, regular co-ordination meetings were held to enable those involved to discuss the broader research and policy-related issues arising out of the work of the Initiative, and an electronic mail system was set up which included all the main documents relating to the work programme to facilitate internal communications between participants. A two-day workshop for RRL staff was also organised at the University of Edinburgh in September 1987 to broaden their range of hands-on experience with respect to various geographic information systems (GIS) software applications. This was timetabled so that staff could also participate in the ESRC/NSF state-of-the-art workshop on GIS that immediately preceded it.

2.2.2 Evaluation

The Laboratory Dimension: The experience of the trial phase vindicated the importance that had been attached to the concept of the well-found RRL by the ESRC (Masser and Blakemore 1988). It demonstrated the extent to which the development of facilities for data resources management involved not only the provision of hardware and software but also human resources in the form of academic and technical staff. As the seed-corn funding provided by ESRC enabled the appointment of only one full-time post in each of the four trial phase RRLs, a high level of commitment of existing staff within the institutions involved was a crucial factor in the creation of these centres. The response of the existing staff was extremely positive to the extent that each RRL had between six and ten staff actively contributing to its research efforts. Consequently, the effective resources at the disposal of the Initiative were in the region of 30 to 40 rather than the handful of ESRC funded posts. Some of this staff support was directly funded by the institutions themselves in order to help the RRLs get started in anticipation of future benefits stemming from RRL designation.

At the start of the trial phase there were considerable variations in the range of experience and the facilities that were available at the four RRLs. Staff at Birkbeck College and the University of Edinburgh were particularly fortunate in being able to draw upon earlier GIS experience with large mainframe packages. During the trial phase the other RRLs pursued various strategies with a view to obtaining extra facilities and extending their range of operational experience. The Northern England RRL, for example, persuaded the University of Newcastle to purchase the Arc/Info package and a number of related hardware facilities as an investment for the University as a whole. In contrast, the Wales and South-west England RRL relied heavily on home-grown GIS software and took the opportunity provided by the trial phase to build up its operational experience through a variety of applications projects.

The development of three of the four trial-phase RRLs as split site operations presented a number of problems with respect to the establishment of well-found laboratories in terms of the location of both staff and equipment. It also imposed an additional tier of management on those involved because of the need to make sure that account was taken of the broader interests of the participating institutions in the decision-making process. In the case of the Northern England RRL physical distance also proved to be a major obstacle to effective collaboration. Despite these problems, however, there were many indications of the benefits to be obtained by institutional collaboration, particularly where there was a complementarity of interests between the sites. In the case of the Wales and South-west England RRL, for example, the data-handling experience and mainframe facilities available at Bath University Computer Services gave a new dimension to the work of the mainly micro-oriented staff in Cardiff.

The Research Dimension: Given the shortness of the trial phase and the emphasis that was given to the production of tangible products, it was inevitable that its research output drew heavily on work that was already in progress before the RRL Initiative began. The Scottish RRL built on its experience in data library and software development with a view to improving links between users and software and between different software systems. The Wales and South-west England RRL drew upon earlier work involving planning applications and user services to mount demonstration projects in the health care and rural planning fields. The Northern England RRL made use of its experience in spatial analysis and software applications to develop techniques for spatial data integration with special reference to the requirements of the users of large datasets, while staff at the South-east RRL built upon earlier work on planning and related GIS applications in the development of an integrated settlement and infrastructure database for South-east England.

RRL designation provided a focus for existing research activities in the institutions involved and stimulated collaboration between research groups from different disciplines. This had a catalytic effect on existing staff who were often forced to rethink their own research priorities within the framework of the Initiative as a whole and the benefits of this were immediately apparent in the emergence of new projects that exploited the facilities being developed at the RRLs.

Because of the geographical and planning background of the staff in the four sites that were finally selected for the trial phase, there was an emphasis on GIS applications within the general field of data resources management from the outset. These activities were given a substantial boost by the publication of the Chorley Report on Handling Geographic Information (Department of the Environment 1987), less than three months after work began on the trial phase. This report and its accompanying video had an immediate impact in raising the overall level of public awareness of the potential presented by GIS technology and its set of 64 recommendations prompted a speedy and generally favourable response from the Government in February 1988 (Department of the Environment 1988).

The overall development of the RRL Initiative was influenced by the Government's decision to reject the Chorley Committee's recommendation to establish a national Centre for Geographic Information to act as a forum for the various interest groups involved and promote the use of GIS technology. The Government rejected this recommendation on the grounds that existing organisations including the RRLs and the Data Archive were already starting to meet these requirements and should be encouraged further to develop their roles in this area in order to cover the whole range of applications (Department of the Environment 1988, 5, para. 64). However, despite the positive tone of this recommendation, the Government did not see an immediate need for additional funding to be made available for these purposes.

The value of focusing on GIS applications as an important subset of large dataset management can be seen in the impact that the outputs from the trial phase had on the wider research community. The tangible products and the operational GIS experience gained during the trial phase enabled the RRLs to provide potential users with much needed information about the relative advantages and disadvantages of particular GIS and database management systems for different types of application fields. This body of experience was felt to be a valuable national resource in a situation where technology is rapidly changing and the amount of public and private sector investment that is involved is considerable by any standards.

The Regional Dimension: The four trial-phase RRLs covered only part of the UK area and the special interests of the regional research communities in the Midlands and Northern Ireland in particular were not directly taken into account at this stage of the Initiative. The view was also expressed by officials at the Department of the Environment that not enough attention was being given to the major urban concentrations that represented key policy-making and decision-making areas for many government departments. Inevitably, therefore, the experience of the trial phase with respect to the regional dimension was partial in nature. Nevertheless, a number of positive points emerged from the contacts between the RRLs and their regional client communities which generally confirmed the importance attached to the regional dimension by ESRC and the extent to which it is this dimension that gives it its distinct character.

First, the experience of the trial phase indicated that there were active regional research communities in all four areas covered by the RRLs. As might be expected, these regional communities were particularly well developed in Scotland and Wales. The extent to which the former acts as an independent entity with respect to the collection of major datasets such as the Census of Population was obviously an important factor in building up such a community. The development of these regional communities was also substantially assisted by the efforts of the Scottish and Welsh Offices in both regions. The Wales and South-west England RRL in particular was given considerable support by the Welsh Office for the development of its research activities. In England, inevitably, the picture that emerged was less sharply focused, given that standard datasets are available for England as a whole. However, even in this case the impact of the North–South divide was clearly apparent in the interests of the local research communities and the priorities that were given to particular fields of research. Nevertheless, there was a tendency for the English RRLs also to develop specialist application fields such as settlement infrastructure or emergency planning with a view to providing national level services alongside purely regional projects.

Second, the experience of the trial phase demonstrated the length of the lead time that is required in situations of this kind to make a noticeable impact upon the regional research community. At the outset, there was a shortage of tangible products at the disposal of the RRLs which could be demonstrated to their potential client communities. Furthermore, it should be borne in mind that RRLs were also new institutions which had to go through some form of collective learning process to establish their own objectives with reference to their regional client communities. As a result, it was not until the later stages of the trial phase that most of them were in a position to discuss their experience and identify the most fruitful areas for client-based research. The experience of individual RRLs in connection with their regional research communities was also paralleled at the national level. It was not until the end of the trial phase that the Initiative as a whole was able to take on the task of presenting its findings to the wider research community through the medium of a special meeting at the Royal Geographical Society and the publication of a special GIS issue of the ESRC Newsletter in October 1988 (ERSC 1988).

2.3 The Main Phase

2.3.1 Overview

The main phase of the RRL Initiative began in October 1988 and will continue until December 1991. As a result of the positive evaluation of the experience of the trial phase, the sum allocated by ESRC for the main phase was substantially increased to £2m and eight RRLs are now in operation covering the whole of the United Kingdom. Within England separate centres have been established to meet the needs of regional client communities in the Midlands, the North East, the North West, the South East and the South West (as before in established to meet the needs of regional client communities in the Midlands, the North East, the North West, the South East and the South West (as before in conjunction with Wales). The RRLs involve research teams based at Birkbeck College at the University of London, Queen's University of Belfast, the University of Wales College of Cardiff, Bath University Computer Services and the Universities of Edinburgh, Lancaster, Leicester, Loughborough, Newcastle and Ulster. A separate RRL involving a consortium from Liverpool and Manchester Universities has also been set up to provide additional expertise on urban research and policy evaluation matters. In addition to these sites, related research projects are being carried out at the ESRC Data Archive and London School of Economics.

The main phase RRLs cover the whole of the United Kingdom. Five of the sites are based on trial centres as Lancaster and Newcastle decided to operate as separate RRLs for the North West and North East respectively in the light of their experience of the trial phase. All three new centres involve split site operations. Two of these are based on a consortium of four different organisations. In addition to Queen's University and the University of Ulster, the Northern Ireland RRL involves two special research groups, the Northern Ireland Economic Research Centre and the Policy Research Institute. Similarly, the Urban Research and Policy Evaluation RRL involves the Greater Manchester Research and Information Planning Unit and Merseyside Information Service as well as staff on the Universities of Liverpool and Manchester.

There are a number of important differences between the nature of the main phase of the RRL Initiative as against the trial phase. The most striking difference is in the resources that have been allocated for the main phase by ESRC in the light of its evaluation of the trial phase. Even allowing for the establishment of eight as against four centres in the trial phase and the fact that the main phase will last for twice the length of the time allocated for the trial phase, the average amount allocated to each RRL per month during the main phase is broadly the same as that allocated per month to the entire Initiative during the trial phase.

Another distinctive feature of the main phase is the special facilities that have been provided for each RRL both through the ESRC sources and as a result of the attraction of support from other bodies. An RRL-wide study contract has been awarded by IBM UK Ltd to provide each RRL with a fully equipped IBM PS/2 80 work station. This has been supplemented by ESRC with a view to enable RRLs to purchase extra equipment and GIS software including PC Arc/Info, PC MAPICS and PC GIMMS. It has been estimated that the market value of this package is over £800k alone.

There is also an important difference between the trial phase and the main phase in the amount of effort that is being given to general corporate activities. The main phase is regarded by ESRC as a tightly coordinated initiative to the extent that all the RRLs are contractually committed to collaboration with respect to matters relating to publicity, image and presentations at conferences and seminars. The resources at the disposal of the Initiative for collective brainstorming and evaluation activities have also been substantially increased and limited funds have been made available to enable special projects to be commissioned in the interests of the Initiative as a whole. The RRL Steering Committee set up by ESRC has also been substantially enlarged in scope and individual members have been asked to maintain regular contacts with specific RRLs. Dissemination-related activities are being given a very high priority. Each RRL has produced its own series of technical reports and ESRC has allocated additional resources to commission a 17-minute video entitled Mapping the Future: The Regional Research Laboratory Initiative (ESRC 1989).

The last important difference between the main phase and the trial phase of the RRL Initiative is that ESRC has made it clear to the RRLs that there will be no further direct funding from them when the main phase ends in December 1991. By this time ESRC anticipate that its seed-corn funding will have enabled the RRLs to get off the ground and that they will be capable of financing their own operations. With this in mind all the RRLs were asked to include business plans which indicated how they intended to achieve self-financing status in their main phase applications.

2.3.2 Evaluation

The main phase of the RRL Initiative has been in operation for only one year, but nevertheless a number of issues have already emerged which merit further consideration under the headings of the laboratory, research and regional dimensions inherent in its title.

The Laboratory Dimension: The resources provided by ESRC for the main phase have enabled each RRL to make up to three full-time appointments in the database management, software development and spatial analysis fields for the duration of the main phase. As a result, the size of each RRL team has increased to between 8 and 12 persons and the total effective resources at the disposal of the Initiative have increased to somewhere between 70 and 80 persons. As in the trial phase, some additional staff have been directly funded by the institutions involved to get the RRL off the ground or to maintain the momentum built up during the trial phase.

Some RRLs have also been successful in obtaining additional facilities to build up their technical resources. The University of Lancaster, for example, has purchased the mainframe Arc/Info package and three Sun 3/60 workstations to build up the north-west RRL's range of services and Birkbeck College has been designated as an Apple Mapping Centre. This had already paid dividends for the latter in the development of a portable Geographical Information Systems Tutor (GIST), which is now being marketed commercially (Raper and Green 1989).

Despite the increase in staffing resources in the main phase there are still severe pressures on both short-term and tenured staff within the RRLs. These pressures have been aggravated by the difficulties faced by some RRLs in holding contract staff with GIS experience in the face of a highly competitive market. With these considerations in mind the first of a series of meetings specifically for RRL contract staff was held in Sheffield in June 1989. The objective of this meeting was to bring them more directly into the overall programme of the Initiative and to promote the exchange of ideas and experience between RRLs as well as fostering the view that these staff will be the seedbed of future GIS developments.

Another issue which has already emerged in the main phase is the amount of time that needs to be devoted to overall management and dissemination activities as against the more traditional research and teaching commitments of academic staff. The pressures that this creates on the Executive Directors of the RRLs was recognised in the evaluation of the trial phase experience and a number of special arrangements were made in several main phase contracts to enable key individuals to be relieved of their normal teaching and administrative duties to devote more time to management issues. In general, these arrangements have proved successful in reducing the pressure on the individuals concerned but the extent to which they must also develop new skills in order to carry out these duties was largely ignored. With this in mind a special workshop on business planning techniques was held at the

University of Stirling in June 1989 for RRL Executive Directors. This not only introduced Directors to business planning methods but also performed a valuable brainstorming function regarding the overall objectives of the Initiative as a whole.

The research dimension: Given the additional resources and the extra time that are available, the RRLs have been able to develop substantial research programmes for the main phase of the Initiative which build upon their existing strengths in data resources management. As in the trial phase, the Scottish RRL is particularly involved in software development-based activities but will also be concerned with the application of GIS methods in the fields of education and labour market entry, rural land use change and locational analysis. Similarly, the Southeast RRL continues to develop its settlements and infrastructure database as the centrepiece for its research efforts but in the main phase it will also use this as a resource for policy-based research and as a means for attracting potential contracts in GIS-related fields. Of the two RRLs constituting the former Northern England RRL, the North-east RRL is consolidating the project-based approach that it developed in the trial phase alongside technical research on policy relevant GIS software and urban and regional development issues where the group has particular research strengths. The particular interests of the North-west RRL lie in fundamental research associated with development of GIS in emergency planning, landscape ecology and pollution control with particular reference to relational database management systems. The strategy of the Wales and South-west England RRL focuses on a range of methodological issues associated with the provision of data management services, spatial-modelling and decision support systems in a number of substantive areas including housing, health and rural resource management.

Of the new RRLs, the Midlands RRL is giving high priority to establishing a Midlands regional database which acts as a core facility for work on data linkage, spatial representation, human computer interfaces and multimedia demonstration systems for specific client groups. The Northern Ireland RRL is adopting a project-based research strategy with an emphasis on policy-related research in economics, environmental studies and social policy with particular reference to the management of a wide range of largescale data sets. The Urban Research and Policy Evaluation RRL, on the other hand, is focusing its attention on a number of broadly defined policy issues associated mainly with urban regeneration and service delivery in the context of Liverpool and Manchester. These include monitoring and evaluation, the construction of needs indicators, land development and housing, and labour markets.

Alongside these individual RRL research programmes, a number of small-scale research activities have been commissioned as part of the overall work of the Initiative. These are mainly exploratory in character and seek to identify emerging issues which are likely to come on the research agenda towards the end of the main phase. These include a state-of-the-art review (Openshaw 1990), which sets out some priorities for future British research efforts in the spatial analysis field. Further state-of-the art papers have been commissioned to explore the potential for applications-related research in the fields of education, environment, geodemographics, health and hazards in organisations, and settlement and infrastructure. Special attention is also being given at an Initiative-wide level to evaluation research, given the unique opportunity that is presented by the installation of identical system configurations in each of the eight RRLs (Plummer 1990). Work has also begun on the issues that must be tackled when evaluating the impact of GIS and similar large-scale database management systems on organisations (Campbell 1990).

Since the main phase of the RRL Initiative got under way a further boost to GIS-related research in Britain has been given by the Advisory Board for Research Council's decision to allocate a further £1.05m to ESRC and NERC for a three-year joint research and training programme in the field of Geographic Information Handling. The majority of this money is being used to commission a wide range of underpinning research projects and GIS applications research which straddle the boundaries between socioeconomic and environmental research. As a result of these developments, the North-east, South-east and Scottish RRLs have already benefited by a significant supplementation of the Research Council funds at their disposal.

A conspicuous feature of all these research activities is the extent to which the development of GIS facilities has blurred conventional boundaries between socioeconomic and environmental applications and led to a substantial degree of collaboration between human and physical geographers. Because of the degree to which the methodological issues that are involved in large-scale datasets management override conventional disciplinary boundaries, RRL staff have found themselves increasingly categorised in terms of their methodological interests rather than their study fields.

The regional dimension: It is still too early in the main phase to be able fully to assess the impacts that the RRLs are likely to have on their respective regional research communities. The experience of the trial phase highlighted the long lead times involved in getting facilities off the ground. This is particularly important for the three new RRLs which did not have the benefit of the trial phase experiences.

The way in which individual RRLs will develop their activities in conjunction with their regional client communities is linked to their plans for future funding, and some additional support is likely to be required if they are to offer the full range of information and advisory services that was envisaged for them in the Government's response to the Chorley Report. Despite these concerns, the RRLs are becoming increasingly prominent within their regions and there is a growing demand for their services. The extent of this demand as well as the growth of current interest in their activities is particularly apparent in the response of both public and private sector organisations to the launches of the Midlands, North-east, North-west and Southeast RRLs during late 1988 and early 1989. As the example of the Wales and South-west England RRL indicates, the importance of agencies such as the Welsh Office cannot be underestimated in this respect and in Northern Ireland this is likely to be a factor of even greater significance because of the unique circumstances surrounding data collection and provision in the province.

2.4 The Long-Term Impact of the RRL Initiative

2.4.1 The Laboratory Dimension

By the end of the main phase in 1991, the RRLs will have built up a strong resource base for research and policy analysis throughout the United Kingdom. They will have not only an impressive range of technical facilities at their disposal but also teams of researchers with a wide variety of operational experience in the GIS and data resources management fields. Together with the ESRC Data Archive, the RRLs will be the vital nodes in the national network for largescale data resources management. In this way they are likely to have a profound effect on the development of British research in this field during the 1990s.

However, some concern must be expressed about the extent to which the RRLs will be able to provide the full range of advisory and information services envisaged for them in the Government response to the Chorley Report without further support from public sources after the end of the main phase. If they are to provide a comprehensive resource for the wider research community which includes not only academic bodies but also public and private sector organisations, they must be recognised as such and given the basic core funding that they will require for this purpose.

2.4.2 The Research Dimension

The resources given to the RRLs together with the opportunities for research that have been created within the Initiative give British researchers a head start in the GIS research field. As a result, they have been able to build up vital operational experience in large-scale data resources management at a relatively early stage in the development of this field. The long-term impact of this experience will be apparent in a variety of ways. It will be evident in the scale of future research operations in the GIS field in the UK. This is already apparent in the 135 outline bids received by the joint ESRC/NERC programme on Geographic Information Handling as a result of its call for proposals in spring 1989. It will also be apparent in the nucleus of skilled researchers that is represented by the staff recruited by the RRLs. By the end of the Initiative, there will be 20 to 30 highly experienced staff who will inevitably be in the forefront of the next phase of GIS research and development. RRL designation is also having its effect on PhD research and curriculum development. The fruits of this will be reflected both in higher levels of general awareness of current developments in data resources management and in a greater degree of involvement at all levels than would otherwise be the case.

The RRL Initiative in the medium to long term is also likely to enhance the richness and vitality of large-scale data resources management research activities as a whole. The long-term impact of between 70 and 80 persons working alongside one another over a three year period in an integrated research initiative cannot be underestimated. The Initiative has also been given the resources to draw up and develop agendas for future research. Such activities are not restricted to RRL staff and a deliberate attempt is being made to involve key individuals from the wider research community in these discussions. In this way the RRL Initiative can be regarded as providing a seedbed for future research efforts.

2.4.3 The Regional Dimension

The RRLs have had a major impact on their regional research communities in raising overall levels of awareness about the issues involved in data resources management. It is already clear that the cumulative impact of the collective efforts is much greater than the sum of a number of individual efforts to raise levels of awareness. They have also had facilities at their disposal for this purpose that would have been impossible other than in the framework of a co-ordinated research initiative.

2.4.4 Value for Money

It is also necessary to ask the question whether the investment of over £2 million by the ESRC in one of its largest research initiatives so far will give value for money. Could, for example, the same impacts have been achieved at a lower cost by funding a number of independent projects rather than through an integrated initiative?

In overall terms, the findings of the above evaluation are very positive from the standpoint of value for money. As a result of the resources that have been placed at their disposal, British researchers have been given a head start in a rapidly developing field. A resource base is being established which will be of considerable value to the whole research community for many years to come. A special feature of this is the extent to which it takes account of the needs of the various regional research communities in the United Kingdom.

Could this have been achieved by other means than a co-ordinated initiative? The answer to this must unequivocally be 'no'. The advantages of a co-ordinated initiative are apparent in relation to the achievement of all three objectives. In the laboratory dimension the high profile attached to the Initiative as a whole has been a crucial factor in negotiations with data suppliers and equipment vendors. An integrated initiative also opens up possibilities for collective learning and comparative evaluation that would not occur to anything like the same extent in a series of independently funded projects. From the research standpoint, it is unlikely that the same opportunities could have been created by other means for strategic thinking about the future research agenda. Finally, the benefits of an integrated initiative from the regional standpoint are probably greatest of all in overcoming the isolation of many researchers and ensuring that the unique features of different regions are taken into account in national research efforts. In this way, the regional research communities have the best of both worlds by being given resource bases which take their special needs into account while linking them into a broader national network of research facilities.

2.4.5 A Comparison of British and American Approaches to GIS Research

In conclusion, it is useful to compare the research strategy that has been developed in the RRL Initiative with the strategies that are being implemented by research councils in other countries. The experience of the US National Science Foundation's National Center for Geographic Information and Analysis is particularly interesting in this respect. In October 1988, the National Science Foundation agreed a contract worth $5.5 million spread over five years with a consortium consisting of the University of California at Santa Barbara, the State University of New York at Buffalo and the University of Maine at Orono to undertake an integrated research programme with respect to 12 priority fields (NCGIA 1989). The most striking differences between the two initiatives are in the style of operations and the character of the research that is in progress. In contrast to the NCGIA, the RRL Initiative is very much a bottom-up initiative which gives individual RRLs a considerable amount of freedom within the overall programme to respond to the specific needs of their regional research communities. This dimension is reinforced by the need for each RRL to achieve some measure of self-financing through its own efforts by the end of 1991. As a result, much of the research that is being undertaken involves collaborative ventures with both public and private sector agencies. These strategic alliances highlight the extent to which GIS is very much an applications-led technology.

2.5 The RRL Initiative – An Evaluation

Due to the interest that had been stirred up by the RRL initiative the UK Economic and Social Council (ESRC) decided to evaluate it a year and a half before it was formally completed and set up a panel chaired by Professor Peter Burrough to carry out this task. Somewhat unusually, a summary of the main findings of this evaluation was published in a leading GIS journal (Burrough and Boddington 1992). The reasons for this step are explained in their report.

> The Regional Research Laboratories generated so much enthusiasm and productivity among the researchers at the various sites that in October 1990, the ESRC commissioned an evaluation of the Regional Research Laboratory Initiative from its newly established Research Evaluation Division (RED), even though the initiative was only two thirds of the way through the programme for the Main Phase.
>
> *(Burrough and Boddington 1992, p. 426)*

Consequently, it was felt that 'Nevertheless, the scale of RRL activities, the outputs to date and the national and international impact achieved, to be reviewed in more detail below, demonstrate that significant results were quickly achieved' (Ibid, p. 430).

The findings of the evaluation were impressive even at this stage in the work programme. Burrough and Boddington's (p. 428) analysis showed:

> The RRLs have been very prolific in generating publications, attributing some 580 printed publications to the initiative and 336 papers presented at conferences. The numbers demonstrate clearly how the initiative has collated together and boosted the production of both work in progress and new work. Such was the excitement and momentum associated with the initiative that many associated researchers were keen to have their work recognized under the RRL brand name.
>
> *(p. 430)*

The initiative was also very successful in attracting resources to carry out its tasks.

> This competitive funding process has been largely successful in getting the RRLs started and in generating extra resources. By September 1990, there were 151 research staff contributing to the RRLs, three times more than the ESRC RRL initiative funds could directly support. The 151 staff represented a time commitment of 76 full-time equivalents (FTEs) (36 per cent of RRL staff were allocated to the RRLs full time) of which 44 FTEs were funded from various ESRC sources – 23 FTEs through the RRL initiative itself.
>
> *(p. 428)*

The initiative also generated a book of essays entitled 'Handling Geographic Information: methodology and potential applications' (Masser and Blakemore 1991). The gives a good overview of the issues that were being tackled by the RRLs. The methodology section contains five papers by authors such as Stan Openshaw, Robin Flowerdew, David Maguire, and Jonathan Raper. The applications section examines applications issues relating to information handling in four potential applications fields: environmental monitoring (David Rhind), managing natural and technological hazards (Anthony Gatrell and Peter Vincent), settlement planning (John Shepherd) and geodemographics (Peter Brown). The final chapter by Heather Campbell rounds off the discussion with an examination of the organisational issues that need to be taken into account in the development and implementation of all GIS applications.

These topics also feature prominently in the statistics collected during the evaluation process.

> About 50 per cent of RRL research projects involve applied research. Of these, most effort was concentrated in geodemographics, employment and education (27 per cent), in health (22 per cent) and in planning and the environment (19 per cent). Just 15 per cent of projects involve basic research into methodology, including fuzzy demographics, object-oriented GIS and knowledge-based spatial database systems. In all, 33 per cent of projects involved technical software development.
>
> *(p. 433)*

Many of these projects had a strong regional dimension, reflecting the overall objectives of the initiative. According to the evaluation report, 'All RRLs are conducting regionally based projects with at least one third of the research projects and nearly half of consultancies being regionally based. Hence the initiative has successfully achieved a regional focus' (p. 432).

Given the overall thrust of the evaluation, it is not surprising to find that the overall conclusion of the evaluators can be summarised in the following terms.

> In launching the RRL Initiative, the ESRC made a bold shot in the dark and scattered small laboratories across the U.K. in the hope that they would become well found and permanent without long-term ESRC support. The investment was timely: it caught the wave of international interest and the main period of technical development. The response from the universities was enthusiastic and many researchers, particularly those just starting their careers, have made strenuous efforts to ensure that their RRL succeeded.
>
> *(p. 440)*

Nevertheless, it is interesting to note that there was a high degree of mobility amongst RRL in the years immediately following the end of the initiative. The findings of an analysis carried out in 1993 (Masser and Hobson 1993) show that 39 ex-RRL contract staff had found new positions during the main phase of the initiative in the 18 months since the end of the main phase. These positions were not restricted to academic posts but included a wide range of public and private sector appointments. In addition to moves by contract staff there had been a surprisingly high degree of mobility among the grant holders themselves. These included Mike Batty (Wales) to become Director of the NCGIA, Buffalo in 1990, Neil Wrigley (Wales) to a chair at the University of Southampton in 1991, Stan Openshaw (Newcastle) to a chair at the University of Leeds, and David Rhind (Birkbeck College, London) to become Director General and Chief Executive of Ordnance Survey, both in 1992.

It should also be noted that the first GIS Research UK (GISRUK) Conference was held at the University of Keele in 1993. Since then the GISRUK international conference series has grown out of the UK's national GIS research conference. GISRUK conferences are primarily aimed at the academic community and attract those interested in Geographical Information Science and its applications from all parts of the UK, together with the European Union and beyond. A selection of the papers presented at the 25th GIS Research UK conference at the University of Manchester in April 2017 can be found at Huck (2018).

2.6 Final Thoughts

Looking back on the RRL initiative more than 10 years later, Flowerdew and Stillwell (2003, p. 381) offer some pertinent reflections on the initiative as a whole. 'At the end of the main RRL funding in 1991, the RRLs remained in existence but have followed different paths, some effectively disappearing while others have grown in size through a mixture of strategies.' With the benefit of hindsight, they feel:

> there were always some ambiguities and contradictions in the roles that they were expected to play, reflected in part in the name. Part of the original idea was that each RRL would be particularly responsible for collecting and providing access to datasets about its region, and for liaison with potential clients in that region. This function worked reasonably well in the RRLs with responsibility for Scotland, Wales and Northern Ireland, where there were government and commercial offices with devolved responsibilities with which they could work. The English RRLs, however, tended to work with datasets and organisations at the national level, and although they may have built up closer relationships with local government or other organisations in their own regions, their roles quickly became demarcated in terms of subject specialisation rather than region.

Flowerdew and Stillwell (2003, p. 382) also highlight the inherent ambiguity regarding the role of RRLs.

The requirement to become self-supporting appeared to demand a concentration on commercial and consultancy work. However, potential 'users' of the research, especially those from the private sector, were very concerned that the RRLs might become subsidised competitors for their own companies.

In practice, this could cause conflicts with the requirement for RRLs to conduct basic research. Consequently, it might be argued that the RRLs ended up trying to be all things to all men.

References

Burrough, P. A. and A. Boddington, 1992, The UK Regional Research Laboratory Initiative 1987–1991, *International Journal of Geographical Information Systems*, 6, 5, 425–440.

Campbell, H., 1990, The Organisational Implications of Geographic Information Systems for British Local Government, paper presented at the European GIS Conference, Amsterdam.

Department of the Environment, 1987, *Handling Geographic Information*, Report to the Secretary of State for the Environment of the Committee of Inquiry into the Handling of Geographic Information, HMSO, London.

Department of the Environment, 1988, *Government Response to the Report of the Committee of Inquiry into the Handling of Geographic Information*, HMSO, London.

Economic and Social Research Council, 1988, Working with GIS, *ESRC Newsletter, No. 63*, ESRC, London, pp. 5–29.

Economic and Social Research Council, 1989, *Mapping the Future: The Regional Research Laboratory Initiative*, video film, ESRC, Swindon.

Economic and Social Research Council/National Science Foundation, 1986, *Largescale Data Resources for the Social Sciences*, ESRC, London.

Flowerdew, R. and J. Stillwell, 2003, Undertaking applied GIS and spatial analysis research in an academic context, in J. Stillwell and G. Clarke (eds.), *Applied GIS and Spatial Analysis*, Berlin: Springer.

Huck, J., 2018, 25 years of GISRUK, *Journal of Spatial Information Science*, 16, 53–56.

Masser, I., 1988, The Regional Research Laboratory Initiative: A Progress Report, *International Journal of Geographical Information Systems*, 1, 2, 11–22.

Masser, I. and M. Blakemore, (eds.), 1988, *The Regional Research Laboratory Initiative: The Experience of the Trial Phase*, ESRC, London.

Masser, I. and M. Blakemore, (eds.), 1991, *Handling Geographical Information: Methodology and Potential Applications*, Longmans Scientific and Technical, Harlow.

Masser, I. and C. Hobson, 1993, Where Are They Now? The Diffusion of RRL Staff, *Area*, 25, 279–282.

National Center for Geographic Information and Analysis, 1989, The Research Plan of the National Center for Geographic Information and Analysis, *International Journal of Geographical Information Systems*, 1, 3, 117–136.

Openshaw, S., 1990, A Spatial Analysis Research Agenda for GIS, *Regional Research Laboratory Initiative discussion paper No. 3*, Department of Town and Regional Planning, University of Sheffield.

Plummer, S.E., 1990, The ESRC Regional Research Laboratories: A Technical Profile, *Regional Research Laboratory Initiative discussion paper No. 2*, Department of Town and Regional Planning, University of Sheffield.

Raper, J. and N. Green, 1989, *The Geographical Information Systems Tutor*, Department of Geography, Birkbeck College, University of London.

3

Data Integration Research

Overview and Future Prospects

Ian Masser

3.1 Introduction

> All GIS experience thus far strongly suggests that the ultimate value is heavily dependent on the association of one data set with one or more others, thus in the EEC's CORINE project, ... the bulk of the success and value came from linking data sets together.

This quotation from one of David Rhind's papers (Rhind 1992, p. 16) is one of many similar statements by leading authorities in the field that highlight the vital importance of data integration for the effective utilisation of modern geographic information handling technologies. For this reason, the original GISDATA programme proposed and identified a number of new research opportunities associated with the increasing need to integrate geographic information collected by a wide range of different national and international agencies for a great diversity of purposes.

In several important respects the work carried out as part of the data integration stream in the GISDATA programme differs from that undertaken in the other two research streams. Most importantly, data integration is a political and organisational as well as a technical matter. This can be seen particularly when data integration is looked at from the standpoint of information sharing. As Onsrud and Rushton (1995, p. xiv) point out:

> The value and social utility of geographic information comes from its use. Sharing of geographic information is important because the more it is shared, the more it is used, and the greater becomes society's ability to evaluate and address the wide range of pressing problems to which such information may be applied. Failure to share geographic information is also economically inefficient and wasteful. The expertise and time it takes to collect and maintain information about land creates a need to share that information.

Consequently, there is a strong policy as well as a research dimension in the work that has been undertaken on data integration in the GISDATA programme, and this is evident in the tangible products that have been produced.

With these considerations in mind, this chapter reviews the work carried out on data integration in this programme and evaluates its achievements with respect to the research agenda that is likely to emerge during the next 5 to 10 years.

3.2 Data Integration in the GISDATA Programme

The work of the GISDATA programme has been largely organised around a series of specialist meetings on particular topics. Four specialist meetings were organised as part of the data integration stream during the life time of the programme. The first of these was held in October 1993 and the last took place in May 1996.

The sequence of these meetings was as follows. The first meeting was essentially a scene setting meeting which explored some aspects of the diffusion of GIS with particular reference to local government in Europe. The primary objectives of this meeting were to put together a comprehensive picture of the level of adoption of GIS in a key applications sector and to identify some of the factors facilitating and/ or impeding adoption in various countries. The second meeting tackled some of the political and organisational issues associated with data availability. Its main objective was to produce a position statement on these matters on behalf of the GISDATA Research Community as an input into the discussions taking place at the European level on the further development of the research resources infrastructure. The third meeting in this steam considered one important facet of the data explosion that is currently taking place as a result of the growing use of digital data and the development of new forms of data collection technologies. It examined the opportunities that are being created for monitoring urban change as a result of the use of new forms of airborne and satellite remote sensing technologies. The final meeting was planned so that it made a direct input into the debates that have been taking place at the European level and the need for a European geographic information infrastructure. This involved the preparation of a further GISDATA position statement on these issues as well as a systematic evaluation of the experiences of the creation of trans-national databases.

Given that there are some overlaps between the subject matter discussed in the second and fourth specialist meetings, the presentation below does not strictly follow this sequence. It discusses the findings of the third meeting on urban remote sensing before those of the two specialist meetings on data availability and data policies.

3.2.1 GIS Diffusion

There is a substantial body of literature on the diffusion of innovations that indicates the degree to which their adoption is governed by social and cultural factors as well as technical considerations (see, e.g., Rogers 1983). Consequently, the study of GIS diffusion processes is of considerable interest for two main reasons. First, it indicates the extent to which the new technologies have been taken up by prospective users. This shows the capacity that has been built up for data integration and sharing information within a particular user community or within a specific country. Second, it also draws attention to the contextual factors that facilitate and/or inhibit the diffusion process in particular circumstances. This points to some of the policy implications arising out of the diffusion process.

GIS diffusion differs from much of the existing literature on diffusion in one significant respect: the technology is not adopted by an individual user in this case but by an organisation or a section in an organisation. Consequently, it is necessary to study not only the adoption of GIS by organisations but also the way in which the technology is subsequently adapted and reinvented to meet the specific needs of each organisation. This implies that the outcomes of adopting GIS technology can be very different even within similar organisations carrying out similar duties (see, e.g., Campbell and Masser 1995).

The processes of GIS diffusion in Europe were explored with reference to the experience of local government in nine European countries during a specialist meeting at Knutsford in October 1993. Local government was chosen as the focus for the discussion because it is a key player in the development of the GIS industry in most countries and also because there is a strong national dimension to local government which is evident from the expectations that people have with respect to local democracy, the tasks it carries out and the professional cultures that have come into being to support them.

A book was subsequently published as a result of this meeting (Masser et al. 1996). This consists of case studies of the nine countries together with two other contributions which explore the broader theoretical issues associated with diffusion research. The case studies of Denmark, Germany, Italy, Portugal, and the United Kingdom make use of similar survey methodology, whereas those of France, Greece, the Netherlands and Portugal present a number of alternative perspectives on the diffusion process. The last section of the book is devoted to a comparative analysis of these experiences. The main findings of this analysis can be summarised as follows. In all the countries surveyed GIS is a relatively recent phenomenon in local government and there are relatively few examples of the adoption of the technology before 1990. Consequently only a limited amount of experience has been built up so far to support the claims that GIS implementation will result in improved information processing and more informed decision making.

Two contextual factors appear to account for most of the differences between the case study countries. One of these measures the links between digital data availability and the level of GIS adoption. The question of digital data availability is not simply a matter of the information rich versus the information poor. It is much more a question of central and local government attitudes towards the management of geographic information. Consequently, countries with relatively low levels of digital data availability (and GIS adoption) tend also to be countries where data sources are fragmented in the absence of central government coordination. Conversely, those countries with relatively high levels of digital data availability (and GIS adoption) tend also to be countries where governments have taken steps to a locate responsibilities and resources for the collection of management of geographic information.

The other factor reflects the professional cultures surrounding GIS applications in local government. This is closely linked to the responsibilities that have been given to local government in different European countries. In countries where local governments play an important role in maintaining land registers and/or cadastral systems, a highly organised surveying profession has come into being to carry out these tasks. Consequently, land information systems rather than geographic information systems tend to predominate in local government applications in these countries. Conversely, in countries such as Great Britain, where land registration and digital topographic mapping are dealt with centrally, there is a greater emphasis on applications which are broadly linked to the various planning activities of local government. In this case the predominant professional cultures tend to be those of the town planner and, to a lesser extent, the transport engineer and geographic information systems rather than land information systems tend to occupy central place.

3.2.2 The Data Explosion

One of the main consequences of recent developments in information and communication technologies has been a considerable increase in the number and type of data sources that are potentially at the disposal of the user. This is largely due to two factors. The first of these reflects the emergence of new kinds of data collection methodologies. These include the widespread use of global positioning systems to collect locational data, the use of video cameras for routine surveillance purposes and the employment of airborne and satellite remote sensing technologies to monitor routine events on the Earth's surface. The amount of data collected over the last two decades by the latter alone is enough in its own right to justify the use of the term "data explosion" (see, e.g., Dureau and Weber 1995).

The second factor reflects the near universal use of electronic data processing techniques for routine administrative tasks throughout both the public and private sectors. Examples of data of this kind include statistics about traffic accidents, criminal offences, purchases of consumer goods or financial products, and medical records. Given that most of this data has a geographic reference attached to it in the form of an address the potential for integrating data from sources of this kind is enormous. It should be noted that this data largely consists of individual personal records collected under varying levels of confidentiality rather than aggregate information. Consequently, these developments have given rise to a growing concern about the need to regulate such activities to protect personal privacy particularly when data of this kind is used for purposes other than those under which it was obtained. This in turn raises important ethical questions regarding the use of data matching technologies in both the public and private sectors to target individuals and/or to reduce the number of fraudulent claims (Data Protection Registrar 1996).

From this brief discussion it can be seen that issues relating to data matching go way beyond the geographic information field and most of them are being debated in general terms rather than with specific reference to geographic information. Consequently, it was decided to concentrate the discussion in the data integration side not on these generic issues but on data which is essentially geographic in nature: i.e., the collection of spatial data using airborne and satellite remote sensing technologies. Given the emphasis on data integration it was also decided to concentrate the discussion on the use of urban remotely sensed data for monitoring urban change and urban analysis. This was the subject of a specialist meeting on remote sensing and urban change in Strasbourg in June 1995. The main objectives of the meeting were to consider the impact of recent technological developments with particular reference to the following questions:

- Cartographic feature extraction and map updating;
- Delimiting urban agglomerations;
- Characterising urban structure, settlement and distribution;
- Urban modelling.

An edited selection of the papers presented at this meeting was published in a book (Donnay and Barnsley 1997). This shows that there have been significant advances both in remote sensing technology and in the price/performance ratio of computer workstations in recent years. Continued enhancements in sensor technology, particularly with respect to spatial resolution mean that remote sensing is increasingly able to provide planners with pertinent information on urban areas, while improvements in the price/performance ratio of computer workstations mean that the capability to process digital, remotely sensed images is now within the reach of national and local planning agencies.

Partly in response to these developments, a number of organisations involved in the collection of statistical information on urban areas have recently shown renewed interest in the operational use of remote sensing at the urban level. The impetus of this interest stems from the need to acquire consistent, objective, accurate, and timely data on urban areas at both the national and European levels.

Remote sensing, as it is in the position to provide multiple, comparable and repeatable answers, can make an essential contribution to the redefinition of concepts connected with urban entities. This makes it possible to define their scope and limits according to scenarios which employ different criteria. There is also considerable scope for integrating various socio-economic data sets with remotely sensed images to model urban morphology over time in a consistent way at both the national and international levels.

3.2.3 Data Availability

Given what has been said above about the data explosion it may seem incongruous to some people that data availability should be regarded as a problem. However, the fact that data exists does not mean that it is readily available to potential users. There are several reasons why this is the case. First, it must be borne in mind that a large proportion of geographic information is collected by central and local government agencies in the course of their administrative duties. In some cases, the terms under which this data is collected preclude its release to other parties. In many other cases it should also be noted that the agencies which collect the data have no mandate requiring them to disseminate this data, nor do they usually have the resources at their disposal to cover the additional costs incurred as a result of dissemination. Consequently, despite frequent exhortations to make government more open in the interests of democratic accountability, only a relatively small proportion of this data is released to the public at large in Europe.

The second reason why data is not readily available to potential users is related to the economics of geographic information. It goes without saying that the data products and services provided by private sector agencies are not free to potential users and that there are also growing pressures on central and local government agencies in Europe and elsewhere to recover some or all of the costs of database creation and maintenance through the sale of data products and services. Such developments inevitably further restrict the availability of data to some groups of potential users.

There is an additional reason for the lack of geographic information at the European level. This is largely because there are no European wide agencies with a mandate to collect data on a pan European scale. The lack of geographic information at the European level presents a major barrier to social science research according to the report commissioned from the European Science Foundation/Economic and Social Research Council (1991, p. 74) by DGXII.

> The benefits of social science to the Community are constrained by the lack of social science data in an appropriate form. On a national level Europe is data rich. In most countries there is a well-established system of market and opinion research institutes, statistical offices, academic social research institutes and, to a varying extent, the social science data service infrastructure ...

> In spite of this data wealth, research with European perspectives is seriously handicapped: the European database is not well integrated, large scale research is hardly coordinated, measurement instruments and data representation lack compatibility, data access and data protection regulations differ, and even information about the availability of data is not always easy to obtain.

Given these circumstances and the extent to which policy and research questions relating to data availability are inter-related, it was agreed that an exploratory meeting to review these issues should be held at Malgrate in Italy at the end of June 1994. Unlike the other specialist meetings in the GISDATA programme, the main objective of this meeting was to prepare a position statement on "Building the European Social Science Resource Base: the Geographic Dimension" for the programme as a whole to provide a basis for further discussions with the main European agencies involved.

The position statement prepared at the Malgrate meeting and subsequently approved by the GISDATA Programme's Steering Committee identifies four key issues that must be taken account of in any discussions regarding the creation of a European geographic information resource bases (Masser and Salgé 1995). These are:

- the need for more coordination of existing efforts at both the national and European levels;
- the lack of consistency with respect to the nature and enforcement of existing legislation concerning copyright and related matters at the national level;
- the impacts of the increasing tendency towards the commodification of geographic information; and
- the need to reconcile the growing demands for geographic information at the small area level with the duty of data providers to protect personal privacy.

Because of these issues, it is argued in the position statement that high priority must be given to the management of data resources at the European level and that the creation of a geographic information resource base should be regarded as an important strategic objective for the research community as a whole. To achieve such an objective, it will be necessary to concentrate efforts on five main strategic tasks:

- developing an overall acquisitions strategy;
- acquiring and preserving strategic data;
- creating, documenting and integrating databases;
- establishing networked catalogues; and
- facilitating wider and more informed use.

3.2.4 Data Infrastructures

During the lifetime of the GISDATA programme, there have been major developments in both North America and Europe with respect to the concept of data infrastructures. These indicate the extent to which governments are having to rethink their roles with respect to geographic information as a result of recent developments in data handling technologies (see, e.g., Masser 1997). It is worth noting that the Malgrate meeting on data availability took place less than three months after the publication of an Executive Order signed by President Clinton himself establishing a National Spatial Data Infrastructure in the United States (Executive Office of the President 1994) and only eight months before the publication of the first draft of "GI 2000: Towards a European Geographic Information Infrastructure" in February 1995 (CEC 1995). These developments gave the position statement produced at the Malgrate meeting a new significance and provided a unique opportunity for the GISDATA research community to contribute to the ongoing policy debates that took place on the GI 2000 document before it entered the Commission machinery in September 1996 with a view to its eventual publication as an EU Communication in late 1997.

The text of the version of the GI 2000 document that entered the Commission machinery in September 1996 (CEC 1996) echoes many of the concerns expressed at the Malgrate meeting and has many features in common with the US National Spatial Data Infrastructure Executive Order. The GI 2000 document presents a policy framework for European geographic information which highlights the need for political action to overcome the barriers identified above:

The major impediments to the widespread and successful use of geographic information in Europe are not technical, but political and organisational. The lack of a European mandate on geographic information is retarding development of joint information strategies and causes unnecessary costs, is stifling new goods and services and reducing competitiveness.

(p. 1)

What is required, then, is "a stable, European wide set of agreed rules, standards, procedures, guidelines and incentives for creating, collecting, exchanging and using geographic information" (p. 2).

The GI 2000 document points out that many elements of a European geographic information infrastructure already exist. There is a growing number of high-quality digital geographic databases in Europe held by local, national, and European providers and users in both the public and private sectors. However, to convert these assets into an effective European geographic information infrastructure requires the following political decisions:

- Agreement of member states to set up a common approach to create European base data, and to make this generally available at affordable rates;
- A joint decision to set up and adopt general data creation and exchange standards and to use them;
- A joint decision to improve the ways and means for both public and private agencies and similar organisations to conduct European level actions, such as the creation of seamless and European data sets;
- Agreement and actions to ensure that European solutions are globally compatible. (p. 2).

Given these circumstances, it is not surprising to find that the last specialist meeting in the data integration stream that took place at Buoux in France in May 1996 was devoted to the subject of "Geographic Information: the European Dimension". Part of this meeting was devoted to the preparation of a position statement on this topic as a direct contribution from the programme as a whole to the ongoing GI 2000 debate (Burrough et al. 1997). This pointed out that it will be necessary to take account of the very different needs of four kinds of GIS activity in the process:

- operational functions associated with routine administrative activities at various levels;
- strategic planning activities;

- pure and applied research; and
- the emerging mass market for geographic information.

A common feeling that underlies the whole of this position statement is the need to raise overall levels of awareness of the need for an overall policy framework of this kind among key decision makers at both the national and European levels.

One feature of much of the current debate on the European geographic information infrastructure is the lack of systematic information on the nature of the problems experienced by different types of GI user and/or supplier in developing databases or products that involve data from more than one country within Europe. Another feature is the relatively limited discussion that has taken place so far on the legal and institutional issues involved at the European as against the individual national level. With these considerations in mind, the Buoux meeting attempted to fill these gaps by the production of a book (Burrough and Masser 1997), which presents the findings of a number of case studies of multinational geographic information projects within Europe and puts forward a number of different perspectives on the legal and institutional issues that must be resolved before a European wide infrastructure can be effective. In the process, the book seeks to make a contribution to the debates that are currently taking place at the European level while also drawing attention to some of the policy research issues associated with them.

The main findings of this book can be summarised in the following terms. The demand for transnational data of all kinds is growing throughout the European Union. However, there are still many problems that must be resolved in the creation of transnational databases. These are primarily political and organisational rather than technical in nature and cover a wide range of legal and institutional issues.

Their impact varies according to the nature of the project. Focused efforts driven by specific applications in both the public and private sectors seem to encounter fewer problems than multipurpose initiatives involving a large number of different interests or the development of products for the mass market. These problems are exacerbated at the present time by a number of short-term factors. These include the absence of clear rules or agreed practices for the acquisition and reuse of data sets, the lack of adequate metadata to help users find the information they need, and the immaturity of the market for transnational geographic information products itself.

3.3 Evaluation

Taken as a whole the tangible products of the data integration stream of the GISDATA programme are substantial. The diffusion book provides a snapshot of the state of GIS in a key application sector in Europe in the early 1990s. Similarly, the remote sensing urban change book highlights the new opportunities that are being opened up by current airborne and satellite remote sensing technology for monitoring urban change. The two position statements on the resource base and the European dimension of geographic information can be seen as both timely and much needed contributions on behalf of the geographic information research community as a whole to the overall policy debates that are currently taking place at the European level. Similarly, the European geographic information infrastructures book draws attention to the problems experienced by different kinds of transnational database developer while also underlining the need for further policy driven research on data integration issues.

One of the most distinctive features of the programme of work that has been undertaken as part of the data integration research stream is the changes that have occurred during its lifetime. These show no sign of slowing up and will probably increase in pace in the immediate future. Consequently, the geographic information research community in Europe faces a formidable challenge just to keep up the momentum that has been built up during the GISDATA programme over the next few years. The importance of meeting this challenge at the European level cannot be underestimated and it is essential therefore that the needs of geographic information research are reflected in the EU's Fifth Framework for Research and Development, which will begin at the end of 1998. For example, the diffusion book should be seen as a useful benchmark in geographic information research in this field, which must be followed up by other studies to keep track of changes in the countries concerned, as well as by further studies of countries which were not included in the original study and studies of diffusion in sectors outside local government. The advent of even finer resolution remote sensing data is likely to present further opportunities for the urban remote sensing community over the next few years while the policy dimension of geographic information is likely to increase in importance as the amount of digital information that could be made available expands exponentially.

The close links that exist in the data integration field between policy and research create both problems and opportunities for the European geographic information research community. They present problems because most of the political and organisational issues involved are not restricted to geographic information and can be regarded as problems arising out of the use of digital information in general. However, this does not imply that the geographic information research community can leave it to others to solve their problems. It is essential that they participate in these broader discussions to ensure that the special characteristics of geographic information are taken into account in them. It should also be noted that many of the research opportunities in this field lie outside the traditional strengths of the geographic information research community and that it may be necessary to take special steps to build up a critical mass of new geographic information researchers who have the appropriate skills to take advantage of these opportunities.

Several research issues relating to data integration are also likely to need much more attention over the next 5 to 10 years. One of the most important of these is the economics of geographic information markets in the digital age. Unlike many types of information, a great deal of geographic information is collected by public agencies primarily for administrative purposes. To facilitate the dissemination of this information and its integration with other data both fundamental and applied research are likely to be needed on the nature of different geographic information markets and the new kinds of public–private partnerships that will be required in the process. Given the extent to which some of these public agencies resemble natural monopolies in their data holdings, it is also likely to be increasingly necessary to consider what forms of economic regulation will be required to ensure maximum data availability.

Alongside these questions there are a number of issues associated with the broader social impacts of greater data integration (see, e.g., National Research Council 1997). The discussion of data matching earlier in this chapter highlighted some of the ethical issues that will need to be debated over the next few years. More generally, it will be increasingly necessary to take account of the negative as well as the positive consequences of the diffusion of GIS. This is particularly important as geographic information research, like most other innovation related research, suffers from a strong pro-innovation bias. To counter this bias there will be growing demands for the political economy of geographic information to be explored in some depth and also for a wide range of alternative scenarios of the brave new GIS worlds that are emerging to be evaluated (see Chapter 4).

3.4 The GISDATA Programme (1993–1997) and Its Impact on the European Research Community

3.4.1 The GISDATA Programme in Perspective

The European Science Foundation (ESF) is a non-governmental organisation which brings together 54 research councils, academies, and institutions devoted to basic research in 20 European countries. Its aims are to help member organisations promote high-quality science in Europe, to advance European co-operation in basic research through collaborative programmes, to promote the mobility of scientists and young researchers, and to encourage the sharing of research experience in order to build scientific communities on a European scale in specific fields.

GISDATA was one of the scientific programmes of the European Science Foundation's Standing Committee for the Social Sciences. It was launched in January 1993 for a four-year period with the support of 14 member organisations which contributed to a budget of just over 5.3 million French Francs (about a million dollars in 1993 prices). In 1996 the programme was extended for one for further year until the end of 1997. The author and Francois Salgé from IGN France were co-directors of this programme.

The term GISDATA aptly summarises the underlying objectives of this scientific programme. The GIS component refers to the need to exploit the research potential of a new technology for capturing, storing, checking, integrating, manipulating, and displaying data which are spatially refer-enced to the earth. The DATA component of the term refers to the need to build up the geographic information resource base at the European level to facilitate the integration of spatial data collected by a wide range of different national and international agencies for a variety of purposes.

The origins of GISDATA go back to January 1991 when the European Science Foundation funded a small specialist meeting at Davos in Switzer-land to explore the need for a European-level GIS research programme. The participants at this specialist meeting felt very strongly that a programme of this kind was urgently needed to overcome the fragmentation of existing research efforts within Europe. They also argued that such a programme should concentrate on fundamental research and that it should have a strong technology transfer component to facilitate the exchange of ideas and experience at a crucial stage in the development of an important new research field. Following this meeting a small coordinating group was set up in 1992 to prepare more detailed proposals for a GIS scientific programme. A central element of these proposals was a research agenda of priority issues grouped together under the headings of geographic databases, geographic data integration and social and environmental applications.

The main GISDATA programme started in January 1993 and was completed in December 1997. Its objectives were:

- to enhance existing national research efforts and to promote collaborative ventures overcoming European wide limitations in geographic data integration, geographic databases and social and environmental applications;
- to increase awareness of the political, cultural, organisational, tech-nical, and informational barriers to the increased utilisation and inter-operability of GIS in Europe;
- to promote the ethical use of integrated information systems, includ-ing GIS, which handle socio economic data by respecting the legal restrictions on data privacy at the national and European levels;
- to facilitate the development of appropriate methodologies for GIS research at the European level; to produce output of high scientific value;
- to build up a European network of researchers with particular emphasis on young researchers in the GIS field.

The GISDATA programme was built around three main activities: a series of 12 specialist meetings on topics largely identified in the original research agenda (Arnaud et al. 1993), a midterm and an end of programme review to evaluate what had been achieved, and two summer institutes for early career scientists which have been organised jointly with the American National Science Foundation through its National Center for Geographic Information and Analysis (NCGIA) (Figure 3.1).

- *Specialist meetings:* Because of the wide range of disciplines involved and the nature of the subject matter of the GISDATA programme, it was decided to organise a series of four specialist meetings within each of the three clusters of research activities in the programme. The basic advantage of the specialist meeting format is that it enables the participants for each meeting to be selected because of their expertise in relation to the specific topic of that meeting. As a result, some 300 scientists from 20 European countries have participated in the twelve specialist meetings that took place between Autumn 1993 and Summer 1996 (Masser and Salgé 1997). Another advantage of this approach was that it was possible to invite leading European academics with specialist knowledge in particular fields to participate in the organisation and planning of these meetings. The topics considered at each of the 12 specialist meetings are listed in Figure 3.1.

- *Strategic reviews:* During the planning stages it was recognised that two strategic reviews would be needed to evaluate progress at the half way and end of programme stages, respectively. This was felt to be particularly important in the GISDATA programme given the large number of participants involved and also because the extent to which the products of the programme are of interest to policymakers as well as academics.

- *Summer institutes:* It was also envisaged from the outset that there would be a strong trans-Atlantic dimension to GISDATA as a result of agreements reached between the European Science Foundation and its American counterpart, the National Science Foundation, and between the Management Committees of GISDATA and the US National Center for Geographic Information and Analysis. The organisation of two summer institutes involving equal numbers of North American and European researchers was seen as a flagship activity in both the GISDATA and the NCGIA programmes. The first summer institute was held at Wolfe's Neck in Maine in the summer of 1995 and the second took place in Berlin during the summer of 1996. Both institutes were targeted towards early career scientists on both sides of the Atlantic who were chosen on the basis of open competitions in Europe and the United States.

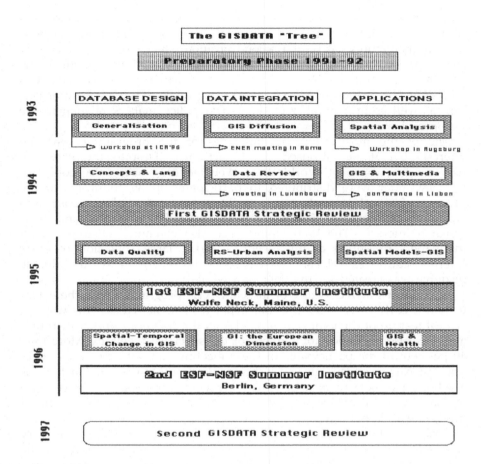

FIGURE 3.1
Research agenda of the GISDATA scientific programme

3.4.2 Its Impacts Om the European Research Community

The specialist meeting format adopted for the GISDATA programme proved particularly successful. It brought together a large number of scientists from a wide range of disciplines extending from information law to computer science on the one hand and from philosophy to hydrology on the other. The products of 9 of the 12 specialist meetings were published in a special GISDATA series by Taylor and Francis (now CRC Press). Further publications included the proceedings of the two summer institutes.

There has also been a strong policy dimension to GISDATA, which reflects its data integration theme discussed above. One specialist meeting was devoted primarily to the preparation of a position paper for circulation to European agencies involved in building the European social science resource base and an important output of the recent meeting on European data infrastructures has been a second position paper which has also been circulated widely among the European agencies involved. Participants in the GISDATA programme have also been very active in the consultations regarding the creation of the European Geographic Information Infrastructure as well as in a number of other data policy matters relating to the European Union.

One of the most gratifying outcomes of the programme was the interest that it has aroused in its activities. Despite the extensive use that was made of World Wide Web facilities, over 1,000 copies of the newsletter were distributed to participants in 35 countries. In the process of developing the Web facilities, a substantial database was created which contained the details of over 700 scientists and their research interests. The Web Home Page also contained abstracts of all the papers presented at the meetings and a large collection of other Internet addresses relevant to GI research. In this way, a valuable resource was created for the field as a whole.

Another distinctive feature of the GISDATA programme was the extent to which early career scientists participated in its activities. This was the central feature of the joint ESFGISDATA/NSF/ NCGIA summer institutes and was also an important element of the specialist meetings. More than 40 per cent of the participants of these meetings fell into the early career scientist category. This clearly indicates the youthfulness of the field as a whole, particularly in areas involving new technology such as multi media where the proportion of early career scientists at the specialist meeting was 75 per cent.

However, it must be recognised that, although GISDATA was very successful in building on existing European research strengths in the field of geography/spatial analysis and cartography/surveying, many potentially interesting fields of social science research were not fully explored within the programme. With this in mind, members of the GISDATA planning group took a leading role in the organisation of four Euro conferences on Socio Economic Research and GIS. These meetings provided additional opportunities to further explore socio economic research issues within the GIS field in Europe. In autumn 2001, a specialist workshop for participants from Europe and the United States was also funded by the European Science Foundation and the US National Science Foundation to assess the current state of research on access to geographic information generally, and also more specifically, to consider participatory approaches surrounding the use of geographic information. This was held in Spoleto, Italy, and the papers presented at the workshop were published in two special volumes in the URISA Journal (Onsrud and Craglia 2003).

The Summer Institute model was modified and adopted by the Vespucci initiative for their annual summer school on geographic information sciences (Gould 2004). In 2015 a further Summer Institute was organised in Bar Harbour, Maine, for participants in the two GISDATA meetings to celebrate the 20th anniversary of these meetings and the presentations at this meeting were published in book form under the title "Advancing Geographic Information Science: the past and next twenty years" (Onsrud and Kuhn 2016).

Last but definitely not least, it must be recognised that possibly the most important outcome of the GISDATA scientific programme was the establishment of an Association of Geographic Information Laboratories in Europe (AGILE) in 1998 to keep up the momentum built up for networking in Europe during the programme (Masser 1999). In 2017 AGILE (https://agile-online.org/) also celebrated its twentieth anniversary with another volume in the Springer series (Bregt et al. 2017). A paper by Hardy Pundt and Fred Toppen (2017) in this volume describes some of the main events in its history which have led to its development "towards a lively, well established, and widely appreciated organization with around 85 member organizations throughout Europe."

References

Arnaud, A. M., M. Craglia, I. Masser, F. Salgé, and H. Scholten, 1993. The research agenda of the European Science Foundation's GISDATA programme, *International Journal for Geographic Information Systems 7, 5,* 463–470.

Bregt, A., T. Sarjokowski, R. van Lammeren, and F. Rip, (eds.), 2017. *Societal innovation: selected papers from the 20th AGILE Conference on Geographic Information Science,* Berlin: Springer.

Burrough, P., M. Craglia, I. Masser, and F. Salgé, 1997. *Geographic information: the European dimension,* position statement on behalf of the European Science Foundation Scientific Programme on geographic information systems: data integration and database design (GISDATA), Strasbourg: European Science Foundation.

Burrough, P. and I. Masser, (eds.), 1997. *European geographic information infrastructures: opportunities and pitfalls,* London: Taylor and Francis.

Campbell, H. and I. Masser, 1995. *GIS and organisations: how effective are GIS in practice,* London: Taylor and Francis.

Commission of the European Communities, 1995. *GI 2000: towards a European geographic information infrastructure,* Luxembourg: DGXIII.

Commission of the European Communities, 1996. *GI 2000: towards a European policy framework for geographic information,* Luxembourg: DGXIII.

Data Protection Registrar, 1996. Data matching: what limits, what safeguards? *Conference held in Manchester,* 6th December 1996.

Donnay, J. P. and M. Barnsley, (eds.), 1997. *Remote sensing and urban analysis,* London: Taylor and Francis.

Dureau, F. and C. Weber, (eds.), 1995. *Télédétection et systèmes d'information urbaines*, Paris: Authro pos.

European Science Foundation/Economic and Social Research Council, 1991. *Social sciences in the context of the European communities*, Strasbourg: European Science Foundation.

Executive Office of the President, 1994. Coordinating geographic data acquisition and access: the National Spatial Data Infrastructure, Executive Order 12906, *Federal Register* 59, 17671–17674.

Gould, M., 2004, 2004 Vespucci summer school on GI science – a success! *Directions Magazine*, August 4th 2004.

Masser, I., 1997. *Governments and geographic information*, London: Taylor and Francis.

Masser, I., 1999, The Association of Geographic Information Laboratories in Europe (AGILE): aims and scope, *Proc 4th EC-GIS Workshop*, Joint Research Centre, Ispra, pp. 235–239.

Masser, I., H. Campbell, and M. Craglia, (eds.), 1996. *GIS diffusion: the adoption and use of geographic information systems in local government in Europe*, London: Taylor and Francis.

Masser I. and F. Salgé, 1995. *Building the European social science resource base: the geographic dimension*, position statement on behalf of the European Science Foundation Scientific Programme "Geographic information systems: data integration and database design (GISDATA)", Strasbourg: European Science Foundation.

Masser, I. and F. Salgé, 1997. *The GISDATA programme: an overview, in geographic information research at the millennium*, Strasbourg: European Science Foundation.

National Research Council, 1997. *The future of spatial data and society: summary of a workshop*, Washington, DC: National Academy Press.

Onsrud, H. and G. Rushton, (eds.), 1995. *Sharing geographic information*, New Brunswick, NJ: Centre for Urban Policy Research.

Onsrud, H. J. and M. Craglia, 2003. Introduction to the special issues on access and participatory approaches in using geographic information, *URISA Journal 15, 1*, 5–9.

Onsrud, H. J. and W. Kuhn, (eds.), 2016. *Advancing geographic information science: the past and next twenty years*, Needham, MA: GSDI Association Press.

Pundt, H. and F. Toppen, 2017. 20 years of AGILE, in Bregt, A., T. Sarjokowski, R. van Lammeren, and F. Rip (eds.), *Societal innovation: selected papers from the 20th AGILE Conference on Geographic Information Science*, Berlin: Springer.

Rhind, D., 1992. Data access, charging and copyright and their implications for geographic information systems, *International Journal for Geographic Information Systems 6*, 13–30.

Rogers, E., 1983. *Diffusion of innovations*, New York: Free Press.

4

Brave New GIS Worlds

Michael Wegener and Ian Masser

4.1 Introduction

Traditionally, diffusion research has a pro-innovation bias. Diffusion is studied as a process by which older, outdated technologies are replaced by more advanced, more efficient and hence more beneficial ways of doing things. The first systematic studies of diffusion processes by Rogers (1962) and Hägerstrand (1968) looked into the way new technologies, such as new farming techniques, the telephone or television, gradually penetrated their markets. The implicit assumption, never discussed, was that adoption of the new technology was in the interest of the adopters.

However, the view that a new technology is always better than the older one it displaces has long been discredited by the dialectic of technological progress – the experience that more often than not a successful technology, once it has become dominant, also displays a destructive dark side. At least in the case of farming, this dialectic has become commonplace. The 'green revolution' helped farmers to multiply their crops by introducing fast-growing plants with higher yields, more efficient fertilizers and more effective pesticides, yet also brought new risks of water contamination and soil erosion, contributed to rural unemployment and depopulation and caused grave imbalances in global food markets.

Diffusion research was not able to visualize these long-term impacts because it was concerned with the early phases of diffusion, in which the new technology was still virgin and innocent. As only the beneficial aspects of the new technology were seen, rapid diffusion was interpreted as success and lack of it as deplorable backwardness. In fact the most frequent motivation for diffusion research has been to identify barriers to the rapid adoption of the new technology and, once these have been identified, to recommend strategies to overcome them (cf. Rogers, 1993).

Diffusion research concerned with the adoption of geographic information systems (GIS) is no exception. The growing volume of studies on the
adoption and use of GIS in the United States and in European countries
have been stimulated not by academic interest but by the well-intended
drive to identify and help remove institutional and technical bottlenecks
to their universal distribution and application – not surprisingly, because
the authors of the studies generally are members of the GIS community,
i.e. individuals with a strong interest in this rapidly growing market (see,
for instance, Masser and Onsrud, 1993). The dangers of a pro-innovation
bias of this kind of research are obvious. Diffusion research which only
sees the positive side of the new technology it is concerned with is
unable to distinguish between backwardness and other more serious
reasons for differences in speed or intensity of adoption. Even where it
subsumes barriers to adoption under the broad but diffuse heading of
'cultural factors', it will always view these as regrettable and something
to overcome, and this will inevitably colour the conclusions drawn from
the research.

Yet there are good reasons to move beyond a naive all-out promotion
of GIS and arrive at a differentiated and balanced stance which carefully
weighs their obvious benefits against their potential risks. As long as
there have been computers, there have been warnings that the information revolution may endanger fundamental human values. The 'data
bank' and 'information system' have always carried the Janus-face of
unlimited knowledge about and control over the individual. There have
always been fears that this kind of knowledge, in the hands of irrespon-
sible bureaucrats, law-and-order police officers, power-hungry politicians,
criminal organizations or unpredictable fanatics, could be used to undermine democracy and individual privacy. However, the recent success of
GIS has given these warnings a new dimension. GIS, with their capability
to localize every conceivable object or activity in Cartesian space, are the
ultimate expression of the rationalist dream of measuring and knowing
everything. In combination with concurrent technologies such as electronic
data interchange, GIS introduces a new powerful threat, that of total
spatial control.

The first concerns that GIS might be far more dangerous than previous
information systems have been expressed not by proponents of GIS, but by
social and political scientists and a few critical geographers (e.g. Smith,
1992; Obermeyer, 1992; Curry, 1993; Lake, 1993; Onsrud et al., 1994;
Pickles, 1994). The GIS community itself has largely remained confined to
an uncritical promotional attitude towards GIS.

GIS journals such as *GIS Europe* are technology- and application-
oriented and rarely deal with the social impacts of GIS. Academic
discussions orbit around epistemological or methodological issues of GIS
or what GIS do to geography (e.g. Openshaw, 1991, 1992, 1993; Taylor
and Overton, 1991; Couclelis, 1993) but hardly touch upon their limits

and risks. However it is time that also GIS experts become aware of the debate on the impacts of GIS on society and develop adequate answers to its serious questions. This chapter tries to contribute to this debate by suggesting a number of scenarios of possible future GIS diffusion which capture the range of perceptions of the impact of GIS on society found in different countries of Europe today.

4.2 GIS in 1996

GIS include a wide range of different applications including automated mapping and facilities management as well as land information systems. As the number of applications grows, the term GIS is used increasingly as shorthand for a great diversity of computer-based applications involving the capture, manipulation, analysis and display of geographic information and the associated services that go with them.

Although many of the basic concepts underlying GIS were developed more than 20 years ago, the computer technology required to manage large amounts of geographic information and display them in graphical form has only been available since the mid-1980s. Since that time the GIS hardware and software industry has dramatically expanded in terms of both the number and range of applications. It is estimated that sales of GIS hardware since 1985 have grown at rates well over 10% per annum, while software sales have increased by 15–20% each year. As a result the volume of hardware sales has doubled every six years since 1985, while that of software has doubled every three to four years.

The pace of technological innovation is still accelerating and the range of applications continues to expand. In fact the number of GIS facilities in operation has grown at an even faster rate than overall sales as an increasing number of budget-price installations come on the market.

The main users of GIS are central and local government agencies and the utility companies. Together these account for well over half the overall GIS market in most countries. Other important application areas are in the field of environmental management and facilities planning. Over the last few years there has been a considerable increase in the number of business applications for sales analysis and marketing. These already account for 8% of the GIS market and it is forecast that their share will rise to at least 15% over the next five years. Other potential fields which are still to be exploited include the use of GIS in vehicle navigation systems.

The utilization of GIS is heavily dependent on the availability of digital topographic data. As a result of variations in national government policies towards data provision, there are considerable differences between countries in terms of the availability of digital topographic data at both the large and small scales. There are also important differences in the cost of information of this kind to users, as the providers in some countries attempt to recover the cost of data provisions. In Britain, for example Ordnance Survey data are protected by copyright, and the agency itself already recovers 70% of its costs, whereas the TIGER files developed by the United States Bureau of the Census are available at minimal cost without copyright restrictions.

The growth of GIS over the last years has stimulated a massive growth in specialist GIS services of all kinds. These range from bureaus specialising in digitizing, automated data capture and customizing spatial databases to management consultancies advising agencies on the benchmarking and implementation of particular GIS packages. An important sub-group of GIS services is associated with the development of customized software for particular application fields. Of particular importance in this respect is the development of decision support systems for commercial marketing operations. As the number of systems in operation has increased, there has been a parallel growth in legal actions regarding the accuracy and reliability of the information provided by them. As a result, litigation and claims for liability and compensation are emerging as an important growth area.

The spatial impacts of these developments are not homogenous. The GIS market is particularly well developed in North America, whereas the European GIS market is still divided by national interests and highly fragmented in character. There is little agreement on common standards, and there are considerable differences in the professional cultures that are involved in GIS applications. Generally the north and west European countries have experienced higher levels of GIS penetration than those of southern and eastern Europe. However, there are also marked variations within countries in the level of GIS penetration, particularly in southern and eastern Europe.

4.3 The Four Scenarios

This is the situation from which the four scenarios start. Each of them is a projection of one possible evolution of the uses of GIS and their impact on society. The first scenario is the *Trend* scenario characterized by incremental diffusion of information systems along the lines experienced in the past. The other three scenarios highlight and exaggerate specific tendencies that can be observed today. The *Market* scenario extends current tendencies towards commodification of information, which restricts access to information to the more powerful. The *Big-Brother* scenario dramatizes the potential of GIS to be used for surveillance and control by fully integrated omniscient systems, which pervade all aspects of life. The *Beyond-GIS* scenario, finally, speculates on how information in the public domain might contribute to more democratization and grassroots empowerment. All four scenarios look 20 years into the future and are expressed as narratives of a person looking back to the 1990s.

4.3.1 The Trend Scenario

The end year for the scenarios was 2015. The 20 years since 1996 have been a period of stupendous technological developments. All of them have been based on innovations made in the 1980s, but nobody at that time would have expected the speed by which they have penetrated their markets. New materials have brought unprecedented levels of miniaturization, memory and computing speed of all kinds of electronic devices. Telecommunications, cable and computer companies have merged into transnational media conglomerates. Fibre optics, cable, cellular radiophony and satellite communications have grown together into an integrated multi-layer network of information superhighways bringing fax, e-mail, smart TV and electronic data interchange to every home and office. Artificial intelligence, multimedia and virtual reality have amalgamated to create new kinds of computer applications that are more user-friendly, entertaining and unobtrusive than ever.

All these developments have had their impacts on GIS. As a result of the immense advances in performance and reduction in cost of both hardware and software, the number of GIS installations, the development of user-friendly interfaces and the range of applications have multiplied to the extent that GIS are now used universally like spreadsheets and database management software before them. One impact of universal GIS is that most users make use of GIS facilities without ever being aware that they are doing so. This is particularly the case with multimedia applications using virtual-reality GIS. Together with GIS installations, the number, size and diversity of spatial databases has grown explosively. Today there is a huge variety of public and private spatial databases for all conceivable purposes, from postcode systems precise to the letter box to multimedia, virtual reality house catalogues, or travel guides. GIS education is now part of the conventional school curriculum. The GIS industry has become increasingly specialized and fragmented in order to meet the great diversity of demands placed on it by different applications groups. The term GIS is used less frequently than during the 1990s, and when it is used, it tends to be prefaced by another term indicating the specific subset of applications that is involved.

Within Europe as a whole there are still considerable differences between countries. Although efforts to promote greater harmonization of GIS by the European Organization for Geographic Information have had some success, there are still considerable differences between the countries in terms of the data that are collected and the extent to which they are made available to users as well as with respect to the data models and data interchange formats used. Many of the differences between professional cultures also remain, particularly with respect to the key GIS users such as local governments and the utility companies. However, considerable progress has been made in reducing regional disparities within Europe, partly as a result of initiatives of the European Union. As a result, the gap between the European and north-American GIS industries in terms of market penetration has largely disappeared. This is also due to technological developments and automated data capture, which have resolved many of the problems previously faced by information-poor countries.

In this scenario therefore, GIS in Europe is both universal in extent and largely benign in operation by 2015, while the applications field as a whole has become highly fragmented and specialized in nature. Variations between European countries still persists despite efforts to promote harmonization. Against this backcloth, the potential of the technology and the capabilities of organizations to manage it are still being constantly tested in practice. Because of the risks involved in such operations, the media contain occasional reports on gross incompetence and inefficiency in public-sector GIS applications as well as about the enormous sums that are being paid out in compensation as a result of court decisions regarding GIS.

4.3.2 The Market Scenario

In this scenario for 2015 the information industry has become the largest and most powerful economic sector. As goods production now largely takes place in the developing countries and in eastern Europe, more than 70% of all economically active persons in western Europe primarily handle information during their daily work. Digital data, text, audio and video, fax, telephone, and electronic data interchange have amalgamated into one integrated multimedia information and communications technology. The desktop computer has given way to a flurry of miniaturized, interconnected electronic gadgets from credit card to hand-held super computer. All individuals and households are part of and connected with thousands of electronic networks putting at their disposal all conceivable kinds of deliveries and services. Every transaction in daily life leaves a trace in these networks: orders, sales, invoices, receipts, itineraries, reservations, inquiries, messages, sounds and images.

A large part of the traffic over the networks is geocoded. Every customer or supplier is associated with a unique address, which not only represents a point in geographic space but also is a node of the transport and telecommunication networks and is linked to a postcode, enumeration district, electoral ward, municipality, and county. Attaching a geocode to an item has become so easy that geocoding is used even where it does not serve any other purpose than identifying an object. Every trip, delivery route, or electronic message represents a spatial interaction between two addresses and can be aggregated to flows of people, goods, and information across the territory. Knowledge about these stocks and flows, about potential customers and the pattern of their activities, is economic power, which can be used to contest or defend a market.

This is why most of the networks and the information they contain is private. In the 1990s, many European governments, following the neoliberal economic doctrine of that time, privatized their postal services, transport, and telecommunications networks and enforced a strict cost-recovery policy for government agencies providing postcoded directories or cartographic or statistical services, which traditionally had been free or could be obtained for a nominal charge. Local governments followed suit by privatizing their utility companies and contracted out mapping and surveying tasks. Privacy legislation, which had been overly constraining the information industry in some European countries, was harmonized between European countries in the late 1990s. Today, it is legal to collect and trade data on individuals as long as the information appears to be correct.

The result was the emergence of an immense market for value-added telecommunications services and geocoded information. During the 1990s, small- and medium-sized firms specializing in digital databases with the associated software mushroomed. There was a proliferation of digital road databases for trip planning and fleet management, of small-area population and household databases for marketing planning, and of large-scale digital city maps for real-estate development and property management and sales. Prospective home buyers could browse in virtual reality through offered houses and their environment without actually going there. The same technology was used by travel agencies, instead of bulky catalogues, to market package tours. Other rapidly growing markets were utility planning, facility management, and vehicle tracking and navigation systems for the rapidly growing intelligent vehicle-highway systems (IVHS) industry.

Besides these applications for commercial and professional users, a booming market for consumer or home GIS emerged. People could download virtual travel experiences as a surrogate for actual travel to far-away countries, cities, or museums – one could even book a trip to ancient Rome. Other popular home applications of GIS were trip planning, geography courses, and spatial computer games. As with today's video games, customers were lured into buying cheap hardware to make them captive to expensive software.

It was a period of creative turbulence and confusion. Every conceivable spatial information of commercial value was digitized over and over again by a multitude of data suppliers. Needless to say that all these proprietary databases were of varying accurateness and reliability and incompatible with each other. As competition became fiercer, prices plummeted. There were real data wars between suppliers; even sophisticated encrypting techniques did not prevent massive reverse engineering of databases resulting in an explosion of litigation about data ownership and copyrights. Because of the proliferation of data suppliers, it became less and less profitable to trade raw data. The real business was to compile customized 'designer information'for the specific purposes of individual clients; and as these were more often than not persuasion and manipulation if not deception, the notion of what was 'correct' information underwent a subtle change. As a consequence, litigation on the liability for damages due to the use of incorrect or distorted data emerged as a second fast growing field of legal disputes. In particular, some spectacular cases of large-scale fraud in international virtual space created a worldwide legal debate about which country's jurisdiction to apply.

In the late 1990s, the market consolidated and many small suppliers of digital geocoded information went out of business. After some spectacular mergers and take-overs, a few big transnational players remained: among them Mitsubishi, Siemens-Bull and Warner-Murdoch, the US-British media giant, who had ventured into the geoinformation business by swallowing EtakMap of Atlanta, Georgia, and by launching its own fleet of imagery satellites. The Warner-Murdoch (formerly Etak) map encoding system became the factual industry standard. More recently, alliances between the geoinformation industry and credit card companies, travel agencies, and telecommunication networks operating worldwide have created giant online data banks capable of tracking not only lost luggage but also travellers or customers with any desired detail.

Governments at all levels, once the sole providers of geoinformation, found themselves at the mercy of the information conglomerates. Their retreat from the information scene in the 1990s, based on short-sighted budgetary considerations, proved to be a costly mistake. Since in most European countries now population and employment censuses have been abolished, governments have to pay the market price for the same kind of information which in former times they had produced themselves. Even worse, they do not get all the information available, as certain kinds of data on property values, household income, consumer preferences or travel patterns are too commercially valuable to be released to the public domain. Ironically, the refusal to sell commercially profitable data to government is often justified by reasons of confidentiality, although everybody knows that no such constraints are observed where that data are used for commercial purposes.

The loss of public control over the geoinformation market has seriously affected the status and effectiveness of public planning. Some kinds of data of potential value for local planning have practically disappeared because their collection or updating is not profitable, such as historical or time-series small-area data or maps of non-metropolitan areas. Other kinds of data have a negative effect on urban development because they are selectively available only to certain groups. Proprietary information on the socio-economic composition of neighbourhoods and on property values, for instance, has been used by real-estate agents to speculatively manipulate land and house prices with the effect of displacement of poor households and reinforcement of spatial disparities. In fact, dealing with manipulated real-estate information has become an important field of activity of organized crime.

Other users of geoinformation, who formerly relied on government services for their information needs, are effectively excluded from access today. University research using geographic information is hardly able to afford privately collected data offered at market price. Users without financial means, such as students or citizen groups or protest movements, who need information for their study or political work, have no chance. But nobody complains; people see that information is a commodity and understand that the market does not produce where there is no demand. Nor is the issue discussed in the media which are controlled by the same multinationals that produce the data. The free information market is not free for everybody but only for the rich and powerful. The consequence is a widening gap between the information-rich and the information-poor: between those who participate in the information society as providers and manipulators of information and those who participate in it only as consumers and have access only to manipulated information.

The information gap is widening also between regions and countries. Developing countries have become dependent on the transnational geoinformation corporations from whom they buy GIS-processed satellite images indispensable for resource exploration and water supply management. In addition, also east European countries, which had not had the time to develop their own geoinformation industry, are victims of this dependency. After their privatization, the statistical offices and mapping agencies of Poland, the Czech Republic, and Hungary were acquired by Mitsubishi, whereas Siemens-Bull succeeded with their bid for designing CISGIS, the distributed GIS for the countries of the former Soviet Union.

4.3.3 The Big-Brother Scenario

In this scenario for 2015 there is a sense of relief that the opposition of the 1990s against the geoinformation networks had not been successful. The European corporate state depends on reliable intelligence to defend itself against crime and subversive activities. Today, it is hard to imagine how the security of residential areas or shopping malls could be guaranteed without efficient spatial surveillance systems. Even driving on highways has become more secure since every vehicle is being monitored by police, although this had been introduced originally merely for accounting purposes.

There had been a time when some people had resented being registered in the new geocoded information systems. There had been even fears that data banks with the capability to track everybody's movements, might endanger basic human rights. Fortunately, these concerns have long been dismissed as exaggerated fabrications of the individualistic liberal period. Today, citizens realize that modern information systems are only to their benefit. They appreciate the convenience and safety of the welfare state and are eager that they are correctly represented in as many data banks as possible as home owners, customers, subscribers, patients or drivers and wherever they go, at home or abroad. Of course, people who are denied the privilege of membership may complain; but they must understand that the exclusion of people without credit line is necessary for the protection of the majority. Also people who find their whereabouts tracked in police data banks, such as narcotics dealers, traffic delinquents, HIV positives, homeless, or people with questionable political views may not like this, but they can only blame themselves for being observed in the interest of a safe society.

The integration of the geocoded information systems started in 1998, when Eurostat and Europol, with the help of Siemens-Bull, the European information giant, were amalgamated into the European Intelligence Agency (EIA). It was the task of the new public-private agency to integrate all hitherto isolated national spatial information systems into a coherent hypernetwork of distributed information interchange following the lead of the 'information highways' programmes in the United States and Japan. It was argued that only by this integration would Europe have a chance to compete with these two rivals in the fight for global economic dominance. One can say that the integration of geoinformation did more for the unification of Europe than the Single European Market in 1993.

The impacts of this restructuring of the geoinformation scene in Europe were dramatic. It ended the chaos of uncoordinated production of geoinformation by small suppliers of the liberal period. Now it was recognized that spatial information which is freely available to everybody is intrinsically dangerous, whereas in the hands of the corporate state it can guide a society to achieving its highest economic potential. Since the European Freedom-of-Information Act of 1998 therefore every collection of geocoded information has to be licensed by the local subsidiary of the EIA, and any collected geoinformation is classified unless explicitly released by the EIA to the public domain as economically not sensitive. This law has greatly reduced the number of unqualified suppliers of geoinformation and the volume of litigation in this area.

From an engineering point of view, the European information network is a marvellous achievement. DESCARTES (distributed European spatial control and real-time early-warning system) represents the latest advance in hypernetwork technology. It is in fact a network of networks, superseding the hotchpotch of formerly separated and incompatible public (police, secret service) and private (commercial data bank and corporate data and transaction) networks in one grand, unified design – a splendid synthesis between German thoroughness and French elegance. Equipped with latest artificial intelligence, DESCARTES is an adaptive, learning, decentralized system. It has therefore no single primary control centre; its alert rooms are virtually distributed over all its levels, in police head-quarters, corporate offices or the various spatial levels of government. In the control rooms planners watch for sensor lights to flash on floor-to-ceiling maps at places where trouble is likely to occur (cf. Wegener, 1987). The EIA has the responsibility of maintaining the network and controlling access to it as well as linking it to similar networks in the United States, Japan and China.

However, DESCARTES was not only an engineering achievement; it also has had a deep influence on the relations between people. Never before had there been such a harmonious society. Violence and street crime have practically disappeared, since all public spaces have been equipped with video surveillance system; without surveillance people would not feel safe. Most people have asked the authorities to link their homes to the circuits to demonstrate that they have nothing to hide, in fact privacy has become associated with something unethical if not illegal. Surveillance is moulding behaviour in many beneficial ways. For instance, neighbourhoods now look much tidier, since remote sensing has enabled police to monitor garden maintenance.

Of course, where there is much light, there must be some shadow. There remains the misery of those who are excluded from the surveillance society, such as illegal immigrants, tax evaders, or subversive elements living in sewers or abandoned underground tunnels. They do not enjoy the benefits of surveillance but are themselves strictly observed by police and, if necessary, ruthlessly attacked. It remains an interesting question why the authorities have tolerated the existence of this underclass in an otherwise perfect society.

4.3.4 The Beyond-GIS Scenario

The final scenario for 2015 begins by highlighting the view that,seen from today, the GIS craze of the 1990 looks like a strange fad. Certainly, there have been some useful applications of GIS in cartography, planning or facility management, but to call this a revolution was a vast exaggeration. More likely it was fuelled by the hope of a fringe discipline for 'the movement of geography to center stage' (Curry, 1993). Today geography already exploits its next revolution, holography in four-dimensional hyper-space, hailed by an elderly Lord as the greatest revolution in geography since the invention of the globe.

The end of the GIS boom in the late 1990s coincided with a major change of values. What had been a minority opinion in the early 1990s, now became a broad movement: that the most advanced countries in the world could not continue to pursue economic growth forever, but needed to move towards qualitative growth in terms of equity and sustainability

(Masser et al., 1992). Political landslides in major European countries brought back the welfare state but also a revival of grassroots democracy.

These developments changed the role of information and by that of GIS in society. People rediscovered that the most important types of knowledge are *not* data and are *not* spatial, but are informal, personal and political, i.e. everything information systems, and GIS in particular, are unable to offer. Some even claimed that the hypothesis that with more and better information all problems could be solved, was itself an expression of a technocratic and functional view of the world (Postman, 1991). All sorts of computerized information systems became associated with everything that was negative: central power, technocracy, the corporate state, police surveillance and organized crime. A wave of violence against computer centres and agencies dealing with data of all sorts swept across Europe. In November 1997 a small group of Luddites set fire to the Eurostat complex in Luxembourg. The fire lasted 5 days, and the smoke trails it generated were recorded by satellites.

Violence cannot solve social conflicts, but in this case it forced the information authorities to radically decentralize and democratize their operations. All cross-links between secret service, police and all sorts of public and corporate data banks were interrupted and put under strict public control. New freedom-of-information acts in many European countries determined that information collected in the public domain had to be made available to the public at no or marginal cost, except where privacy constraints precluded it.

Paradoxically only on first sight, the anti-GIS movement benefited local government GIS. As local self-governance and local planning reemerged as a central forum of political debate, local government GIS became even more important decision support systems for local land use, transport and environmental planning. In particular the need to redirect urban development towards sustainability gave an unexpected boost to local government GIS as it became apparent that environmental analysis in fields such as air pollution, noise propagation, vegetation, wildlife or micro climate required a more disaggregate spatial scale than conventional aggregate methods.

However, the relationship of local government GIS to power changed. Whereas they were originally designed for the use of the authorities, they now became a public good explicitly designed for public use in an open and participatory process of social experimentation and grassroots decision making. This, of course, required a different type of GIS, one especially designed to be used by non-experts. Therefore a new generation of GIS designed as 'expert systems for non-experts' emerged. Public libraries and institutions of adult education were given a new responsibility as mediators between non-experts and GIS in order to reduce the information gap between the authorities and the public. The result was a revival of public participation in local decision making, in particular in matters of urban planning, and a surge of self-organized user groups exchanging databases and analytical techniques (cf. Wegener, 1987).

Some say that local planning has become more difficult as public inquiries are more thorough and hence more time-consuming. It is also true that there have been periods of public disinterest when political apathy seemed the ultimate barrier to participatory planning. Moreover, the democratization of knowledge has not solved, but rather acerbated the problem of how to cope with the flood of largely irrelevant information. There even have been instances of deliberate misinformation in the open information arena, and it must be recognized that without the former comprehensive surveillance police work has become less effective. However, most people agree that these are small problems compared with the gain in civic culture.

4.4 Conclusions

Which of the four scenarios is likely to become reality? One view is that there could be different scenarios for different countries. The benign Beyond-GIS scenario, for instance, might have a chance in the mature democracies of north-west Europe, whereas countries with less developed political checks and balances might be at risk of moving into the directions of the Market or Big-Brother scenarios. An opposite view holds that the global competition will bring convergence rather than polarization between countries. In any case it is likely that the future will contain some facets of each of the scenarios. Low-cost GIS software will be widely available and used like spreadsheets and companies will use GIS to increase their profits as in the Market scenario; government agencies will use GIS to process personal spatial data for their purposes as in the Big-Brother scenario and local planning will be changed by access to spatial planning information for everybody as in the Beyond-GIS scenario. Each country can choose to which degree each scenario will come true.

What can be done to enhance the benefits and minimize the dangers of the GIS revolution? The first and most important task is to promote computer literacy and mature and responsible use of GIS through information and education for social consciousness. Like all strategies built on the principle of the enlightened and competent citizen, this may not sound very convincing vis-à-vis powerful economic interests not constrained by moral principles. Therefore good legislation in the area of information is essential. Even though many European countries have made substantial progress towards efficient privacy protection, all are sadly lacking in legislation guarding the right of citizens to have access to information collected in the public domain. Recent tendencies to force public agencies to recover the cost of data collection from their users seem to be steps in the wrong direction. Lastly it remains to be seen whether the forthcoming harmonization of privacy and freedom-of-information legislation within the European Union will settle for the lowest common denominator or will bring genuine progress.

These political considerations should, however, not distract from the more fundamental philosophical questions concerning GIS. These questions have hardly found an answer. For instance, it needs to be asked in how far the data model of GIS implies a certain perception of the world and, if applied, will impose that perception on its users. It has been said that because of their US origin many existing GIS not only require their users to communicate in English but also reflect American cultural values (Campari and Frank, 1993; Wegener and Junius, 1993). Lake (1993) claims that the relationship between spatial units of reference and attributes in GIS is essentially positivist, and Curry (1993) points out that current GIS embody the principles of a property-based society. If this is true, GIS would secretly have a conservative and system-stabilising effect – the direct opposite to their desired innovative and emancipatory role in planning. Under this perspective, the ESRI slogan 'geography organizing our world' takes on an insidious double meaning.

4.5 What Actually Happened during the Last 20 Years?

The Brave New GIS Worlds paper attracted quite a lot of attention at the time of its publication. For example, it was used at a workshop organised by the Mapping Science Committee of the US National Research Council on 'The future of spatial data and society' (National Research Council, Mapping Science Committee, 1997). In 1995, the Committee decided to organise workshop to explore a series of alternative long-term visions and identify some of the societal forces and changes that would make them more or less likely. If successful, it was argued that the workshop would provide a framework for thinking about the future of the national spatial data infrastructure.

Thirty five leaders in the spatial data field from the public and private sectors as well as academia identified forces affecting the future of spatial data, discussed the four Brave New GIS World scenarios and made suggestions to the spatial data community based on these futures. This formed the focus of its activities in the second day of the workshop when the participants were divided into working groups and each of these groups was consider one of these scenarios in the light of the outcomes of the previous day's discussions.

The final chapter of the report considers the lessons learned from this exercise with reference to future strategic planning given that 'Organizations and stakeholders within the spatial data community will be making many strategic decisions and choices in the coming years' (National Research Council, Mapping Science Committee, 1997, p. 40). In the process they concluded that particular attention should be paid to those involved in partnership initiatives, the activities of data suppliers and the directions of education and research.

Twenty years have passed since these scenarios were published and lots of imporrtant developments have taken place in terms of GIS technologies as well as in terms of administrative and organisational environment in which they operate. As a result technical terms such as GPS (Global positioning system) and Satnav (Satellite navigation) have become commonplace in society. Consequently it is worth asking how far do developments over the last 20 years bear out the assunptions underlying the original four scenarios.

A recent paper by the original aurhors (Masser and Wegener, 2016) compared their speculations of 20 years ago with actual developments. Looking back, with the benefit of hindsight, they decided that, regarding the Technology scenario, 'we correctly anticipated most of the unprecedented levels of miniaturisation, memory and computing speed of all kinds of electronic devices possible today and we predicted the emergence of transnational media conglomerates integrating telecommunications, cable and computer companies.' A good example of the impact of technological innovation is the academic research that has begun on geographically referenced virtual environments (GRVE) (Lin et al., 2015). It should also be borne in mind that remotely piloted aircraft sytems (drones) were largely a feature of science fiction in 1996 and that the increasing use of such systems brings with it a number of legal problems as well as operational benefits.

The Market Scenario assumed that by 2015 the information industry would be the largest and most powerful economic sector. However, Masser and Wegener (2016, p. 1158) felt that that may be a bit exaggerated. However, the latest geospatial industry outlook and readiness index report from Geospatial Media and Communications (2018) estimates that with an estimated market size of around 300 billion US\$ the geospatial industry will grow between 15% and 20% a year over the next few years to reach almost 500 billion US\$ by 2020. By this time the size of the Asian market will probably be equal that of the North American market.

The Big Brother scenario foresaw that the potential of geoinformation networks is exploited by private companies and the corporate state to protect themselves against crime and subversive activities. However, it was also felt that they had not fully appreciated the extent of the widespread invasion of personal privacy that follows on from the enormous data holdings created by large private multinational companies such as Facebook and Google. A number of regulatory measures such as the European Union's General Data Protection Regulation (GDPR) (CEC, 2016) have been formulated to deal with these problems but it is not yet possible to evaluate their likely outcomes (Chakravarty, 2018).

The Beyond GIS scenario correctly reflected the broad movement towards grass-root democracy which has led to a more critical attitude regarding the information society. This has also encouraged public participation in GIS and partcipatory mapping projects (see, for example, Brown and Kyatta, 2014). Nevertheless, they did not anticipate the explosive growth in the volume of volunteered geographic information (VGI) and the universal diffusion of mobile phones with GPS location capability. This has given rise to a number of issues about VGI, public participation and citizen science (see, for example, Sui et al., 2017).

On the basis of their analysis Masser and Wegener (2016) felt that the four scenarios would not change much if they were defined now as the reality of today contains elements of all of them and may even underestimate the changes lying ahead. Comparisons of their speculations of 20 years ago with actual developments are the point of departure for some speculation about trends over the next 20 years. On the basis of this analysis it is felt that four important questions still need to be considered:

- What new, still unknown technological advances appear possible?

From past experience this is probably the hardest question of all. Who could have forecast the spectacular development of the internet or the emergence of Satnav or GPS as universal products 20 years ago. However, according to a report on the spatial industry industry prepared for the Australian Cooperative Research Centre for spatial information (Coppa et al, 2018) there is no shortage of potential candidates in the enabling infrastructure and technology field. These include sensor networks, visualisation technologies including augmented reality, mapping and scanning systems such as 3D scanning and mobile mapping, and auotnomous transport such as driverless cars, drones and flying cars.

- How may they be exploited by private stakeholders and public institutions?

It must be recognised that a spatial data infrastructure (SDI) is an enabling platform for data sharing between stakeholders from both the public and private sectors as well as academia and local community groups in the spatial data community. In recent years the distinction between public and private has changing as a result of a movement away from national small-scale data to more people relevant large-scale information, generally derived at a sub-national level (Rajabifard et al., 2006). It can also be argued that it is increasingly neccesary to consider several different types of SDI in practice given the evolution of SDIs in response to crowd sourcing and ubiquitous mobile technologies (see Chapter 14).

- Will this lead to further growth of the global spatial services economy?

As yet there is no sign of a slow down in growth and there are still considerable differences between wealthier countries such as the United States and Germany and the rapidly developing countries such as Bangladesh and Vietnam. Consequently, the Geospatial Media and Communications report (2018) shows that the geospatial market (comprising Global Navigation Satellite Systems, GIS, Earth Observation, and 3D Scanning) is growing steadily with the 2018 market worth USD $339 billion and forecast to grow to USD $439.2 billion by 2020.

- Will there be a solution to the conflict between the goals of open access and privacy?

The simple answer to this question is yes but in reality the situation is much more complex especially where big data in concerned. Mayer-Sch önberger and Cukier (2013) have discussed both the promises and dangers of Big Data: the huge gains in efficiency of shopping, travel, and health as well as the benefits of improved security and crime prevention, but with risk of a loss of privacy. 'However, dazzling we find the power of big data to be, we must never let its seductive glimmer blind us to its inherent imperfections' (p. 197).

Given these questions, Masser and Wegener (2016, p. 1159) conclude:

> Already, there is a wide range of reflections and speculations about these questions and more definitive answers to some or most of them may emerge during the next twenty years. For example, in 1998 The US Vice President Al Gore proposed a vision of a Digital Earth as a virtual representation of the earth accessible to all citizens. Since then several virtual globes, such as Google Earth, have been implemented. In 2012 a team of experts (Goodchild et al., 2012) has recommended that the next generation of Digital Earth will consist of multiple connected infrastructures for different needs, be problem-oriented, allow search through time and space, ask questions about change, enable open access to data, services, models and scenarios, support visualisation and be engaging, interactive and exploratory.

Another interesting development occurred in 2012 when the International Federation of Surveyors proposed the concept of the Spatially Enabled Society (Steudler and Rajabifard, 2012). Spatial enablement is a concept that adds location to existing information. Societies and their governments need to become spatially enabled to take the right decisions. The United Nations Committee of Experts on Global Geospatial Information Management (UN-GGIM, 2015) has also envisaged an integration of official statistics, geospatial information, satellite data, Big Data, crowd-sourced data and building information modelling (BIM) to facilitate the creation of Smart Cities based on the Internet of Things. Despite these developments it is worth noting that, a former Federal Bureau of Investigation (FBI) expert has painted a frightening picture of global cybercrime making everyone connected vulnerable to attacks by criminals and terrorists who are more innovative than the public authorities responsible for protecting them (Goodman, 2015). The consequences of these kinds of activities can also be seen in the *New York Times* (2018) report that a voter profiling company, Cambridge Analytica, was able to harvest personal information from 50 million Facebook users without their permission to use in its electoral profiling activities. This kind of activity reflects the threats to society as a whole that are associated with unprecedented power free from democratic oversight as a result of what Zuboff (2015) has termed 'surveillance capitalism'.

References

Brown, G., and M. Kyatta 2014, Key ssues and research priorities for public participation GIS (PPGIS), *Applied Geography, 46,* 122–136.

Campari, I. and Frank, A.U. 1993. Cultural differences in GIS: A basic approach, in Harts, J., Ottens, H.F.L. and Scholten, H.J. (Eds) *EGIS '93 Conference Proceedings,* Vol. I, 10–16, Utrecht/Amsterdam: EGIS Foundation.

Chakravarty, S. 2018, How GDPR impacts location data. *Geospatial World, 9,* 1, 10–11.

Commission of the European Communities (CEC) 2016. Regulation (eu) 2016/679 of the European Parliament and of the Council of 27 April 2016 on the protection of natural persons with regard to the processing of personal data and on the free movement of such data, and repealing Directive 95/46/EC (General Data Protection Regulation). *Official Journal of the European Union, L119,* 1–88.

Coppa, I., P.W. Woodgate and Z.S. Mohamed-Ghouse 2018. *Global outlook 2018: Spatial information industry,* Melbourne: Australaia and New Zealand Cooperative Research Centre.

Couclelis, H. 1993, The last frontier, *Environment and Planning B: Planning and Design, 20,* 1–4.

Curry, M.R. 1993. Producing a new structure of geographical practice: On the unintended impact of geographic information systems, Mimeo, Los Angeles: Department of Geography, University of California at Los Angeles.

Geospatial Media and Communications 2018. *GEOBUIZ: Geospatial industry outlook and readiness index,* 2018 edition, Noida, India: Geospatial Media and Comunications.

Goodchild, M., F.H. Guo, A. Annoni, L. Bian, K. de Bie, F. Campbell, M. Craglia, M. Ehlers, J. van Genderen, D. Jackson, A. J. Lewis, M. Pesaresi, G. Remetey-Fülöpp, R. Simpson, A. Skidmore, C. Wang, and P. Woodgate 2012, Next generation digital earth. *Proceedings of the National Academy of Sciences, 109,* 28, 11088–11094. www.pnas.org/content/109/28/11088.full.pdf (last accessed 16 February 2019).

Goodman, M. 2015. *Future crimes: Everythingis connected, everyone is vulnerable and what we can do about it,* New York: Doubleday.

Hägerstrand, T. 1968. *Innovation diffusion as spatial process*, Chicago: Chicago University Press.

Lake, R.W. 1993, Planning and applied geography: Positivism, ethics, and geographic information systems, *Progress in Human Geography, 17*, 404–413.

Lin, H., M. Batty, S, Jorgensen, B.Fu, M. Konecny, A. Voinov, P.Torrens, G. Lu, A-X. Zhu, J.P.Wilson, J. Gong, O. Kolditz, T.Bandrova and M. Chen 2015, Virtual environments begin to embrace process based geographical analysis. *Transacttion in GIS, 19*, 4, 493–498.

Masser, I. and H. J. Onsrud (Eds) 1993. *Diffusion and Use of Geographic Information Technologies*, Dordrecht: Kluwer.

Masser, I., O. Sviden and M. Wegener 1992. The *Geography of Europe's Futures*, London: Belhaven Press.

Masser, I., and M. Wegener 2016, Brave new GIS worlds revisited, *Environment and Planning B, 43*, 1155–1161.

Mayer-Schönberger, V., and K. Cukier 2013. *Big Data: A revolution that will transform how we live, work and think*, London: John Murray.

National Research Council, Mapping Science Committee 1997. The future of spatial data and society: Summary of a workshop, Washington DC: National Academy Press.

New York Times, 2018. How Trump consultants exploited the Facebook data of millions, New York Times March 17th 2018.

Obermeyer, N.J. 1992. GIS in democratic society: Opportunities and problems. Mimeo, Terre Haute, Indiana: Department of Geography and Geology, Indiana State University.

Onsrud, H.J., J. P. Johnson, and X. R.Lopez 1994, Protecting privacy in using geographic information systems, *Photographic Engineering and Remote Sensing, 60*, 1083–1095.

Openshaw, S. 1991, A view on the GIS crisis in geography, or, using GIS to put Humpty-Dumpty back together again, *Environment and Planning A, 23*, 621–628.

Openshaw, S. 1992, Further thoughts on geography and GIS: A reply, *Environment and Planning A, 24*, 463–466.

Openshaw, S. 1993, GIS 'crime' and GIS 'criminality', *Environment and Planning A, 25*, 451–600.

Pickles, J. (Ed) 1994. *Grand Truth: The Social Implications of Geographic Information Systems*, New York: Guildford Press.

Postman, N. 1991. *Technopoly*, New York: Alfred Knopf.

Rajabifard, A., A. Binns, I. Masser and I. Williamson 2006, The Role of Sub-National Government and the Private Sector in Future Spatial Data Infrastructures, Int Jour Geographic *Information Science, 20*, 727–741.

Rogers, E.M. 1962. *The Diffusion of Innovations*, New York: The Free Press.

Rogers, E.M. 1993. The diffusion of innovations model, In Masser, I. and Onsrud, H.J. (Eds.) *Diffusion and Use of Geographic Information Technologies*, 9–24. Dordrecht: Kluwer.

Smith, N. 1992, History and philosophy of geography: Real wars, theory wars, *Progress in Human Geography, 16*, 257–271.

Steudler, D., Rajabifard, A. 2012. *Spatially Enabled Society*, Copenhagen: International Federation of Surveyors (FIG).

Sui, D., S. Elwood and M. Goodchild 2017. *Crowd sourcing geographical knowledge: Voluteered Geographic Information (VGI) in theory and practice*, Dordrecht: Springer.

Taylor, P.J. and M. Overton 1991, Further thoughts on geography and GIS, *Environment and Planning A, 23*, 1087–1232.

United Nations Committee of experts on Geospatial Information Management (UN-GGIM), 2015. *Future trends in Geospatial Information Management: The five to ten year vision*, 2nd edition, http://:ggim.un.org/docs/Future-trends.pdf (last accessed 16 February 2019)

Wegener, M. 1987. Spatial Planning in the Information Age, in Brotchie, J.F., P. Hall, and P. W. Newton (Eds.) *The Spatial Impact of Technological Change*, pp. 375–392. London: Croom Helm.

Wegener, M. and H. Junius 1993. 'Universal' GIS versus national land information traditions: Software imperialism or endogenous developments? in Masser, I. and Onsrud, H.J. (Eds.) *Diffusion and Use of Geographic Information Technologies*, pp. 213–228. Dordrecht: Kluwer.

Zuboff, S. 2015, Big other: Surveillance capitalism and the prospects of an infomation civilisation, *Journal of Information Technology, 30*, 75–89.

5

Geographic Information

A Resource, a Commodity, an Asset or an Infrastructure?

Robert Barr and Ian Masser

5.1 Introduction

The development of an economy which is increasingly based on trading information rather than material goods requires a re-examination of fundamental economic principles. This re-examination is far from complete: 'How knowledge behaves as an economic resource, we do not yet fully understand, we have not had enough experience to formulate a theory and test it' (Drucker, 1993, p. 183).

Negroponte (1995) sums up this transformation as the shift from an 'atom-based economy' to a 'bit-based economy' and re-emphasizes the fact that information – bits behaves differently from atoms. In particular, the economics of atoms are based on value being derived from rarity. Material goods arc inevitably rationed by availability, and where the goods are manufactured, or extracted, the price of material goods depends on the balance between supply and demand. Geographical variations in the costs of material goods are based, at least in part, on the effort required to transport those goods. In contrast, information, once created, can be reproduced, distributed and reused at negligible cost. To make information behave in a way analogous to material goods, it is necessary to limit access to it and create an artificial scarcity. This can be achieved either by embedding information in a material medium, such as a book or a map, or by pricing policies and the stringent assertion of intellectual property rights. It is necessary to treat information in this way in order to achieve the financial returns necessary to ensure that information continues to be created, refined and maintained. Without such a return information may cease to flow.

As one of the first major information markets, the market for geographic information has found itself embroiled in the controversies that arise from the paradox that information is costly to produce but can effectively be distributed to consumers at virtually no cost.

With these considerations in mind, this chapter considers geographic information from four different standpoints and explores the implications of each of them with respect to the role of governments:

- geographic information as a resource: this emphasizes the fact that geographic information is only of value when steps are taken to exploit its potential for users;

- geographic information as a commodity: this emphasizes the degree to which geographic information can be bought and sold like any other commodity;

- geographic information as an asset: this introduces the concept of custodianship and distinguishes between matters relating to access and those related to ownership;

- geographic information as an infrastructure: this highlights the extent to which the high costs of the creation and maintenance of geographic information requires government involvement to protect the public interest.

The discussion draws upon the general literature surrounding the debate on information in order to highlight the main issues underlying these four viewpoints, as well as on literature from the geographic information field.

This chapter is organized as follows. Section 5.2 defines the special significance of information and geographic information in the context of the emerging global information economy. This is followed by four sections dealing with the resource, commodity, asset and infrastructure viewpoints. Finally, Section 5.7 summarizes the main findings that emerge from this analysis, while Section 5.8 considers some more recent views on the nature of information infrastructures.

It should be noted that the discussion focuses on different conceptual standpoints with respect to geographic information rather than current practice in different parts of the world. As a result, it is hoped that the findings of the analysis will provide a better context for future discussions relating to national geographic information strategies throughout the world.

5.2 The Significance of Information and Geographic Information

Dictionary definitions of information are not entirely helpful because of the degree of circularity in the relationships between knowledge, information and data.

In a geographical context, it is perhaps easiest to think of a hierarchy. At the bottom of the hierarchy comes data. Data can be thought of as the symbolic representations of some set of observations. Data comprise the raw characters and digits that occupy computer storage media. Data alone have little value but incur significant storage costs and some transmission costs. At the next level in the hierarchy comes information. Information implies both data and context. The context may be the definitions of the variables that the data represent, their reliability and timeliness. These additional characteristics of data are usually termed metadata. In simple terms, information = data + metadata.

Knowledge, in turn, implies understanding. For information to become knowledge, an element of understanding of the significance and the behaviour of that information is required. In the academic world, data and information alone are considered to be of little value except as steps on the path to knowledge. However, in everyday life information alone is usually sufficient. It is enough to know that a postal address, or a house number and a post code, refers to an addressable property. No deeper significance needs to be sought.

As raw data are of no value, and true knowledge is seldom attained, much of our argument revolves around information. It is useful to consider the concept of information in the context of the information economy in general terms. Goddard (1989, pp. xvi–xvii) argues that this contains four propositions:

- Information is coming to occupy centre stage as a key strategic resource on which the production and delivery of goods and services in all sectors of the world economy will depend.

- This economic transformation is being underpinned by technological transformation in the way in which information can be processed and distributed.

- The widespread use of information and communications technologies is facilitating the growth of the so-called tradable information sector in the economy.

- The growing internationalization of the economy is making possible the global integration of national and regional economies.

These propositions provide a useful starting point for the general discussion in that they emphasize the linkages between information as such and the communications technologies that have recently come into being. They are also valuable in that they draw attention to some of the implications of these developments for economic restructuring and globalization. General support for these propositions can be found in the work of Cleveland (1985, p. 185): 'information (organised data, the raw material for specialist knowledge and generalist wisdom) is now our most important and pervasive resource'.

Geographic information must be seen as a special case of information as a whole. It can be defined as: 'information which can be related to a location (defined in terms of point, area or volume) on the earth, particularly information on natural phenomena, cultural and human resources' (Department of the Environment, 1987, p. 131). This definition makes the important distinction, albeit subtly, that there are two elements to geographic information location and attribute. Location is clearly essential in order to make information geographic. However, locational information without attribute information is sterile and is of little inherent interest.

The economic significance of geographic information lies in the general referencing framework that it provides for integrating large numbers of different data sets from many application fields in both the public and private sectors. For this reason, items of geographic information such as topographic maps, standardized geographic coordinate references such as the National Grid, standardized geographic referencing systems such as street addresses, and standardized areas such as administrative subdivisions and postcode sectors are particularly valuable in that they make it possible to link different data sets, and thereby to gain additional knowledge from them. An early example of adding value in this way was the work of the London doctor John Snow, who demonstrated in 1854 that there was a strong association between the home locations of cholera victims and the water pump in Broad Street by combining the two sets of data on deaths and water pumps on a single map (Gilbert, 1958).

Although the value of linking such data has long been recognized, it was not until recently that computer technologies came into being which are capable of manipulating large quantities of geographic data in digital form. With the arrival of geographic information systems (GIS) handling technology during the 1980s, the potential for linking geographic data sets increased dramatically and the modern geographic information economy came into being. The subsequent growth of the tradable information sector and the globalization of the geographic information industry are now part of history.

It is important to note that those elements of geographic information that are of special significance are also those that require a considerable measure of consistency in order that they can be used effectively. For example, in practice, map projections may vary from one country to another and in different applications. Similarly, the administrative areas used for local government are not necessarily the same as those used for other purposes. Furthermore, subdivisions such as these vary over time as modifications are made for administrative purposes. Consequently, some measure of standardization is needed if the potential for linking data sets is to be fully exploited. This means that units such as postcodes, which were originally intended to facilitate the operations of the postal services, acquire new significance as a national geographic information resource.

Standardization also implies a criterion for correctness. The literature on data quality identifies many dimensions to this concept, but for the topographic template and for fundamental referencing information such as the postal address file, it is possible to specify precisely what constitutes a correct and up to date version of the data set. While the ideal correct set of geographical references may be hard to achieve, a standard provides a target to aim at. This in turn implies that there is little point in attempting to produce several versions of the same referencing data set. In fact, the production of multiple versions of a data set, such as a comprehensive address file, for varying purposes reduces the general value of each version. It is preferable that a single standard version of such a file should exist. This creates a difficulty with geographic data because it implies that a natural monopoly should exist for fundamental referencing data and that, paradoxically, the more similar data sets there are, the lower the aggregate value and quality of the result.

These features of geographic information make it a special case not only of information in general but also of the new information economy that is transforming society.

5.3 Geographic Information as a Resource

Both Goddard and Cleveland refer to information as a resource which has many features in common with other economic resources such as land, labour and capital. However, Cleveland (1985) argues that information possesses a number of characteristics that make it inherently very different from these traditional resources. As a result, the laws and practices that have emerged to control the exchange of these resources may not work as well (or even at all) when it comes to the control of information.

Cleveland identifies six unique and sometimes paradoxical qualities of information which make it unlike other economic resources:

1. Information is expandable; it increases with use.
2. Information is compressible, able to be summarised, integrated, etc.
3. Information can substitute for other resources, e.g. replacing physical facilities.
4. Information is transportable virtually instantaneously.
5. Information is diffusive, tending to leak from the straightjacket of secrecy and control, and the more it leaks the more there is.
6. Information is shareable, not exchangeable; it can be given away and retained at the same time. (Cleveland, cited in Eaton and Bawden, 1991, p. 161).

Given the significance attached to these six qualities in much of the literature on the economics of information, it is useful to consider them in some detail with particular reference to their applicability to geographic information.

5.3.1 Information Is Expandable

Mason et al. (1994, p. 42) point out that information tends to expand with its use as new relationships and possibilities are realized. Cleveland (1982, p. 7) also draws attention to the synergetic qualities of information: 'the more we have, the more we use, and the more useful it becomes'. There are close parallels between these views and those of leading writers on geographic information. For example, Rhind (1992, p. 16) concludes that all GIS experience thus far strongly suggests that the ultimate value is heavily dependent on the association of one data set with one or more others, thus in the EEC's CORINE (and in perhaps every environmental) project, the bulk of the success and value came from linking data sets together.

However, Rhind also emphasizes that this depends very much on the unique properties of the geographic information referencing system referred to above:

> Almost by definition, the spatial framework provided by topographic data is embedded in other data sets (or these are plotted in relation to it, or both); without this data linkage, almost no other geographical data could be analysed spatially or displayed.
>
> *(p. 16)*

Paradoxically, geographic data often expands, while being degraded. Census data, for example, are made anonymous by aggregation, but the process of aggregation and tabulation actually increases the size of the data set and this increase is potentially almost infinite. A small number of census questions posed to individuals and households, yield over 8000 counts when aggregated up to the level of the enumeration district and tabulated. The tables which are produced are only a carefully selected subset of all those that could be produced. Thus, the process of geographical analysis and reaggregation itself has the potential to expand information.

Similarly, the editors of a major work on sharing geographic information (Onsrud and Rushton, 1995, p. xiv) point out that 'sharing of geographic information is important because the more it is shared, the more it is used, and the greater becomes society's ability to evaluate and address the wide range of problems to which information may be applied'.

5.3.2 Information Is Compressible

Another unique feature of information is the extent to which complex data sets can be summarized. Cleveland (1982, p. 8) points out:

> We can store many complex cases in a theorem, squeeze insights from
> masses of data into a single formula [and] capture lessons learned from
> much practical experience in a manual of procedure.

The enormous flexibility of information and the opportunities it opens up
for its repackaging in different ways to meet the demands of particular
groups is clearly evident in the current geographic information practice.
A good example is the field of geodemographics, which makes extensive use
of a wide range of lifestyle classifications derived from a mass of small area
census statistics. This reverses the process of expansion described above.

In addition to the compression of attribute information, referencing
information can also be compressed. For example, given a suitable look-
up table, a house number and a postal code in the UK almost always
uniquely identifies a postal delivery point. It is important to distinguish
between application-specific means of compression, where geographic
characteristics determine the form of compression used, and generic
compression that can be used on any digital data.

5.3.3 Information Is Substitutable

In essence, this property of information means that it can replace labour,
capital or physical materials in most economic processes. According to
Mason et al. (1994, p. 44), substitutability is one of the main reasons for the
power of information, and it can be used to harm as well as to improve the
human condition.

Once again there are many examples of substitutability in the geographic
information field. A typical example is the requirement placed on UK local
authorities by the 1991 New Roads and Streetworks Act to maintain
a computerized street and roadworks register. The geographic information
contained in such a register makes it possible to locate underground facilities more precisely than before, thereby saving the time needed by workers
to carry out essential maintenance and repairs, as well as reducing the
number of disruptions caused by the works.

5.3.4 Information Is Transportable

Given the digital networks that have been created by modern technology, large
quantities of information can be transferred from one place to another in the
world almost instantaneously. This means that information users can create
their own virtual databases by accessing data they need from other databases
as and when it is needed. A good example of a virtual database for geographic
information is the national land information system pilot project for Bristol,
which operates on the Land Registry's mainframe and provides online access
to live databases held by the Ordnance Survey at Southampton, the Valuation
Office at Worthing, the Land Registry at Plymouth, and Bristol City Council.

Technical developments will have an enormous impact on the transport-ability of information because there are generally only two reasons for holding information. One is to enforce intellectual property rights by acting as the holder and gatekeeper to a set of information. The second is to overcome the difficulties in transporting and reusing the data. In the next few years, the development of very reliable high-speed wide-area networks based on ATM (asynchronous transfer mode) technology will make it just as fast and easy to access a data set residing on a distant disk attached to a distant computer as it is to recover data from one's own hard disk. At that stage holding data in multiple locations becomes a relatively inefficient way of using it. It makes more sense to access it from source when necessary. While still relatively slow, the holders of the US Census CD-ROM based archives at the University of California Berkeley library already advise users to mount one of their CD-ROM drives as a remote disk rather than transferring data in bulk

5.3.5 Information Is Diffusive and Shareable

These two properties have a number of common features and are best dealt with at the same time. However, the idea that information is diffu-sive reflects the intangible qualities of information that distinguish it from material resources. The idea that intonation is shareable draws attention to the extent to which information can both be given away and retained and does not wear out as a result of being used. Both these ideas have been given new significance by recent technological developments which make it possible to copy information at near-zero cost. Furthermore, both raise fundamental problems about the ownership of digital information. For this reason, Branscomb (1995) argues that the traditional ways used to control the flow of information, such as copyright, need to be rethought in the context of digital information. This is because the roots of the concept of copyright are embedded in the printing press and the notion that there is an artefact that can be copied: 'In the computer environment it is access to organized information which is valuable and everything is copied. It is impossible to use a computer program without copying it into the memory of the computer' (p. 17). Because of its diffusability and shareability, information has many features of a public good in economics, i.e. its benefits can be shared by many people without loss to any individual, and it is not easy to exclude people from these benefits.

Similar views have also been expressed in the debates regarding geo-graphic information. For example, Rhind (1992) notes that geographic information does not wear out with use, although its value may diminish over time due to obsolescence, and also that it can be copied at near-zero cost although this may not be the case in situations where currency is important. Rhind also identifies two conflicting tendencies on the interna-tional scene which reflect these properties: the increasing commercialization of geographic information supply and the free exchange of data among scientists working on global problems.

Onsrud (1995) considers this situation from the standpoint of power:

> Because geographic information has potential value to those with effective access to it, this realization gives rise to the desire to exercise ownership rights over this information ... the desire to control information is in direct tension with recent technological realities that make the copying, dissemination and sharing of information very inexpensive ... As technology improves, this tension will only increase.
>
> *(p. 293)*

A particular problem with geographic data is that in addition to being diffusive and shareable, it is also very easily transformed. Some transformations are reversible, for example, a change of projection can be reversed and the original product reproduced, within the limits of machine accuracy. However, a wide range of operations such as generalization, random perturbation, translation or selection can make it very difficult to identify the origins of a particular data set. This causes difficulties in the assertion of intellectual property rights. Suppliers are obviously keen that their rights in a data set should remain regardless of the series of transformations it has been subjected to. They also want to be able to prove that a derived data set is based on their original. One way of achieving this is to place intentional errors, patterns or perturbations into their data which can be identified as an electronic 'watermark', regardless of subsequent transformations. In contrast, users often assume that once they have carried out a number of operations on raw data, their addition of value and intellectual property should override the original claim. It is the mutability of geographic information that raises many of these issues in a way that does not apply to other types of data.

When drawing the analogy between geographic information and other resources, it is also important to distinguish between ubiquitous resources, such as land, air or water, and rare resources such as particular minerals or resources that arise as a result of invention, such as a particular type of drug. The economic and legal nature of each of these resources is different. Certain resources are seen to be part of the public domain and their ownership and exploitation is tightly regulated. Others are seen as private resources and ownership alone determines most of the rights associated with the resource. Likewise, geographic information can and should be differentiated. For example, a company's client list is clearly a vital resource (and can be seen as an asset; see below). But are the addresses of the clients, once deprived of the attribute information that these are clients for particular goods, part of that resource? Clearly not, because anyone can hold that set, or an overlapping set of addresses. So, the geographic information contained in the client list has two parts: a public geographic key, the address; and a private attribute, the client details. The debate that ensues concerns the extent to which public geographic keys can legitimately be seen as private resources.

The above discussion highlights some of the parallels that exist between the issues that are being debated in the geographic information field and those identified by Cleveland as unique properties of information as a whole. The conclusion to be drawn from such a discussion, however, is that, because of these properties, information cannot be treated as a resource in the same way as land, labour or capital. Nevertheless, as Eaton and Bawden (1991, p. 165) point out, information must still be regarded as a resource in the sense that it is vital to organizations because of its importance to the individuals within them. The task for managers is therefore to exploit the potential of information as a resource while taking account of its singular qualities.

5.4 Geographic Information as a Commodity

The concept of the information economy assumes that information can be bought and sold like any other commodity. Yet, as was the case with the notion of information as a resource, its unique features make it significantly different from other commodities. This is particularly apparent in the public good dimension that information possesses because of its shareability and its diffusiveness. Unlike other commodities, information remains in the hands of the seller even after it has been sold to a buyer, and its inherent leakiness makes exclusive ownership of information problematic. In addition, the compressibility of information makes it difficult to define what constitutes a unit of information, and its expandability raises questions about how to gauge its value given that this depends heavily on its context and its use by particular users on particular occasions (Repo, 1989). In summary, it is fashionable to speak of information as a commodity, like crude oil or coffee beans. Information differs from oil and coffee, however, in that it cannot be exhausted. Over time certain types of information lose their currency and become obsolete, but, equally certain types of information can have multiple life cycles. Information is not depleted by use, and the same information can be used by, and be of value to, an infinite number of consumers (Cronin, 1984).

The discussion of information as a commodity is further complicated by the fact that large quantities of information are collected by government agencies in the course of their administrative duties, and most of this information is not made available to the public. This is particularly the case with respect to geographic information. In the UK, for example, over 500 separate data sets held by government agencies are listed in the Ordnance Survey Spatial Information Enquiry Service (SINES), but a large proportion of these are not available to external users.

There are a number of reasons why information collected by government is not made generally available. First, it can be argued that some of this information is obtained on the condition that its confidential nature will be respected. Second, it can also be argued that personal information acquired for the performance of a specific statutory duty should not be made available for other purposes. This principle is embodied in British data protection legislation. A case in point, quoted by the British Data Protection Registrar (Jones, 1995), was when the Department of Education was not allowed to use a list of child benefit claimants assembled by the Department of Social Security as a mailing list to send out leaflets to all parents, on the grounds that the use of this information was restricted to the statutory purpose for which it was collected. Finally, it can be argued that the primary task of government agencies is to carry out their statutory responsibilities and that the dissemination of information collected in the course of these duties may be both a burden on resources and a distraction from their primary responsibilities.

These arguments have been criticized on the grounds of both the needs of the information economy and the need for open government. With respect to the former, Openshaw and Goddard (1987), for example, draw attention to the emerging market for geographic information in the private sector and urge the public sector to make more data holdings available to promote the expansion of this important market. With respect to the latter, Onsrud (1992, p. 6) claims, 'All other rights in a democratic society extend from our ability to access information. Democracy can't function effectively unless people have ready access to government information to keep government accountable.'

Given these criticisms there has been much debate about how the costs of collecting, maintaining and disseminating these data sets to the public at large should be paid for. The traditional view is that governments collect information primarily for their own administrative purposes and that the costs they incur in the process are offset by the benefits obtained by the public in the form of the delivery of better public services. Under these circumstances it is argued that, as the public has already paid for the collection of this information through taxes, it should be made available to the public at no more than the marginal cost of reproduction. It is claimed that such a policy would not only increase the accountability of government departments to the public but also stimulate the growth of the information economy.

The cases for and against these arguments have been widely debated over the last few years (see, e.g., Maffini, 1990; ALIC, 1990b; Blakemore and Singh, 1992). The arguments against public domain databases have been summarized by Rhind (1992, p. 17) in five main points:

> First is that only a small number of citizens may benefit from the free availability of data which has been paid for by all and this is not fair. The second argument partially follows on from the first and is that any

legal method of reducing taxes through the recouping of [public] expenditure is generally welcomed by citizens. The third is that the packaging, documentation, promotion and dissemination of data invariably costs considerable sums of money ... thus nothing can be free. The fourth argument in favour of charging is that putting a price on information inevitably leads to more efficient operations and forces consumers to specify exactly what they require. Finally, making data freely available is liable to vitiate cost sharing agreements for its compilation and assembly.

On the other hand, opponents of cost recovery, such as Onsrud (1992, p. 6), claim that it runs counter to the principle of public accountability:

> Because of the need to maintain confidence in our public adminis-trators and elected officials and to avoid accusations that they may be holding back records about which citizens have a right to know, we should not restrict access to GIS data sets gathered at tax payer expense simply because that information is commercially valuable to the government.

Onsrud also argues that cost recovery will also increase bureaucratic costs and discourage the sharing of geographic information. Further-more, he claims that cost recovery will result in the creation of government-sanctioned monopolies and that once these monopolies come into being, there will be little incentive for them to improve their services.

It should also be noted that the notion that geographic information can, and should, be treated as a commodity is used largely to justify pricing policies that are based on the transfer of a measurable quantity of informa-tion for a given sum of money and, often, an annual service fee. By treating data as a commodity, it is implied that more data is better and should cost more, even though the client may be entitled to some bulk purchase discount. While this fiction could justify some pricing regimes, we know that geographic data aren't really like that.

Because information is expandable and compressible, its volume cannot be measured sensibly. It varies in quality over space and over time. While data is sometimes intentionally degraded to justify selling it at a lower price to clients who are not prepared to pay for the full quality, price is seldom related to the achieved quality of any particular data set (although this argument is often articulated, the reduction in quality is usually simply a device to ensure that premium customers, requiring high quality, continue to pay a premium price). It is difficult to think of other commodities which are delivered in uncertain quanti-ties, are of uncertain quality, yet continue to be sold at an apparently fixed price per unit.

True commodities should also be fully substitutable; sugar or corn sold on the commodity exchanges in Chicago can be traded internationally simply because there are fixed standards for the product regardless of the producer. This substitutability is seldom the case for geographic information, in particular geographic referencing information. A standard has been established in the UK (BS7666) for the creation of a master address file. If that standard is fully adhered to, any authority should, in theory, be able to create a standard address file. However, because such an authority has certain powers to name and allocate identifiers, only one authority can be responsible for the operation. In this case, as is often the case with geographic information, a natural monopoly exists in the creation of geographic references.

Topographic mapping is more problematic. Although surveys are expensive, and a creative element exists in mapping, the rapid development of digital orthophotography, the availability of accurate and inexpensive GPS receivers, and the higher standards of specification for topographic maps are all removing the uniqueness of the product. In one respect this could be considered to be a good thing because competition between alternative mapping agencies could lead to better products at lower prices. However, in practice, outside major metropolitan areas where the demand for maps is high, this is unlikely to happen and natural monopolies would emerge.

Arguments such as these highlight some of the problems associated with treating information as a commodity, particularly where government agencies are involved. This presents particular problems in the field of geographic information because of the large number of data sets involved and the vital importance of some data sets – such as the topographic data collected by national mapping agencies – for linking together other data holdings. It should also be noted that national mapping agencies differ from most other government agencies in one significant respect. They do not collect data in order to carry out their administrative duties. Their administrative duties are primarily to collect the data needed to maintain the national topographic database (Rhind, 1991). In this case the arguments for cost recovery may be stronger than those for most other government departments.

In summary, then, information has some of the features of a conventional commodity that can be bought or sold. However, it also has a number of unique qualities which must be taken into account in the process. These are closely linked to the previous discussion regarding the notion of information as a resource. However, when it comes to information collected by government agencies, it is necessary to add a number of additional qualifications to the concept of information as a commodity. This includes the recognition that much information is not primarily collected for resale as a commodity, and that the costs of much of this information are met through taxes on the public in the interests of good governance.

5.5 Geographic Information as an Asset

At the outset, it must be recognized that the concept of information as an asset introduces a new dimension into the discussion. As Branscomb (1994, p. 185) points out:

> We are in the process of designing a new paradigm for our information society, one that offers room for great economic, intellectual, social and political growth. It must be based upon the recognition that information is a valuable asset whether the claimant is an individual, a corporation, a national entity or humanity at large.

For this reason, she argues that it will be necessary to recognize the rights of individuals to withhold personal information and also to redefine the responsibilities placed upon the custodians of public information.

Similarly, the Australian Land Information Council (ALIC, 1990a, p. 5) has argued that 'the principle of custodianship lies at the core of efficient, effective and economic land information management'. In simple terms,

> all data collected by a State Government agency forms part of a State's corporate data resource. Individual agencies involved in the collection and management of such land related data are viewed as custodians of that data. They do not own the data they collect but are custodians of it on behalf of the State.

The concept of custodianship is important in that it highlights the importance of defining rights of access to databases and of specifying the responsibilities of custodians for database maintenance. In essence then, 'the distinction between ownership and access is quite clear. The issue is how to ensure effective access to information of importance to the community for a variety of purposes' (Epstein and McLaughlin, 1990, p. 38).

Once geographical information is recognized to be an asset rather than a commodity, it becomes feasible to sell access rights rather than the asset itself. Few fishermen are keen to have all the problems involved in owning a stretch of river, but they are keen to have the rights to fish in that river. Likewise, most users of geographic information have no real interest in owning, or having to maintain, copies of that information. They want rights of access and rights of use.

Another characteristic of geographic information that makes it better to consider it an asset rather than a commodity is that at any point in time a user only requires a very small subset of the total asset. Motoring organizations, and some route planning software, produce customized strip maps of routes. This is a customized, highly selective view of an underlying geographic data set that contains a small amount of 'just in time' information for a specific purpose. It is easy to forget that the comprehensive nature of many topographic maps, the full coverage of motoring atlases and the large volume of postcode, zip code or telephone number data is necessary because in the past it has proved impractical to give users access to the underlying asset at reasonable cost.

The debate over how to create a national extensible and accessible base of geographic information has been addressed in various ways. In the United States it has been proposed that a 'national spatial data infrastructure' should be built which comprises all the geographic information assets that are required for the effective operation of the federal government, and that inexpensive access to this asset should be provided for the nation as a whole (and if the rest of the world also wants to take advantage, so much the better). In the UK it has been proposed that a distributed network of geographic resources should be assembled by the private and public sectors, which would collectively produce a coherent asset base and consistent charging mechanisms (Nanson et al., 1995).

Consequently, it can be argued that information must be seen increasingly as an asset that acquires value only at the point of use and at the point of transformation into a product a customer wants to buy. However, this still leaves the issue of who should own this asset. Conventional property law relating to both real property and intellectual property is as yet weak in defining not where we are in relation to treating information as an asset, but where we should be in order to maximize the usage of that asset for the common good.

5.6 Geographic Information as an Infrastructure

Infrastructure has been defined as 'the basic facilities, services, and installations needed for the functioning of a community or society, such as transportation and communications systems, water and power lines, and public institutions including schools, post offices, and prisons' (*American Heritage Dictionary of the English Language*, 3rd end., Houghton Mifflin).

Once it is accepted that geographic information, and in particular geographic referencing information, is an asset, the question of ownership, management and funding appears. A simplistic monetarist approach would be to argue that there is no clear case for the production of geographical referencing data where the market does not demand it, and it is the case that many advanced capitalist economies, such as the United States and many parts of Europe manage their affairs without either large-scale mapping or comprehensive cadastral or address-based gazetteers. It can be argued that the cost–benefit case for geographic information has not been made. However, when such situations are examined more closely, a much less clear picture emerges of patchy coverage, or of duplications of effort in profitable markets, with no concomitant improvement in the product, its quality and coordination (see, e.g., National Research Council, 1993, pp. 20–27). It is very difficult to assess the expenditure, both by government and by the private sector on maintaining such a mess.

> Technology also plays an important part in determining what can be done. With traditional survey methods and clerical means of handling data there were few if any economies of scale to be gained from handling data consistently in different areas. Most large-scale geographical data are only of interest to users in the area covered and the duplication of manual efforts mattered little. But as technology is increasingly determining how we handle geographical information, standards are becoming a great deal more important.

The division of responsibility between the private and public sectors is also important. In a strict monetarist environment, it is clear that 'society does not exist'; there are solely buyers and sellers (and those without the money to buy are invisible). However, such a radical scenario has not emerged and governments have found it difficult to shrink their area of operations. Osborne and Gaebler (1992) recognized many collective responsibilities of governments, even if they felt that governments should shift from being providers to being enablers of privately provided services.

It follows from a concern for the common good, and for the efficient management of affairs for the whole of society, that a basic core of geographic referencing data must be considered part of the national infrastructure. It then becomes the responsibility of government to ensure that the infrastructure is produced, financed and maintained as efficiently as possible and that access to it is provided equitably. Consequently, governments are increasingly recognizing that a basic framework of topographic data, address files, administrative and statistical boundaries, and cadastral information is required for the efficient management of a modern economy. It is hard to argue that such information is anything other than infrastructure. Directly or indirectly, it is required by all the people. It needs to be captured only once and there are no efficiency savings, rather the reverse, in duplicating that effort.

To argue that such data form an infrastructure is not necessarily to argue for funding from taxation, usage fees are common for most other elements of infrastructure and a geographic infrastructure is no different from these elements in this respect. It is also not an argument for the government to carry out the work directly, since private or semiprivate agencies can be employed to do so. However, it is an argument for regulation. One of the principal characteristics of other infrastructural elements in the economy is that they are regulated for the common good and not left to the vagaries of the market or to exploitation by monopoly suppliers. An element of regulation and standardization is essential for the building of spatial data infrastructure, and it will be important to track how these processes develop around the world.

5.7 Conclusions

It is clear from the above analysis that information in general and geographic information in particular has some of the features of a resource, a commodity, an asset and an infrastructure. However, much of the discussion has focused on the resource and commodity standpoints and relatively little attention has been given to the asset or the infrastructure standpoints. This is particularly important with respect to geographic information because of the significance of its public interest dimension.

Considering geographic information from the asset standpoint enables a critical distinction to be made between questions relating to access and questions relating to ownership.

It also introduces the notion of custodianship into the discussion with respect to the role of governments in relation to information. The special characteristics of geographic information as such, together with the high costs that are involved in the development and maintenance of core data sets in the public interest, also make it necessary to take account of the infrastructure standpoint in the formulation of national geographic information strategies. It is clear that the nature of these strategies will vary from country to country because of the different institutional contexts that govern information and geographic information policymaking. This will be reflected in the choices that are made about the mix between public and private sector involvement and between public interest and cost recovery in each case:

The legal line between public and private domains is not fixed. Information is a valuable asset, a treasured resource and archive of knowledge. What use we make of it is up to us to decide (Branscomb, 1995, p. 18).

5.8 Some Notions of Information Infrastructures

At the time that the chapter was written, the authors were unaware of contemporary research on thinking about the notion of information infrastructures in Norway that clarify some of the most important concepts underling the notion of geographic information as an infrastructure.

The section remedies these deficiencies with references to contemporary and more recent sources.

A useful starting point for this discussion is an unpublished manuscript by Ole Hanseth and Eric Monteiro (1998) entitled 'Understanding information infrastructures'. This contains 11 chapters discussing the basic concepts together with a number of case studies illustrating their application in the Norwegian health care and related fields.

The first chapter starts with the assertion that these infrastructures are different from other kinds of information system.

> If one is concerned with the development of the 'information systems of the future' the phenomena we are dealing with should be seen as information *infrastructures* – not systems. This reconceptualization is required because the nature of the new ICT solutions are qualitatively different from what is captured by the concepts of information underlying the development of IT solutions so far.
>
> *(p. 4)*

In simple terms then, the concept of information infrastructures is characterised by six key aspects (pp. 10–11)

1. Infrastructures have an *enabling* function.
2. An infrastructure is one irreducible unit *shared* by a larger community (or collection of users and user groups).
3. Infrastructures are *open*. They open in the sense that there are no limits for the number of stakeholders, vendors involved, nodes in the network etc.
4. 'IIs' are more than pure technology; they are rather *socio-technical networks*.
5. Infrastructures are *heterogenous*. They are different in different ways.
6. Building a large infrastructure takes time. The whole infrastructure cannot be changed instantly – the new has to be connected to the old – the *installed base*.

With this in mind, the main information infrastructures have been described by Monteiro et al. (2014, p. 1) as follows:

> IIs are characterised by openness to number and types of users (no fixed notion of 'user'), interconnections of numerous modules/systems (i.e. multiplicity of purposes, agendas, strategies), dynamically evolving portfolios of (an ecosystem of) systems and shaped by an installed base of existing systems and practices (thus restricting the scope of design, as traditionally conceived). IIs are also typically stretched across space and time: they are shaped and used across many different locales and endure over long periods (decades rather than years).

This description broadly corresponds to the generally accepted definition of a spatial data infrastructure from more than 20 years ago as

> the means to assemble geographic information that describes the arrangement and attributes of features and phenomena on the Earth. The infrastructure includes the materials, technology, and people necessary to acquire, process, and distribute such information to meet a wide variety of needs.
>
> *(National Research Council, 1993, p. 16)*

Georgiadou et al. (2005, table 2) have tried to summarise the key features of information infrastructures from an SDI perspective. This highlights three particular features of IIs: the installed base and lock in effects, reflective standardisation, and the cultivation approach to design.

They view the implications of the concept of the installed base and lock-in effects as follows:

> History cannot be ignored, for example the existence of paper maps. Understanding how existing and embedded technologies influence the new, for example how the existing focus on remote sensing may influence *the* kind of data that is stored (of vegetation cover). While IIs need to evolve, they are constrained by the inertia of the installed base, such as the lack of map culture and institutional mechanisms and standards to enable sharing. The installed base is both technical (for example, scales of existing maps) and institutional (the ownership of maps within the defence department for example) in nature.

On the subject of reflexive standardisation, they feel that universal standards are a utopian quest, as SDI is a rapidly changing domain, both with respect to new technologies and changes in institutional actors and practices. Standards beget the need for new standards, as more users will continue to join (network externalities), which will place new demands on SDI. The use of gateways to develop interfaces between different parts of the network (say different user groups – e.g., forestry and roads), while preserving local autonomy. Developing 'hierarchy of standards' to provide flexibility for local levels (for example, the health department) to have their standards while ensuring the core standards (defined by the planning department for example) are maintained. Standardization is restricted not only to technical artefacts but also to management practices, for example, related to the incentives for data sharing. Developing and using standards involve a political negotiation of different stakeholders including scientists, user groups, vendors and policymakers. Standards should evolve bottom up and create 'local universalities', implying that standards should suit local needs first and foremost, but these should conform to more global frameworks.

With respect to the cultivation approach to design Georgiadou et al. (2005) argue:

> Design should be in small steps and incremental, for example, the focus could be on particular sectors (like forestry) and developing their applications within a wider SDI defined framework. SDIs can never be designed from scratch, and it is important to build upon existing inventories of spatial databases that exist. Grand, and top-down designs, such as those defined by central ministries or apex scientific institutions, are liable to failure. Small parts of networks (particular user groups) should be changed while keeping consideration of the dynamics of the whole network. SDIs should be flexible to absorb unanticipated effects, such as through new global standards being introduced or new user departments joining or leaving the network.

Aanested et al. (2007) have also argued that information infrastructures such as spatial data infrastructures must be treated as a public good in that they benefit many or all (i.e. they are non-excludable), and their consumption by one person doesn't prevent consumption by another (i.e. they are non-rival in consumption).

Four important considerations underpin this focus. First, they point out that it will be necessary to take a social-technical approach to implementation as 'designing information infrastructures is ... not simply a technical venture, but equally (and in some instances more importantly), a project of enrolling other actors through aligning their interests and practices' (p. 15). Second, it will be important to recognise that 'they are never built rather from scratch but rather they are building on, extending, and enhancing existing structures. Thus, information infrastructures are evolving and will inherit both the strengths and weaknesses of what already exists' (p. 16). Third, it will also be necessary to take account of the politics of representation which is reflected in the organisational, economic and political circumstances surrounding their development and implementation. 'They do not only have to be related to this wider context to work and grow, but they are also a product of it' (p. 18). Finally, successful implementation must also allow for the multiplicity of networks involving the participants in such programmes.

With these considerations in mind, the authors consider their implications for the concept of SDIs as public goods.

> Information infrastructures are not provided by one provider for certain purposes but are built as a joint activity and provide information upon which a range of activities can take place. In this process, information systems are no longer under central control, but become parts of networks on a different scale, where government(s), other institutional actors, and market actors influence to a varying degree the involved components and the overall development.
>
> *(p. 22)*

Hanseth and Monteiro (1998, p. 160) also argue that cultivation rather than design is the best strategy for implementing information infrastructures.

> Acknowledging the importance of the installed base implies that traditional notions of design have to be rejected. However, denying humans any role at all is equally, at least, wrong. Cultivation as a middle position captures quite nicely the role of both humans and technology.

References

Aanested M., E. Monteiro and P. Nielson 2007. Information infrastructures and public goods: Analytical and practical implications for SDI, *Information Technology for Development, 13,* 7–25.

Australian Land Information Council (ALIC), 1990a. Data Custodianship/ Trusteeship, *Issues in Land Information Management* Paper No. 1, Australian Land Information Council, Canberra

Australian Land Information Council (ALIC), 1990b. Access to Government Land Information: Commercialisation or Public Benefit? *Issues in Land Information Management Paper No 4,* Australian Land Information Council, Canberra.

Blakemore, M. and G. Singh 1992. *Cost Recovery Charging for Geographic Information: A False Economy?* London: GSA.

Branscomb, A. W. 1994. *Who Owns Information? From Privacy to Public Access,* New York: Basic Books.

Branscomb, A. W. 1995. Public and private domains of information: Defining the legal boundaries, *Bulletin of the Association for Information Science and Technology,* Dec/Jan, 14–18.

Cleveland, H. 1982. Information as a resource, *The Futurist,* 16, 34–39.

Cleveland, H. 1985. The twilight of hierarchy: Speculations on the global information hierarchy, *Public Administration Review,* 7, 1–31.

Cronin, B. 1984. Information accounting, In A. van der Laan and A.A, Winters (Eds.) *The Use of information in a Changing World,* Amsterdam: Elsevier.

Department of the Environment 1987. *Handling Geographic Information, report of the committee of enquiry chaired by Lord Chorley,* London: HMSO.

Drucker, P. 1993. *Post-Capitalist Society,* New York: HarperCollins.

Eaton, J.J. and D. Bawden 1991. What kind of resource is information? *International Journal of Information Management,* 11, 156–165.

Epstein, E.F. and J.D. McLaughlin 1990. A discussion of public information: Who owns it? Who uses it'? Should we limit access'?, *ACSM Bulletin,* Oct, 33–38.

Georgiadou, Y., S. Puri and S. Sahay 2005. Towards a research agenda to guide the implementation of spatial data infrastructures, *International Journal of Geographical Information Science,* 19, 1113–1130.

Gilbert, E.W. 1958. Pioneer maps of health and disease in England, *Geographical Journal,* 58, 172–183.

Goddard, J. 1989. Editorial preface, in: M. Hepworth (Eds.) *Geography of the Information Economy,* London: Belhaven.

Hanseth, O., and E., Monteiro, 1998. *Understanding information infrastructure*, unpublished manuscript. heim.ifi.uio.no/~oleha/Publications/bok.pdf (last accessed February 16 2019)

Jones, P. 1995. Presentation to the AGI/Government Round Table, London: AGI.

Maffini, G. 1990. The role of public domain data bases in the growth and development of GIS, *Mapping Awareness*, 4(1), 49–54.

Mason, R.O., F. M., Mason, and M. J., Culnan 1994. *Ethics of Information Management*, London: Sage.

Monteiro, E., N. Pollock and R. Williams 2014. Innovation in information infrastructures: Introduction to a special issue, *Journal of the Association for Information Systems*, 15, i-x.

Nanson, B., N. Smith, and A. Davey 1995. What is the British National Geospatial Database? *Proc. AGI '95 Conf.* London: AGI.

National Research Council 1993. *Toward a Coordinated Spatial Data Infrastructure for the Nation*, Washington, DC: National Academy Press.

Negroponte, N. 1995. *Being Digital*, Cambridge, MA: MIT Press.

Onsrud, H.J. 1992. In support of open access for publicly held geographic information, *GIS Law*, 1, 3–6.

Onsrud, H.J. 1995. The role of law in impeding and facilitating the sharing of geographicinformation, In H.J. Onsrud and G. Rushton (Eds.) *Sharing Geographic Information, Center for Urban Policy Research*, Brunswick, NJ: Rutgers University, 292–306.

Onsrud, H.J. and G. Rushton 1995. Sharing geographic information: An introduction, In H.J. Onsrud and G. Rushton (Eds.) *Sharing Geographic Information, Center for Urban Policy Research*, Brunswick, NJ: Rutgers University, xiii-xviii.

Openshaw, S. and J. B. Goddard 1987. Some implications of the commodification of information and the emerging information economy for applied geographical analysis in the UK, *Environment and Planning A*, 19, 1423–1439.

Osborne, D. and T. Gaebler 1992. *Reinventing Government*, Reading MA: Addison-Wesley.

Repo, A.J. 1989. The value of information approaches in economics, accounting and management science, *Journal of the Association for Information Science*, 40, 68–85.

Rhind, D. 1991. The role of the Ordnance Survey of Great Britain, *Cartographic Journal*, 28, 188–199.

Rhind, D. 1992. Data access, charging and copyright and their implications for geographic information systems, *International Journal of Geographical Information Science*, 6, 13–30.

Practice

6

The Development of Geographic Information Systems in Britain

The Chorley Report in Perspective

Ian Masser

6.1 Introduction

By any standards, the publication of the Chorley Report (1987) must be regarded as an event of major importance in the discussion of geographic information systems in Britain. To appreciate its full significance, it is necessary to consider the events that led up to the establishment of the Committee of Enquiry in April 1985 and also to take account of the views expressed in the Government's response to the recommendations which was published in February 1988 (DOE, 1988).

6.2 Background

The immediate origins of the Chorley Report lie in the recommendation made in the report of the House of Lords Select Committee on Remote Sensing and Digital Mapping (1984) for the establishment of a high-level committee of enquiry into the handling of geographical information. In its response to the Report, the Government (DTI, 1984) accepted that a more general discussion of these issues was required and it also recognised the need for a forum of users to be established to coordinate their mutual interests. Consequently, it was argued that: 'Only with the private and public user's standpoint clearly represented will it be possible to achieve a practicable programme which ensures that the potential of the data is realised, and that the costs of the programme are in step with the expected benefits' (paragraph 37).

As a result, an 11-person committee was appointed in April 1985 'to advise the Secretary of State for the Environment within two years on the future handling of geographic information in the United Kingdom, taking account of modern developments in information technology and of market need'. The Chairman of the Committee, Lord Chorley, is a partner in Coopers and Lybrand, the accountants and management consultants. His prior experience in this field included memberships of the Ordnance Survey Review Committee (Serpel, 1979) and the House of Lords Select Committee on Remote Sensing and Digital Mapping (1984).

In its call for evidence, the Committee invited views on a wide range of issues relating to the current state of the art and future developments. These included questions associated with the collection and handling of information, the release of this information to other users, and anticipated changes in technology. The Committee also asked for views on the nature of research and development work being carried out on geographic information handling and the organisational arrangements required to coordinate development in the field.

The extent of public interest in these issues is evident in the 400 submissions that the Committee received from organisations and individuals. In addition, the Committee took oral evidence from 26 organisations and commissioned a number of reviews of specific topics. Over 30 pages of the published report are devoted to a very interesting summary of this evidence and the papers commissioned from Dr Openshaw on spatial units and locational referencing and from Thomlinson Associates on the North American experience are also published as separate appendices.

6.3 The Report

The main body of the Report is divided into two more or less equal parts. The first part consists of three chapters reviewing recent developments in geographic information handling in general terms, whereas the second deals in more detail with the specific issues involved. As might be expected, the reasoning behind the Report's 64 recommendations is to be found in the second part.

The opening chapters of the Report highlight the significance that is attached to recent developments in geographic information handling technology by a wide variety of users, while drawing attention to the obstacles that inhibit its take-up. The Committee makes no bones about its enthusiasm for the new technology. In its view, the development of geographic information systems (GIS) is 'the biggest step forward in the handling of geographic information since the invention of the map' (paragraph 1.7).

The importance it attaches to GIS technology is a result of the degree to which it facilitates the convergence of three previously largely separate fields of applications. The first of these fields involves the development of spatial databases in digital form to enable quick and easy access to large volumes of data. Ordnance Survey topographic information is an example of data currently undergoing a massive conversion process from paper map to digital format. The second field is associated with the relational database-management technology that enables the integration of spatially referenced socioeconomic and environmental data drawn from different sources and their manipulation by a variety of users. The opportunities opened up by these means are reflected in the growing commodification of information in the economy, which can be seen in the marketing of specialised services by both public and private sector agencies (Openshaw and Goddard, 1987). The last application field relates to the new and flexible forms of output produced by computers in the form of maps, graphs, address lists, and summary statistics, which can be tailored to meet particular kinds of user requirement. These have had profound effects on the display of spatial information.

In looking to the future, the Committee expects rapid developments on all three fronts. However, it regards this as 'a necessary, though not sufficient, condition for the take-up of geographic information systems to increase rapidly' (paragraph 1.22). For a rapid take-up to occur, it will also be necessary to overcome a number of important barriers to development.

These include the need for greater user awareness of technological change and the extent to which data collectors and holders can be persuaded to provide data in digital form. The Committee also recognises that the diversity of users with very different technical needs makes it difficult to develop a consistent view of priorities.

These issues are explored in greater depth in the seven chapters that make up the second part of the Report. These cover digital topographic mapping, availability of data, linking data, awareness, education and training, research and development (R&D), and the role of government, respectively.

The Committee's most important recommendations are those relating to digital topographic mapping, the linking and availability of data, and the role of government. The Ordnance Survey is singled out for particular attention because of its role in the establishment of a national topographic database, and 24 of the 64 recommendations relate to its activities. The Committee expresses its concern about the slow pace of digital conversion and recommends a much-accelerated programme, using a simplified specification for conversion purposes.

With respect to data linking and availability, the Committee calls for the release of unaggregated data held by government departments, provided that the users are willing to bear the costs and there are no overriding security considerations. It is also in favour of an increased use of franchising arrangements to enable outsiders to act as distributors of government data. To facilitate the linking of data sets, the Committee recommends that address and unit post codes should be utilised wherever possible, and it suggests that the findings of major public data-collection exercises such as the 1991 Census of Population should be made available in this form, subject to the need to preserve confidentiality.

In the light of the issues raised by the House of Lords Select Committee on Remote Sensing and Digital Mapping and the Government response to them, the discussion of the role of government in the Chorley Report is of particular importance. The Committee recommends that a Centre for Geographic Information should be set up with a clear remit to carry out the following functions:

> 1. To provide a focus and forum for common interest groups, or clubs;
>
> 2. To carry out and provide support for promotion of the use of geographic information technology, including promotional activities carried out within and outside the general education and training process;
>
> 3. to oversee progress and to submit proposals for developing national policy in the following areas: the availability of government spatial data; the operation of data registers and arrangements for archiving of permanent data; the development of locational referencing, standard spatial units for holding and releasing data, the operation of the postcode system and the development of data exchange standards (cartographic and non-cartographic); the assessment of education and training needs and provision of opportunities to meet them; and the identification of R&D needs and priorities, including advice to government on bids for R&D funds.
>
> *(paragraph 10.2)*

In the view of the Committee, the Centre for Geographic Information should be independent of government but should be closely linked to it through membership and funding arrangements. This reflects the Committee's view that developments in this area will be primarily determined by user demands.

Like its immediate predecessor, the House of Lords Select Committee on Remote Sensing and Digital Mapping (see Rhind, 1986), the Chorley Report tends to play down the cost implications of its recommendations. It argues, for example, that any short-term increase in costs resulting from the implementation of its recommendations on the conversion of Ordnance Survey data should be more than outweighed by the subsequent increase in revenue. In the case of education and training provision, the Committee argues that no additional funding will be required 'if the money can be found by some modest redirection of funds in existing programmes' (p. 124). The only case where extra funding will be needed is for the launch of the Centre for Geographic Information. Even in this case, however, it is argued that, although the Centre would initially require £250,000 a year from membership subscriptions, income-generating activities, and launch finance from the government, the launch-finance element should taper to zero after five years.

6.4 The Government Response

The Government's response to the Chorley Report was published in February 1988 (DOE, 1988). It consists of a detailed response to each of the Report's 64 recommendations prefaced by an introductory section outlining the broader issues involved.

In its response the Government indicates that it shares the Committee's view of the potential that exists for the rapid spread of technological applications of geographic information handling and agrees that they offer very considerable benefits. The Government also makes it clear that it regards the Chorley Committee's Report as a major step in drawing attention to its potential and raising overall levels of user awareness.

The Government's response is most positive in respect of the activities of the Ordnance Survey. It points to the progress that has been made in producing a new specification for conversion purposes and notes that plans to speed up the conversion programme are already at an advanced stage. These are likely to be facilitated by the agreements that have been reached between the Ordnance Survey and British Telecom and British Gas whereby the latter agencies will digitise sections of their areas to an agreed standard, and the data that are obtained by these means will be incorporated into the Ordnance Survey database.

The Government is less forthcoming on the subject of data linking and availability, although it accepts the spirit of the Committee's recommendations regarding the release of government-held data. In this respect, it merely draws attention to the role that its Tradeable Information Initiative (DTI, 1986) is likely to play in making government data holdings accessible to commercial users who are willing to pay the full market rate for the privilege. On the subject of unit post codes, the Government's response is also cautious. It is only willing to consider producing results for future censuses of England and Wales in combinations of post codes if it can be demonstrated that there is a clear need for results, even though the results of the 1981 Census of Scotland are already available in this format.

Predictably, the most disappointing feature of the Government's response is its rejection of the proposal to set up a Centre for Geographic Information. The Government argues that it is important to build upon the strengths of existing organisations in this field and cites as examples the efforts of the European Division of Automated Mapping and Facilities Management (AM/FM) International in bringing together users and suppliers over the last few years. It also expresses the view that the Economic and Social Research Council's (ESRC) Regional Research Laboratories (RRLs) have the potential to develop into highly effective and locally based resources which users can draw upon for advice and consultancy work.

In the light of these developments, the Government argues that 'it is therefore unnecessary, and indeed could be harmful, to use public funds to set up an additional organisation which would compete with these developing organisations' (paragraph 14). Consequently, it wishes to 'encourage such organisations to continue to develop their roles in this field' and hopes that they will 'expand to cover the full range of applications' (paragraph 15).

6.5 Discussion

There can be little doubt that the Chorley Report will be of considerable value in raising the overall level of awareness of the potential opened up by information technology in the geographic information handling field. There are also encouraging signs in the Government's detailed response to its recommendations that some of the obstacles inhibiting the take-up of the new technology are being removed. However, these positive features cannot offset the negative impression that is given by its rejection of the proposal to establish a Centre for Geographic Information.

As noted at the outset of this paper, the need for a forum of users to be established which would coordinate their interests was recognised by the House of Lords Select Committee on Remote Sensing and Digital Mapping and their case was accepted in the Government's response to that report. The case for a Centre for Geographic Information that is put forward in the Chorley Report goes a considerable way beyond this initial recommendation. It sees the proposed Centre not only as providing a forum for users but also as a way of giving the sharp boost to existing efforts that is required to overcome the barriers to the rapid take up of the new technology which were identified by the Committee.

In simple terms, then, although the development of the field is likely to be determined largely by users, a clear lead from Government is also required to raise general levels of awareness and set in motion the massive education and training programme that are needed to enable a rapid take-up of the new technology. Most of the crucial decisions in these respects lie in Government hands. They alone can ensure, for example, that the ESRC and the Natural Environment Research Council are given the additional resources they will require to set up their activities related to training and research on geographic information handling to meet the growing demands from users.

Questions such as these appear to be largely ignored in the Government's response to the Chorley Report. For this reason, it must be recognised that the Government's favoured option, the establishment of the Association for Geographic Information (AGI) (AM/FM UK), although providing a useful forum for users, is unlikely, in the short term at least, to have more than a very limited effect on increasing the level of resourcing for training and research activities.

In retrospect, it may turn out that the diagnosis of the problems and the proposed course of treatment that are contained in the Chorley Report will be admired despite the Government's failure to accept their full implications. However, the Committee's tendency to play down the costs involved must also be seen as contributing to the Government's attitude to the Report's recommendations. There can be little doubt that substantial investment is taking place in this field at the present time and it is also clear that there are considerable benefits to be obtained by exploiting new technology. Given the scale of current operations, something more than a modest redirection of resources is required on the part of Government if it is to ensure that adequate provision is made for training and research in the geographic information handling field. On this count, then, the Chorley Report may come to be regarded as a missed opportunity which may make things even harder for those involved to secure the level of resources they will need to satisfy the demands of users.

6.6 What Has Happened in the UK over the Last Thirty Years?

Despite the reservations noted above, the publication of the Chorley report prompted a flurry of activities in the UK. The most important of these in the UK is the creation of the AGI in 1989, which effectively took on the task of becoming a de facto alternative to the national Centre for Geographic Information that was envisaged in the Chorley report.

One month after the Government's response was published, a meeting of interested parties was organised by the AM/FM organisation in Nottingham in March 1988. One outcome of this was that it was agreed that a fully independent body wholly based in the UK should be set up and in January 1989 the AGI was formally launched as a self-supporting independent organisation with a mission 'to maximise the use of geographic information for the benefit of the citizen, good governance and commerce' (Corbin, 2002). This has been delivered to its members through three kinds of activities:

- Informing: The membership is kept up to date through a regular newsletter, e-mails, mailings, seminars, publications and articles.
- Influencing: Many AGI members work in government, allowing access to decision-makers at the highest level.
- Acting: Whenever necessary, the AGI takes decisive action to ensure that its members' legitimate interests are promoted and, by consulting with members, the AGI can give the GI industry a respected voice at the appropriate level (Corbin, 2002).

The first AGI conference was held at the National Motor Cycle Museum in Birmingham on October 1989. Over 600 delegates attended the Conference sessions and well over twice that number visited the exhibition of GIS equipment and services. The theme of the Conference was 'GIS – a corporate resource'. In his opening remarks, the president of the AGI, Lord Chorley, said:

> The UK is in the forefront of developing and applying this new tool to manage spatially referenced data. Many observers have noted the parallel between GIS today and Computer Aided Design (CAD) and Computer Aided Engineering ten or fifteen years ago. In those fields then, the UK had a technically leading role but the opportunity was lost. The AGI has a vital role to play in explaining the value of GIS to British management in the wide range of organisations to which it is relevant.
>
> *(AGI, 1990, 1)*

The AGI has retained the mission statement listed above during its nearly 30 years of operation and has been very active in the development of the geographic information and geospatial industry. In May 1997, it also hosted a symposium at the Royal Society on 'The future for geographic information: 10 years after Chorley', which provided an opportunity to discuss likely future developments at a time when the market for geographic information is broadening every year. Lord Chorley's own reflections ten years after his report was published are also particularly interesting. He noted:

> The timing was about right. I suspect that had we been set up a few years earlier we would have been premature; the IT revolution in this field was at too early a stage for its shape, its problems, and in a practical sense, its potential to be tolerably clear. We were ... at a stage when one could take technical progress for granted and think more about other barriers.
>
> *(Heywood, 1997, 80)*

Prior to this date the AGI also organised a series of round tables between key public data providers such as Ordnance Survey and the Office of Population Censuses and Surveys to discuss key policy issues and supported efforts to create a National Geospatial Data Framework (NGDF) (Nansen et al., 1996). This preceded by a whole decade the UK government's commitment to building a National Spatial Data Infrastructure (Geographic Information Panel, 2008).

One outcome of the AGI's involvement in the NGDF is its longstanding commitment to UK GEMINI. This is a specification for a set of metadata elements for describing geospatial data resources for discovery purposes. It is compatible with the INSPIRE metadata requirements, and also contains additional elements considered to be useful in the UK context. UK GEMINI has been produced and is maintained by the AGI Standards Committee for widespread use in the GI community, as a free of charge resource. It is fully supported with detailed guidance written by AGI and by UK-INSPIRE (see, e.g., AGI, 2012).

On the research front the AGI played a pivotal role in supporting the initial establishment of the GISRUK conference series from 1988. This is an international conference which has grown out of the UK's national GIS research conference, since 1993. GISRUK conferences are primarily aimed at the academic community, but also welcome delegates from government, commercial and other sectors. The conferences attract those interested in Geographical Information Science (GIS) and its applications from all parts of the UK, together with the European Union and beyond. The disciplinary range is broad including, but not limited to, Geography, Environmental Science, Ecology, Computer Science, Planning, Archaeology, Geology, Geomatics and Engineering.

One of the most interesting developments in recent years has been the preparation of the two AGI Foresight Reports for 2015 and 2020 (Coote et al., 2010; Kemp, 2015), which explore some of the issues that are likely to have a significant impact on our economy, environment and society over the next five years. The 2015 report highlights five key themes that are felt to be of relevance not only to the GI industry, but also to anyone with an interest in how technology and information will the world and businesses over the next five years. These five themes – Open, Big Data, BIM and Future Cities, Innovative Technologies and Policy – form the backbone of the report, and bring together a large number of separate contributions from experts across industries and disciplines.

6.7 Conclusions

The previous section contained a brief description of some of the main activities of the AGI since its foundation in 1989. This shows that it has achieved a great deal over the last thirty years with limited resources despite the failure of the UK government to fund and support a National Centre for Geographic Information that was recommended by the Chorley Report in 1987. It might be interesting to speculate what more might have been achieved if the Government had accepted its original recommendation.

References

Association for Geographic Information (AGI), 1990. *The Shape of Things to Come*, AGI Newsletter 5, London: Association for Geographic Information.

Association for Geographic Information (AGI), 2012. *UK GEMENI: Specification for Discovery Metadata for Geospatial Data Resources v 2.2*, London: Association for

Geographic Information. www.agi.org.uk/agi-groups/standards-committee/uk-gemini (last accessed 18 February 2019).

Coote, A., S. Feldman and R. McLaren (eds), 2010. *AGI Foresight Study: The UK Geospatial Industry in 2015,* London: Association for Geographic Information. www.agi.org.uk (last accessed 18 February 2019).

Corbin, C., 2002. *AGI Submission to GINIE Project on the Analysis of National GI Organisations in Europe,* London: Association for Geographic Information.

Department of the Environment (DOE), 1987. *Handling Geographic Information Report to the Secretary of State for the Environment of the Committee of Enquiry into the Handling of Geographic Information,* Chairman Lord Chorley, London: Her Majesty's Stationery Office.

Department of the Environment (DOE), 1988. *Handling of Geographic Information: The Government's Response,* London: Her Majesty's Stationery Office.

Department of Trade and Industry (DTI), 1984. *Remote Sensing and Digital Mapping: The Government's Reply,* Cmnd 9320, London: Department of Trade and Industry (HMSO).

Department of Trade and Industry (DTI), 1986. *Government Held Tradeable Information: Guidelines for Government Departments in Dealing with the Private Sector,* London: Department of Trade and Industry.

Geographic Information Panel, 2008. *Place Matters: The Location Strategy for the United Kingdom,* London: Communities and Local Government.

Heywood, I., 1997. *Beyond Chorley: Current Geographic Information Issues: A Report Compiled on Behalf of the Association for Geographic Information,* London: Association for Geographic Information.

House of Lords Select Committee on Science and Technology, 1984. *Remote Sensing and Digital Mapping,* Reports 11983-84, London: Her Majesty's Stationery Office.

Kemp, A. (ed), 2015. *The AGI Foresight Report 2020,* London: Association for Geographic Information. www.agi.org.uk (last accessed 18 February 2019).

Nansen, B., N. Smith and A. Davey, 1996. A British National Geospatial Database, *Mapping Awareness 10,* 3, 18–20 and 10, 4, 38–40.

Openshaw, S. and J. Goddard, 1987. Some implications of the commodification of information and the emerging information economy for applied geographical analysis in the United Kingdom, *Environment aid Planning A 19,* 423–1439.

Rhind, D. W., 1986. Remote sensing digital mapping, and geographical information systems: The creation of national policy in the United Kingdom, *Environment and Planning C: Government and Policy 4,* 91–102.

Serpel, D., 1979. *Report of the Ordnance Survey Review Committee,* London: Her Majesty's Stationery Office.

7

The Impact of GIS on Local Government in Great Britain

Ian Masser and Heather Campbell

7.1 Background

Technological progress in the last few years has removed many of the barriers that inhibited the development of GIS. It is generally agreed that the potential of this technology to store, manipulate and display spatial data is considerable. However, the introduction of GIS technology involves the complex process of managing change within environments that are typified by uncertainty, entrenched institutional procedures and individual staff members with conflicting personal motivations. Given these circumstances, personal, organizational and institutional factors are likely to have a profound influence on the extent to which the opportunities offered by GIS will be realized in practice (Audit Commission for Local Authorities in England and Wales, 1990; Campbell, 1990a, 1991; Department of the Environment, 1987; Willis and Nutter, 1990).

One of the most important groups of users of GIS is local government. The range of potential applications in this field is considerable, extending from property registers and highways management to emergency and land-use planning. It is often assumed that the cost of equipment and data preparation, when combined with the capacity of GIS to integrate data sets from a wide variety of sources, makes the development of departmental systems inappropriate (Bromley and Selman, 1992; Gault and Peutherer, 1989; Grimshaw, 1988). Strategic and efficiency benefits associated with increased levels of information sharing between departments suggest that the implementation of GIS will be accompanied by an extension of corporate activities, which in turn has significant implications for the development of administrative practices in local government.

Given the theoretical advantages of adopting a corporate approach, there are an increasing number of accounts indicating that the realization of these benefits in practice may be more difficult than generally envisaged (Cane, 1990; John and Lopez, 1992; Openshaw et al., 1990; Van Buren, 1991; Winter, 1991). It is clear from these descriptions that users are encountering mixed results even in similar types of organizations. Furthermore, research that has examined the implementation of other forms of information technology suggests that realizing the potential of such systems is often a highly problematic process (Eason, 1993; Moore, 1993; Rogers, 1993). There is no reason for assuming that successful implementation of GIS technology will prove more straightforward than that of other computer-based systems. However, despite the interest GIS has provoked, particularly in local government, little systematic analysis has been undertaken that provides information on the extent of the diffusion of this technology or on the impact that GIS applications are having on the organizations in which they are being implemented. As a result, there is a need to evaluate the underlying assumptions concerning the processes of adopting and implementing GIS, to provide a benchmark against which the isolated accounts of GIS development can be assessed.

7.2 Objectives

The research examines the impacts of GIS on local government in Great Britain. Particular attention has been paid to the extent to which institutional, organizational and personal factors are impeding the take-up and implementation of GIS technology. The research therefore provides factual information on one of the key issues cited in the Chorley Report (Department of the Environment, 1987), and points to possible strategies for resolving the difficulties that are being encountered in practice. The objectives of the research were:

1. To assess the problems and benefits associated with the implementation of GIS in local government.
2. To evaluate the impact of GIS on existing organizational practices in local government with particular reference to the development of corporate activities.

These objectives were operationalized by splitting the study into two complementary parts. The first provided an overview of the take-up of GIS technology throughout British local government and examined the key benefits and problems associated with the implementation of this technology. Given this general context, the second part investigated the detailed impact of GIS technology on the activities and procedures of local authorities and the extent to which organizational issues affect the outcome of the implementation process. The research findings discussed later indicate that these objectives were achieved, although in doing so, many further questions were raised.

7.3 Methods

A combination of methods was used, reflecting the differing emphasis of the two objectives. These included a comprehensive telephone survey of all 514 local authorities in Great Britain and twelve in-depth case studies. A telephone-based survey approach was adopted because of the large number of postal questionnaires that were being circulated to local authorities and the resulting concern about the level of response. The 100% response rate supports the adoption of this method, in addition to respondents giving valuable subsidiary information during the interviews. This method also removes the ambiguity that exists in some surveys over the respondents' perceptions of the definition of GIS, because the capabilities of the software and the precise nature of its use in the host authority can be related to the operational definition. The research adopted a broad interpretation of GIS, which included automated mapping and facilities management type systems but excluded thematic mapping and computer aided design (CAD) packages. One respondent, generally the project manager, was interviewed with respect to each separate system present within a particular authority. The survey took place over the period from February to June 1991. The database developed as a result of the survey was updated in September 1992 by conducting a further round of telephone interviews with the forty-four authorities that had been identified by the initial survey as intending to purchase GIS technology within the coming year.

The selection of the twelve case studies was based on a preliminary analysis of the findings of the telephone survey. The choice of authorities was based on three criteria. First, each authority must have had GIS technology present within the organization for at least two years (i.e. before June 1989), to avoid the research findings being limited to the initial teething problems associated with the introduction of computer technology. Second, a range of organizational contexts were chosen based on a typology of styles of approach to implementation that were developed from the survey findings. These ranged from systems in which all departments in the authority were participating, to situations where GIS software was being developed by a single department. This criterion also ensured that the full range of types and styles of local authority were included. Third, it was essential that the authority was willing to cooperate.

The twelve case studies were undertaken between May 1991 and February 1992 and built on the methodology developed by Campbell in a study examining the use of geographic information in local authority planning departments (Campbell, 1990b). To prepare further for the case studies, exploratory fieldwork was undertaken in New England to examine the extent to which time affects the experiences of organizations in implementing GIS. Given the longer experience of local authorities in handling GIS in the United States, these investigations proved instructive in highlighting the crucial role of organizational factors in achieving the successful utilization of such systems (Campbell, 1992). As a result of this experience, interviews were undertaken with all the staff connected with the introduction of the GIS, including senior management, potential users, computer specialists, project managers and technical support staff. This was a crucial aspect of the methodology because much of the existing research in this area is based solely on a single interview with either the project manager or computer specialists. The extent of the fieldwork required for the case studies varied considerably, ranging from one day to two weeks, depending on the scale and complexity of the project. In several cases, a preliminary discussion with the project manager was supplemented by a second period of fieldwork. The interviews were supported by direct observation of the various organizational contexts and a review of the existing documentation.

7.4 Results

The main results of the research will be presented in two parts, reflecting the organization of the study. The first part provides an overview of GIS adoption and implementation in British local government based on the main findings of the comprehensive survey, while the second highlights some of the key issues raised by the case studies with regard to the impact of GIS on the activities of British local government.

7.4.1 Overview of GIS Adoption and Implementation

The extent of current interest in GIS in local government in Great Britain can be seen in Table 7.1. This shows that 356 out of the 514 local authorities (69.3%) surveyed were considering introducing GIS in one form or another in April 1991. However, the introduction of GIS was at an early stage in most authorities. Only eighty-five authorities (16.5%) had already acquired a GIS, while a further forty-four authorities (8.6%) had firm plans to purchase one in the twelve months following the survey.

The findings indicated that there was a high level of awareness of GIS in local government in Britain, even among authorities that had no plans to invest in such systems. However, there were marked variations between local authorities of different types, and also between different regions

TABLE 7.1

Plans for GIS in local authorities in Great Britain

Plans for GIS	Number	
Already have GIS facilities	85	16.5
Firm plans to acquire GIS within one year	44	8.6
Considering the acquisition of a GIS	227	44.2
No plans to introduce a GIS	158	30.7
Total	514	100

within Great Britain, with respect to the extent of both current and anticipated GIS availability.

The first two columns of Table 7.2 show the number of authorities that had already acquired a GIS, by local authority type. Around 59.3% of all county- and regional-level authorities in Great Britain already had GIS facilities. In contrast, only 7.2% of shire districts and 7.5% of Scottish districts had GIS facilities. The proportion of metropolitan authorities with GIS fell mid-way between the two extremes, at 31.9%. These findings reflect differences in size, with larger authorities much more likely to adopt GIS than the small shire and Scottish districts.

In addition to the variation in take-up between different types of local authorities, the survey also indicates a distinct north/south divide in the level of GIS acquisition. Nearly three-quarters of all authorities that already have GIS are located in the southern half of Great Britain, and the overall proportion of authorities with GIS in the south is nearly double that of the north (see Table 7.3). The difference is most apparent with respect to metropolitan authorities, where 47.5% of all authorities have GIS facilities compared with only 10.3% in the north. Although some allowance

TABLE 7.2

Plans for GIS by type of local authority

Local authorities	Already have GIS		
	Number		Total
Shire Districts	24	7.2	333
Metropolitan Districts	22	31.9	69
Shire Counties	32	68.1	47
Scottish Districts	4	7.5	53
Scottish Regions	3	25.0	12
Total	85	16.5	514

TABLE 7.3

Percentage of authorities who already had GIS facilities, by local authority type and region

Local authorities	South	North	Great Britain
Shire Districts/Scottish Districts	8.6	5.2	7.3
Metropolitan Districts	47.5	10.3	31.9
Shire Counties/Scottish Regions	71.0	46.4	59.3
All authorities	20.1	11.4	16.5

must be made for those authorities that make use of the facilities provided for them by agencies such as the Merseyside Information Service, the contrast between north and south is striking.

With these considerations in mind, the survey went on the investigate the types of GIS systems being implemented, including the time of adoption, software acquired and the main benefits and problems being experienced during implementation. This analysis is therefore based on systems rather than authorities. A system is regarded as a distinct piece or combination of software, which one or more departments within a local authority are implementing. For instance, a situation where several departments are developing separate applications based on the same software is considered as one system. Table 7.4 indicates that a total of ninety-eight systems had been purchased by eighty-five local authorities. Shire counties were the most likely to have more than one system within an authority, with forty-four systems in thirty-two counties. Evidence from subsidiary information provided by respondents suggested that the presence of several systems within one authority was more likely to reflect the desire of certain departments to have complete control of the form and speed of system development than the perceived inability of GIS software to perform a variety of functions.

The findings indicated that there has been a gradual growth since the early 1980s in the number of GIS being purchased, with nearly 40% of all systems acquired in 1990. Because the survey was conducted over the period from February to June 1991, the figures for 1991 are incomplete, but there is a slight indication of a declining rate of take-up. It is probable that the recession, uncertainty over the future structure of local government and changes in the internal management of authorities have contributed to this situation. Monitoring will be necessary to ascertain whether 1990 represents the peak period of system take-up. Analysis of the length of time authorities have been implementing GIS suggests growing experience amongst authorities, with the average length of time since system take-up being slightly over two years. This increasing experience is important as some authorities are beginning of replace

TABLE 7.4

Level of GIS adoption by authorities and systems

Local authorities	Number of authorities possessing GIS	Number of GIS
Shire Districts	24	24
Metropolitan Districts	22	22
Shire Counties	32	44
Scottish Districts	4	5
Scottish Regions	3	3
Total	85	98

The survey identified four further systems, one adopted by a joint board and three by metropolitan research and information units.

their existing facilities. Among these are a small number that decided to evaluate a microcomputer-based system prior to committing more substantial resources, and a number where the original software proved unable to satisfy their requirements. This discussion focuses on the configurations of equipment currently in use.

Table 7.5 provides a breakdown of the software packages introduced by local authorities. It should be noted that systems with GIS capabilities that are essentially being used to perform other activities such as CAD have been omitted from this exercise. In addition, in instances where packages were purchased to provide specialist GIS facilities to supplement an existing system such as SPANS, it is the main system that is recorded in the analysis. A striking feature of the survey findings is the market share of around 22% held by ARC/INFO (ESRI), with Alper Records (a system developed by a small British company based at Cambridge), and GFIS (IBM) holding a further 12% and 10%, respectively. These three packages therefore account for around 45% of the systems adopted by local authorities, with the remaining 55% accounted for by seventeen packages and four home grown systems. The range of products purchased by authorities tends to indicate the overall immaturity of the market, with as yet only tentative indications of specialization on a few core products.

Analysis of these findings in relation to time highlights a number of issues with respect to the diffusion of GIS software. These results indicate that the market share held by ARC/INFO has been growing from around 17% of all systems in 1987 to 30% in 1990. At the same time, the adoption of GFIS has declined from around 35% in 1988 to 2% in 1990. It is also clear that there has been a steady increase in the variety of systems being acquired by local authorities. For instance, in 1986 four different systems were purchased, in 1988, eight and in 1990, sixteen. A number of more detailed issues are raised by examining the take-up of GIS software by local authority type. It is clear from this that while ARCANFO is dominant

TABLE 7.5

GIS software adopted by British local authorities

Software	Shire and Scottish Districts (%)	Metropolitan Districts (%)	Counties and Regions (%)	Total
ARC/INFO	3.4	9.1	40.4	22.4
Alper Records	24. I	9.1	6.4	12.2
GFIS	3.4	9.1	14.9	10.2
Hoskyns G-GP	3.4	22.7	2.1	7.1
Axis Amis	13.8	9.1	2.1	7.1
McDonnell Douglas GDS Maps	6.9	9.1	2.1	5.1
Coordinate	10.3		2.1	4.1
Wings			8.5	4.1
Other [I]	34.5	31.8	21.3	27.5
	(n = 29)	(n = 22)	(n = 47)	(n = 98)

I Other systems purchased include: SIA Datamap, PC ARC/INFO, Planes Spatial, Viewmap, Tydac Spans, LaserScan Metropolis, GDMS, Siemens Sicad, Pafec Dogs, Maps in Action, Atlas GIS, Hoskyns PIMMS and four home-grown systems.

within the counties and regions, accounting for 41% of all systems, this pattern is not repeated throughout local government. In the metropolitan districts, Hoskyns G-GP (a system developed by a small British company) represents nearly 23% of the total number of systems introduced. This reflects the original development of the software by the Research and Intelligence section of the now abolished Greater London Council and the subsequent interest this provoked on the part of a number of London boroughs. As a result, five of the seven systems purchased are in London. A similar presence in Scotland is enjoyed by Alper Records, but the reasons behind this are less clear. The striking feature of the take-up of software in the shire districts is the broad range of systems and the absence of a clear market leader. In more general terms, these findings indicate a greater demand on the part of shire counties for software that provides a variety of GIS facilities, while the other authorities tend to favour systems with more limited capabilities, such as automated mapping and facilities management.

Given the characteristics of the systems being implemented, respondents were asked to identify the main benefits and problems associated with GIS implementation. Table 7.6 shows that the main group of benefits was perceived to be improved information processing facilities. This includes such advances as improved data integration, increased

TABLE 7.6

The most important group of benefits associated with the implementation of GIS in local government

Benefits	Shire and Scottish Districts (%)	Metropolitan Districts (%)	Counties and Regions (%)	Total
Improved information processing facilities	68.4	70.4	51.1	60.5
Better quality decisions	27.0	13.6	42.5	31.5
Savings	4.6	11.4	4.2	5.9
Other		4.5	2.1	2.0
	(n = 29)	(n = 22)	(n = 47)	(n = 98)

speed of data provision, better access to information and an increased range of analytical and display facilities. It is striking that these benefits were regarded as most important by 60% of the respondents, with 31% stressing better quality decisions, including managerial, operational and strategic considerations, and only around 6% linking the main benefits of GIS to achieving savings. Given the work on cost–benefit analysis that has been undertaken, the very low level of importance attached to the role of GIS in achieving time, staff and financial reductions is particularly interesting. Respondents suggested that while financial justifications are important in gaining agreement from elected members and senior management to purchase a GIS, they perceived that in practice these savings were likely to be limited. Furthermore, there is no evidence to suggest a strong link between GIS and improved decision-making. In instances where better decisions were regarded as the prime benefit to be gained from GIS implementation, just over 38% linked these advances to routine operational decisions, while the figures for strategic and managerial activities were 28.8% and 25%, respectively.

The findings shown in Table 7.7 demonstrate that a range of technical, data-related and organizational difficulties is being experienced by those implementing GIS in local government. Most respondents emphasized that the introduction of GIS into their organization did not prove a straightforward process. Only in four cases was it stated that no problems had been encountered. Overall, there was a slightly greater emphasis on data-related issues, such as the lack of compatibility between existing data sets, problems of maintaining up-to-date data sets, the cost of data capture and the cost and availability of Ordnance Survey digital data, than on the other sets of problems.

TABLE 7.7

The most serious group of problems associated with the implementation of GIS in British local government

Problems	Shire and Scottish Districts (%)	Metropolitan Districts (%)	Counties and Regions (%)	Total
Data-related	21.8	31.8	42.5	34.0
Technical	47.7	27.3	17.0	28.4
Organizational	27.0	27.3	29.8	28.4
Other		4.5	6.4	4.1
No problems	3.4	9.1	2.1	4.1
System not sufficiently developed	(n = 29)	(n = 22)	2.1 (n = 47)	1.0 (n = 98)

Given these general trends, particular difficulties were raised by each category of local authority. For instance, technical problems were most pronounced in the shire and Scottish districts, whereas the counties and regions had fewer technical difficulties, but data-related matters were proving more serious. It is probable that the decreasing level of technical problems encountered by the counties and regions through the metropolitan districts to the shire and Scottish districts is a reflection of the greater resources and experience available within the larger authorities. It is noticeable that organizational difficulties are relatively constant regardless of local authority type.

The additional comments of respondents gave some more detailed insights into the difficulties being experienced by local authorities implementing GIS. Around 20% mentioned resource issues such as cost and a shortage of staff time as inhibiting the introduction of GIS, while a significant number were highly critical of the activities of vendors. In particular, there was concern about the extravagant claims of vendors prior to software purchase, their lack of appreciation of the local government context, the limited scope for linkage between systems and inadequate post-sales support. The importance, as well as the difficulty, of establishing the financial viability of vendors was also stressed. Criticism was also levelled at the Ordnance Survey. In particular, dissatisfaction was focused on the incomplete digital coverage for Britain, which leads to partial data sets for most local authority areas, and the quality of some of the data produced, with regard to the speed of update and edge matching of maps. However, in addition to these external influences, many of the comments of respondents related to the specific organizational circumstances in which GIS were being implemented. These included lack of previous technical experience in GIS, limited awareness among the technical specialists of user needs and lack of appreciation by senior staff that the implementation of GIS involves the complex process of managing change. A number of respondents were also concerned about the tension between developing a corporate system, which tends to involve a long lead time and the imposition of standards, and the immediate service needs of departments. More specifically, given the computer storage space required for handling geographic information, some respondents were concerned about the introduction of direct charging for mainframe time. Such circumstances were often associated with an internal reorganization into business units, leading to a number of less wealthy departments such as planning, feeling that they would have to pull out of some of the mainframe-based GIS projects.

The survey provides a useful overview of the diffusion of GIS in British local government. Overall, the results indicate considerable general awareness about the potential of such systems, with GIS software already present within 16% of authorities. The introduction of GIS has been most striking in the counties, regions and the larger district authorities, especially in the southern part of Britain (see Masser, 1993). It has also been found that within some of the counties there may be more than one GIS. Furthermore, in terms of the systems being adopted there has been an increasing move towards the ARC/INFO software and workstations as the supporting hardware. Important issues were also raised in terms of the benefits and problems associated with GIS implementation. Given the emphasis that has been placed on the contribution of GIS to improve strategic decision-making and cost savings through advances in efficiency, it is striking that these respondents associate the main benefits with the more straightforward activities of improved information processing facilities. The findings also indicate that a range of technical, data and organizational problems are experienced during the implementation of GIS technology. These findings, however, provide only a broad overview of GIS development in local government; the detailed impact of these systems was examined in the case studies.

7.4.2 The Impact of GIS on British Local Government

The twelve case studies included a range of local government environments, with four of the systems examined in shire districts, two in metropolitan districts, a further two in a large Scottish district and the remaining four in shire counties. Of these systems, three were being implemented departmentally, while the development of the other nine involved several departments. The systems had also been present within the authorities for at least two years. The following discussion provides a brief summary of the main case study findings. The overview concentrates, firstly, on the types of application being developed and the level of utilization that has been achieved and, secondly, given these findings, on the processes that appeared to influence whether GIS are successfully implemented in local government.

The case study findings indicate that the vast majority of GIS applications being developed aim to assist with operational activities, regardless of the more sophisticated facilities that may be available within the software. Furthermore, despite the common use of the same GIS software by several departments in nine of the case studies, most of the applications concentrate on the needs of no more than one department. The main departments involved with the development of GIS applications are concerned with technical service type activities such as planning, highways, estates, architecture and surveying, assisted in some cases by the central computing department of the authority. The key application area for most authorities has been to exploit the digital data available from the Ordnance Survey to develop automated mapping facilities. The emphasis given to this application probably explains the significant role played by planning departments, as they have traditionally had responsibility for meeting the cartographic needs of authorities. In addition to automated mapping, the other main areas of application include grounds and highways maintenance and estate management. There is very little current or planned use of complex spatial analysis techniques, with most local authorities only perceiving a need for basic display and query facilities. It should also be noted that active interest in GIS has not yet permeated through to the often large community service type departments of housing, education and social services. This is probably a reflection of the considerable administrative pressures that these departments have faced, partly as a result of frequent legislative changes by central government, as well as the limited resources for additional activities such as new data handling initiatives.

The case study findings indicate the tendency for GIS technology to be employed differently even in apparently similar environments developing similar applications. No two case studies, or in many ways users, were utilizing these facilities in exactly the same manner. In general, staff in local government regard GIS simply as a mapping system or a query facility. These general perceptions of the technology are at variance with the accepted understanding of other groups, such as researchers and vendors. These results highlight an important theoretical issue concerning the tendency for each of those involved with a technology to 'reinvent' the exact nature of that technology (Rogers, 1986, 1993). As a result, there is a need for further work to examine the extent to which users reinvent GIS technology and also to what extent there is a shared understanding of its nature and purpose, because misconceptions are likely to lead to the development of technology and facilities that nobody wants.

A striking finding of the research is the limited impact GIS has had on most of the authorities, even after two years' experience. Only three case studies had reached the stage where at least one application was being employed by end users. The remaining seven were either still developing the system or had achieved an operational application but it was not being utilized by users. These results indicate that the lead time to the development of a working application is often considerable because of the time-consuming nature of data capture. Very few of the authorities already had complete data sets or their existing information in a suitable format for input. This means that the implementation of GIS must be sustained over several budgetary cycles. Two of the case studies had abandoned the development of the software they had originally purchased and were reconsidering whether GIS facilities could provide them with any real benefits. The research also demonstrates that a technically operational system will not necessarily be employed by users simply because it is there. Overall, these findings raise an important theoretical issue about the nature of success and failures in relation to the implementation of computer-based systems. In terms of the research, use was regarded as the basic indicator of success.

With these considerations in mind, factors that appear to account for the successful implementation of GIS were explored in greater depth. In terms of the basic features of the systems, there were no similarities between the more or less successful in terms of technical characteristics, such as software and hardware, or the organizational structures adopted, for example, the involvement of several departments or just one. In addition, the technological problems faced by the less successful authorities seemed to be no greater than the others. Previous work in this area suggests that there are three necessary and generally sufficient conditions for the effective implementation of computer-based systems (see Campbell, 1990b; Masser, 1992). These are:

1. An information management strategy that identifies the needs of users and takes account of the resources at the disposal of the organization.
2. Commitment to and participation in the implementation of any form of information technology by individuals at all levels of the organization.
3. A high degree of organizational and environmental stability.

This framework was used as the basis for exploring the experiences of the case studies. The key question, therefore, was 'What factors appear to increase the probability of successful GIS implementation?'

The findings of the research indicate that the implementation of GIS in local government is as much social and political as technical in nature. Technical problems tend to reinforce existing organizational difficulties rather than be responsible for the failure of the process of implementation. The findings of the analysis suggest that there are some organizational cultures that are inherently receptive to the development of innovations, such as GIS. Out of the twelve case studies, two appear to have the capacity to take on the organizational changes implied by the introduction of GIS and sustain the process in the presently highly dynamic context of British local government. It was evident that the skill levels and expertise among the staff in both these authorities were higher than in other similar environments. However, while the all-round round expertise of these individuals was undoubtedly important, the existing culture was in many ways responsible for gathering these individuals into its service. As a result, it appears that innovative environments attract innovative individuals, and presumably the reverse is also the case. In both these authorities, it was evident that there was a long tradition of being at the forefront of new innovations in the local government sector, including a wide range of policy areas as well as information processing. The outward characteristics of the projects in the two authorities were very different, but both demonstrated a fundamental capacity to treat change as an opportunity. In both cases, mistakes were made, but rather than this outcome tending to thwart initiative, it seemed to be regarded as a process of education from which the authority would move forward. It therefore appears probable that the long-term utilization of GIS seems most likely to be sustained in such an environment.

However, while very few organizations are inherently innovative, it was possible to identify four factors that appeared to enhance the chance of success. These are:

1. Simple applications, producing information that is fundamental to the work of potential users.
2. User-directed implementation, which involves the participation and commitment of all the stakeholders in the project.
3. An awareness of the limitations of the organization in terms of the range of available resources.
4. A large measure of stability with respect to the general organizational context and personnel, or, alternatively, an ability to cope with change.

The features contributing to successful implementation overlap to a considerable degree. For instance, it is unlikely that the key needs of users will be identified without the users themselves taking a leading role in system implementation. It was also evident from the case studies that short-term success can be achieved for a relatively small project, based on the expertise and political skills of an enterprising individual, sometimes referred to as a 'champion'. The critical issue in this case is whether the organization can take the innovation and sustain it. Experience suggests that this is often doubtful, as even a most expert individual cannot ensure successful implementation in a vacuum. It is also possible that organizations will achieve technical success without fulfilling the four factors. In other words, operational GIS applications are developed, but they do not become a routine part of the work of users. The analysis of the case studies suggests that this is most likely to occur in situations where system development is controlled by the computer specialists. A further factor that may thwart development is instability, as changing circumstances alter priorities, with organizations that have set goals that are marginal to the needs of users likely to see the system rejected. It is also often difficult for organizations to alter priorities once a course of action has been established. Overall, the results suggest that departmental systems can fail as easily as any of the various forms of corporate development. However, by involving several departments it is likely that the number of variables will increase.

The findings of the case studies generally confirm the usefulness of the three conditions that provided the framework for the analysis. It is important, for instance, that an information management strategy is devised that identifies the information needs of the users and the types of service they require, as well as considering the resources at the disposal of the organization. The form of this initiative, that is to say, whether it has been formally ratified or not, appears to be largely irrelevant. Of far more importance is that the process has been undertaken. Commitment and participation by staff throughout the organization are crucial and must be set within a user-centred framework. The relationship between stability and utilization is perhaps the most complex, with the impact of such forces appearing to depend on the capacity of the organization to withstand change.

This research provides the first systematic attempt to examine the impact of GIS on a group of organizations. Given these results, there is clearly a need for far more theoretical and practical research if many GIS are not to become redundant and the technology devalued. The findings of this work highlight similar issues to those noted by researchers exploring the implementation of other forms of computer technology (see, for example, Barrett, 1992; Danziger and Kraemer, 1986; Danziger et al., 1982; Eason, 1988, 1993; Hirschheim, 1985; Kanter, 1983; Mumford and Pettigrew, 1975; Pfeffer, 1981). In particular, the crucial impact of organizational cultures on the effective implementation of computer-based systems has been identified. Furthermore, these findings indicate the need for greater understanding of the process of diffusion of an innovation, including the importance of the concept of reinvention in relation to GIS (Rogers, 1986, 1993). In Section 5, specific areas requiring further research are examined.

7.5 Future Research Priorities

The findings of this research suggest that GIS is at a crossroads in British local government. The recent update of the initial survey indicates that while one in four authorities now have a GIS, the pace of take-up has slowed considerably by comparison with 1990 (Campbell, Craglia and Masser, 1993). This poses the question as to whether this simply reflects the current recession and uncertainty about the future of local government or is a more profound comment on the value of GIS. It will therefore be important to monitor the future diffusion of this innovation and explore the reasons behind the actions of the various organizations involved with GIS technology.

There is also a need for further research that examines a number of detailed theoretical and practical questions concerning the impact of organizational issues on the successful utilization of GIS. These are:

1. Investigation of the concept of data sharing in relation to the geographic information needs of users. With the exception of Ordnance Survey maps, this research has found it difficult to identify a latent demand for data sharing within local government. What are the experiences of other contexts?

2. Improving understanding of the role of information in the decision-making processes of organizations. This area of concern must underpin any work seeking to identify user requirements, both in terms of the form of information needed and the type of service best equipped to meet these requirements.

3. The nature of the organizational contexts in which GIS are being implemented needs to be explored. This research has indicated that not all environments are the same. Greater understanding of the social and political processes that influence the particular characteristics of individual organizations would facilitate system implementation.

4. The roles of individuals as against the organizational culture needs detailed consideration. This research suggests environments that are inherently innovative can sustain change. Is there more general evidence for this? What is the role of 'champions' in the process?

5. Investigations into the impact of instability and uncertainty on system implementation need to be undertaken, particularly in the context of the current changes in local government.

Research into many of these issues will have implications for the successful implementation of a wide variety of forms of information technology. However, given the current prominence of GIS, it is vital that further research is undertaken in these areas if resources are not to be wasted and if GIS is to avoid becoming yet another failed technology.

7.6 GIS Diffusion and Implementation Research at Sheffield

Ian Masser's interests in the role that computers play in urban planning go back to the late sixties when urban modelling first came into being. These were reflected in his work at Liverpool and Utrecht during the seventies and fostered a concern for the organisational implications of technological innovations in the eighties. His secondment from Sheffield as Coordinator of the Economic and Social Research Council's Regional Research Laboratory Initiative in 1987 allowed him to develop these interests in greater depth together with two of his PhD students, Heather Campbell and Max Craglia. As a result of contacts with the US NCGIA, Harland Onsrud and Ian applied for and were awarded a contract by NATO to organise an Advanced Science Institute in Greece in April 1992 on 'Modelling the Diffusion and Use of Geographic Information Technologies' (Masser and Onsrud, 1993). In July 1990, Heather Campbell and Ian Masser received a two-year grant from the ESRC/NERC Joint Programme on Geographic Information Handling to carry out an in-depth survey of local authority use of GIS, which provided strategic insight into the behaviour and aspirations of this diverse market. By this time, the group at Sheffield had expanded with the arrival of three more PhD students, Dimitris Assimakopoulos, Ian Smith and Steven Capes.

Given that Max Craglia came from Italy, Dimitris Assimakopoulos from Greece and Ian Smith spoke fluent French, it was at all surprising that there was a strong international comparative dimension to our research. This was strengthened by our participation in several projects involving East European countries as well as a number of extended visits from academics in Denmark, Germany and Italy.

In the years after the publication of this chapter, the research described above developed in various ways. These included a second survey of all the original authorities during the summer of 1993 that was funded by the Local Government Management Board (Campbell et al., 1994), the production of a book entitled 'GIS and Organisations: How Effective Are GIS in Practice' (Campbell and Masser, 1995), support for comparative studies in other European countries that formed part of a specialist meeting on the diffusion of GIS in local government in Europe in October 1993 that formed part of the GISDATA scientific programme (Masser, Campbell and Craglia, 1996), and further work on the theoretical underpinnings of computer implementation research in the planning field by Campbell (see, for example, Campbell, 1996). The GISDATA programme has already been described in Chapter 3 in general terms, and the findings of the specialist meeting on the diffusion of GIS will be discussed in Chapter 9. The rest of this section describes the outcomes of the other components of the programme of GIS diffusion and implementation research that was undertaken a Sheffield during the 1990s.

The interviews of the 514 local authorities that formed the basis of the original research took place in during the spring of 1991. From the outset it was recognised that rapid changes were under way in the diffusion of GIS technologies in British local government at that time. This prompted the Local Government Management Board Group to commission the Sheffield researchers to carry out another round of telephone interviews with all the authorities in the summer of 1993 (Campbell et al., 1994). The questionnaire used in this round also included two additional questions regarding the acquisition of automated mapping facilities and the impact of the comprehensive digital mapping agreement reached between the Local Authority Associations and the Ordnance Survey reached in March 1993.

The findings of 1993 survey indicated the extent to which the number of GIS in British local government had changed in the two and a half years since the 1991 survey. By 1993, 149 local authorities (29% of the total) had acquired GIS and the overall level of adoption would rise to 41% if authorities that had acquired automated mapping facilities were included in the survey. The biggest change was in the take-up by shire districts which had risen from twenty-four in 1991 to sixty-one in 1993. As a result, the level of GIS adoption was approaching 100% in the shire counties, around 50% in the metropolitan districts and 20% in shire districts. In the process, the north–south divide in take-up throughout Britain that was noted in 1991 had been substantially reduced. It is also worth noting that 63% of authorities with GIS had already taken advantage of the digital data option under the Service Level Agreement reached between the Ordnance Survey and the Local Authority Associations in March 1993.

The 1993 survey illustrated the need for comparative research based on a common methodology to track the diffusion of GIS in one of the largest potential user groups in local government in Great Britain. The spectacular findings also encouraged other researchers to carry out similar studies in Germany, Italy, Portugal and Denmark. The findings of these surveys, together with related research in France, Greece, the Netherlands and Poland were presented at a specialist meeting on the diffusion of GIS in local government in Europe in October 1993 as part of the GISDATA scientific programme (Masser, Campbell and Craglia, 1996). There is a distinct Sheffield flavour to this book in that seven out of the thirteen chapters are authored or co-authored by members of the Department of Town and Regional Planning. The findings of this cross-national comparative analysis are discussed in detail in Chapter 9 and suggest that the length of local government experience in most cases is very similar between countries. However, it appears that two main dimensions can be identified which account for a large number of the differences observed between countries. The first dimension appears to be related to the overall extent of diffusion and the level of digital data availability in these countries. This essentially measures the links between data infrastructure and diffusion. The second dimension is associated with the kind of GIS applications that are undertaken and the level of government at which they are carried out in the different countries. This dimension largely measures the professional cultures associated with GIS applications.

The underlying assumption behind the book on *GIS and Organisations: How Effective Are GIS in Practice?* (Campbell and Masser, 1995) is the need for an enhanced understanding of the social and political processes that affect the relationship between organisations and technologies such as GIS which affect the performance of GIS in practice. In the process it examines the relationship between a technological innovation, namely GIS, and organisations, in this case in the context of the framework of British local government.

The process that is responsible for transforming a collection of people, equipment and data into a taken-for-granted part of the everyday activities of an organisation is implementation. This process has therefore been at the heart of this investigation. The very essence of implementation is change. It is impossible to envisage circumstances in which implementation, be it that of a new technology or policy, does not imply change. As a result, the term 'implementation' is increasingly becoming synonymous with the management of change; a phrase that might be more accurately expressed as the cultivation or nurturing of change.

With this in mind a detailed analysis of GIS implementation is undertaken in twelve carefully selected local authorities that cover the whole range of local government. These include authorities that have adopted classically corporate approaches to implementation, authorities adopting more pragmatic corporate approaches and some authorities that have adopted a fiercely independent style of GIS implementation.

The findings of the analysis are considered from three theoretical perspectives: technological determinism, managerial rationalism, and social interactionism. They show that, 'Implicit within the technological determinist and managerial rationalist approaches to GIS implementation are assumptions about the role of spatial data operational, managerial and strategic decision-making and therefore the needs of the user.' (p. 156). Consequently, it is argued that the 'overall the findings of this research tend to accord most closely with the explanatory framework provided by the social interactionist perspective' (p. 157). In other words, 'GIS implementation appears to be essentially a social and political process rather than simply a technological matter' (p. 157).

In later work, Campbell (1996) extended these ideas into a more general discussion of computer implementation in the planning field. In the 1996 paper, for example, she argued that changing the assumptions about the nature of technology and organisations alters the planner's understanding of the processes that influence the outcomes of technological innovations.

The analytical framework provided by the social interactionist perspective shifts the whole focus of the debate on the relationship of computer technologies to planning practice, from the likely impact of technology on planning to how individual planning agencies can mould technologies to meet their needs and organisational cultures.

(p. 104)

From this perspective, local government will shape technology to their agenda not the other way around.

By 2002 the staff position at Sheffield had changed as a result of Ian Masser's move to the International Institute of Aerospace Survey and Earth Observation in Enschede in the Netherlands in 1998, while Max Craglia took up a position at the EU Joint Research Centre at Ispra in Italy in 2001.

References

Audit Commission for Local Authorities in England and Wales 1990. *Management Papers: Preparing an Information Technology Strategy: Making IT Happen.* HMSO: London.

Barrett, S. 1992. *Information Technology and Organisational Culture: Implementing Change, EGPA Conf.*, Maastricht, August.

Bromley, R. and Selman, J. 1992. Assessing readiness for GIS. *Mapping Awareness and GIS in Europe, 6(8)*, 9–12.

Campbell, H., 1990a. *The use of geographic information in Local Authority Planning Departments.* Doctoral Thesis, Department of Town and Regional Planning, University of Sheffield.

Campbell, H., 1990b. The organisational implications of geographic information systems in British Local Government. *Proc. 1st European GIS Conf. (EGIS'90)*, Amsterdam, 10–13 April, 145–157.

Campbell, H. 1991. Organisational issues in managing geographic information. In Masser, I. and Blakemore, H. (eds.), *Handling Geographic Information.* Longman: Harlow, 259–282.

Campbell, H. 1992. Organisational issues and the implementation of GIS in Massachusetts and Vermont: some lessons for the UK. *Environment and Planning, B, 19* (1), 85–95.

Campbell, H. 1996. A social interactionist perspective on computer implementation. *Journal of the American Planning Association, 62(1)*, 99–107.

Campbell, H., M. Craglia and Masser, I. 1993. GIS in British Local Government: an up-date. *BURISA Newsletter, 107*, 2–5.

Campbell, H., and I. Masser 1995. *GIS and Organisations: How Effective are GIS in Practice?* Taylor and Francis: London.

Campbell, H., I. Masser, J. Poxon and L. Sharp 1994. *GIS: The Local Authority Scene.* Local Government Management Board: Luton.

Cane, S., 1990. Implementation of a corporate GIS in a large authority. *Proc. 1st European GIS Conf. (EGIS'90)*, Amsterdam, 10–13 April, 158–166.

Danziger, J.N., W.H. Dutton, R., Kling, and K.L. Kraemer 1982. *Computers and Politics: High Technology in American Local Government*. Columbia University Press: New York.

Danziger, J.N. and K.L. Kraemer 1986. *People and Computers: The Impacts of Computing on End Users in Organisations*. Columbia University Press: New York.

Department of the Environment, 1987. *Handling geographic information*. Report of the Committee of Enquiry chaired by Lord Chorley. HMSO, London.

Eason, K.D. 1988. *Information Technology and Organisational Change*. Taylor and Francis: London.

Eason, K.D. 1993. Gaining user and organisational acceptance for advanced information systems. In Masser, I. and Onsrud, H.J. (eds.), *Diffusion and Use of Geographic Information Technologies*. Kluwer: Dordrecht, 27–44.

Gault, I and Peutherer, D. (1989) Developing GIS for local government in the UK. Paper presented at the European Regional Science Association Congress, Cambridge.

Grimshaw, D.J. 1988. The use of land and property information systems. *International Journal of Geographical Information Science*, 2, 57–65.

Hirschheim, R.A. 1985. *Office Automation: A Social and Organisational Perspective*. John Wiley: Chichester.

John, S.A. and Lopez, X.R. (1992) Integrating data derived from within the British Local Government Institutional Framework. Proc. Ass. Geog. Inf. Conf., Birmingham, 27–29 November. Westrade Fairs Ltd., Rickmansworth, 1.18. I–1.18.8.

Kanter, R.M. 1983. *The Change Masters*. Simon and Schuster: New York.

Masser, I. (1992) Organisational factors in implementing urban information systems. Proc. Urban Data Management Symposium., Lyons, 16–20 November. UDMS, Delft, pp. 17–25.

Masser, I. 1993. The diffusion of GIS in British Local Government. In Masser, I. and Onsrud, H.J. (eds.), *Diffusion and Use of Geographic Information Technologies*. Kluwer: Dordrecht, 99–115.

Masser, I., H. Campbell and M., Craglia (eds) 1996. *GIS Diffusion: The Adoption and Use of Geographical Information Systems in Local Government in Europe*. Taylor and Francis: London.

Masser, I and H. J. Onsrud (eds) 1993. *Diffusion and Use of Geographic Information Technologies*, NATO ASI series, Kluwer: Dordrecht.

Moore. G.C. 1993. Implications from MIS research for the study of GIS diffusion: some initial evidence. In Masser, I. and Onsrud, H.J. (eds.), *Diffusion and Use of Geographic Information Technologies*. Kluwer: Dordrecht, 77–94.

Mumford, E- and Pettigrew, A. 1975. *Implementing Strategic Decisions*. Longman: Harlow.

Openshaw, S., Cross, A., Charlton, M., Brunsdon, C. and Lillie, J., (1990) Lessons learnt from a post mortem of a Failed GIS. Proc. Ass. Geog. Inf. Conf., Brighton, 22–24 October. Westrade Fairs Ltd., Rickmansworth, 2.3.1–2.3.5

Pfeffer, J. 1981. *Power in Organisations*. Pitman: Boston.

Rogers, E. 1986. *Communication Technology: The New Media in Society*. Free Press: New York.

Rogers, E. 1993. The diffusion of innovations model. In Masser, I. and Onsrud, H.J. (eds.), *Diffusion and Use of Geographic Information Technologies*. Kluwer: Dordrecht, 9–24.

Van Buren. T.S., 1991. Rural town geographic information systems: Issues in integration. *Urban and Regional Inf. Sys. Ass. (URISA) Proc*, 3. 136,15 1.

Willis, J. and Nutter, R.D. 1990. A survey of skills needs for GIS. In Foster, M.S. and Shand, P.S. (eds.), *AGI Yearbook 1990*, London: AGI. 295–303.

Winter, P. (1991) Selling a corporate GIS. Proc. Ass. Geog. Inf. Conf., Birmingham, 20–22 November. Westrade Fairs Ltd, Rickmansworth, 2.6.1–2.6.5.

8

Information Sharing and the Implementation of GIS

Some Key Issues

Ian Masser and Heather Campbell

8.1 Introduction

Data integration and information sharing are becoming recognized as increasingly important issues in the development and implementation of GIS. In February 1992, the National Centre for Geographic Information and Analysis (NCGIA) organized a special workshop on this topic (Onsrud and Rushton, 1992) and the subject also featured prominently in the discussion at the NATO Advanced Research Workshop on Modelling the use and diffusion of GIS Technology (Masser and Onsrud, 1993). What emerged from these meetings is that there are pronounced differences between those who regard information sharing as an essential and inevitable stage in the universal implementation of the new technology and those who argue that there are important impediments as well as incentives which are likely to lead to marked differences in practice. With these considerations in mind this chapter considers some of the key issues that need to be taken account of in the discussion.

It is often argued that the greatest benefits from GIS implementation come as a result of the pressures that are placed on organizations to share information. The findings of the joint Nordic Project, for example, suggest that the returns on investment in GIS range from 1:1 where the automation of basic mapping functions are involved, to 2:1 where GIS is applied to planning activities and are as high as 4:1 where they involve the collective use of databases common to all users (Tveitdal and Hesjedal, 1989). It can also be argued that these benefits will be accompanied by considerable cost savings particularly where a lot of time is spent on inter-agency map exchanges. For example, a study of six agencies in Santa Clara county estimates that overall savings of between 42 and 54 per cent of current costs could be obtained by the development of decentralized and centralized shared AM/GIS facilities, respectively (Finkle and Lockfield, 1990).

Nevertheless, the theoretical assumptions behind arguments such as these can be questioned from both an economic and a political standpoint. In economic terms, they assume that geographic information can be treated as a public good: i.e. something accessible to everybody which can be shared without loss to any individual. Yet in many respects geographic information must be regarded as a private good, as a commodity which is traded at a price with access being determined by ability to pay (Bates, 1988; Openshaw and Goddard, 1987). It is important also to note that the commodification of geographic information is not restricted to the private sector given the increasing pressures on public sector agencies to recover their costs (Bryan, 1992; Rhind, 1992).

Similarly, these arguments do not take account of the political economy of geographic information: i.e. the extent to which geographic information is used to maintain the existing balance of power within late capitalist society (Mosco and Wasko, 1988). In this perspective geographic information must be regarded as a tool to be used in turf battles between different interests. Moreover, differences in access to information are likely to result in a growing divide between the information-rich and the information-poor within cities and regions while creating circumstances which lead to the emergence of surveillant societies (Pickles, 1991).

Even where the public good dimension of geographic information is paramount and there are no conflicts between the agencies involved, the success of information sharing is likely to depend on a wide range of organizational and institutional factors. There have been a number of attempts in recent years to set down maxims or conditions that promote effective implementation (Croswell, 1991; Levinsohn, 1990; Masser and Campbell, 1991; Wellar, 1988). However, these are largely directed towards single organizations and must be modified in the context of information sharing.

Some of the key organizational issues involved are highlighted in the reviews by Croswell (1991) and Masser and Campbell (1991, 1992). Drawing on the work of the Irvine group (Danziger et al. 1982; King and Kraemer, 1985) on the use of computers in local government, Masser and Campbell (1992) contend that three necessary and generally sufficient conditions must be met for the effective utilization of GIS in organizations:

1. the existence of an overall information management strategy based on the needs of users in the agency and the resources at its disposal;

2. the personal commitment of individuals at all levels in the organization with respect to overall leadership, general awareness and technical capabilities; and

3. stability with respect to personnel, administrative structures and environmental considerations.

The effects of moving from a single organization to a multi-organization environment is particularly marked with respect to Masser and Campbell's three conditions. Effective utilization will depend, first on the existence of an overall information management strategy in each of the organizations concerned which takes account of the needs of each set of users and the resources at the disposal of the respective organizations. This assumes a high degree of compatibility between the priorities that are given to these operations where sharing is required and the resources that are set aside by each organization for these purposes. The second condition presupposes a high degree of personal commitment at all levels in all the organizations. For example, given the importance attached in the literature to the role of particular individuals or champions in ensuring effective implementation (Beath, 1991), it will be necessary to think in terms of supra organizational champions in the context of information sharing. However, the biggest obstacle to effective implementation is likely to be with respect to the third condition regarding the need for organizational and environmental stability where information sharing is concerned given that the failure of even one organization in this respect may threaten the success of the whole operation.

It may be argued that the above discussion is unduly negative in tone and that information sharing in most cases involves largely technical operations which are affected to only a very limited extent by economic, political or organizational considerations. It may also be argued that in many cases the benefits from information sharing are sufficiently great to justify special efforts to satisfy the conditions set out above. With these considerations in mind, this chapter examines the effect of GIS on British local government with particular reference to information sharing. It examines a case where the circumstances are especially favourable for information sharing: i.e. the local authority environment where information sharing takes place between different departments within the same organization rather than between different organizations. Under these circumstances, given the common corporate objectives of the organizations to which the departments belong and the number of common factors associated with organizational stability it may be argued that the adoption of GIS should be associated with corporate information management strategies which maximize the benefits to be obtained from information sharing.

This chapter considers the thesis that is implicit in such a standpoint. Then it examines the antithesis which draws attention to the strengths of a departmental standpoint. The third section presents some of the findings from a comprehensive survey of the impact of GIS on local government in Great Britain which are discussed in relation to the assumptions of both the thesis and the antithesis. Finally, a number of general conclusions are drawn from the evidence with respect to information sharing as a whole and a number of suggestions are made regarding the further research that is required.

8.2 Thesis and Antithesis

8.2.1 Thesis

The basic thesis underlying the case for information sharing in local government revolves around the need to regard information as a capital asset. To exploit this asset an integrated approach is required which maximizes the potential for resource sharing amongst users. This in turn creates economies of scope in the form of multiple applications.

The case for a corporate approach to information resource management in British local government can be developed in a variety of ways. For example, Hepworth (1990) calls for greater efforts by local authorities to mobilize their information capital to improve the territorial management public services. Similarly, Gault and Peutherer (1989, p. 2) argued that 'to manage change in the future, local authorities need to develop and make better use or the information sources at their disposal'. They also argue that GIS represents a useful tool for strategic information management which adds value by integrating diverse data sets. In their view 'a corporate view is essential if the synergy is to be realized which will assist both the management of change and increase operational efficiency' (p. 13). The Audit Inspectorate (1990) also note the importance of data integration in their advice to local authorities on preparing an IT strategy. They view with concern the disorganized and uncoordinated approach to acquisition that exists in many local authorities as a result of a preoccupation with short-term considerations. 'Users have invariably been able to demonstrate the short-term benefits of providing their own solutions and have only later come to appreciate such disbenefits as the inability to share corporate data and the consequence of need for expensive duplication' (p. 4).

The advantages of a corporate approach are highlighted in many empirical studies. For example, Coulson and Bromley's (1990) detailed assessment of user needs in Swansea district demonstrates 'the impracticability of introducing an isolated departmental GIS' (p. 215). In their view four types of benefits would accrue to the local authority as a result of adopting a corporate approach:

1. Maps: 'access to up-to-date urban mapping is a bonus to any organization for reasons of completeness and common usage with other members of the organization'.
2. Data: 'the main benefits of GIS are due to better access to data ... the corporate approach encourages a steady reduction of the artificial barriers that have grown up between departments over the years'.
3. Operations: 'the combination of better mapping and data improvements brought about by a GIS should result in better operations

within the local authority, in particular better resource targeting and better forward planning'.

4. General benefits 'the most significant is the idea of great coordination, but only if a corporate GIS is adopted. Whilst the management of corporate GIS is clearly more difficult than a departmental GIS the benefits far outweigh the disadvantages'. (p. 216)

8.2.2 The Antithesis

It is also necessary to consider the antithesis to these claims. The case embodied in the antithesis revolves largely around matters relating to control and continuity. It can be argued that those most directly concerned with the provision of specific local authority services, that is the individual line departments, need to retain control wherever possible over the technical facilities and the information that they need to carry out their tasks. This presents local government with a fundamental dilemma: 'maintaining sufficient centralization of control in city government that enables oversight, efficiency and accountability, while providing sufficient decentralization to deal adequately with the problems of size and complexity and the demands of neighbourhoods and the special interest groups' (Kraemer and King, 1988, p. 28).

It can also be argued that the importance of continuity in terms of administrative routines and practices is often underestimated in practice. British local government is highly departmentalized and many departments have their own professional cultures which are reflected in the distinctive character of their administrative practices. Under such circumstances movements between departments within local authorities are relatively rare and it can be argued that the most successful technological innovations are likely to be those which reinforce existing departmental practices.

8.3 Evaluation

The arguments underlying the thesis and antithesis can be expressed in terms of pressures towards more corporate approaches and more departmental approaches, respectively. The main advantages and disadvantages of these two approaches are summarized in Table 8.1, which is based on Campbell (1992). This shows that the respective advantages of corporate approaches in terms of information sharing, more informed decision-making and increased efficiency which were incorporated in the thesis are often offset by corresponding disadvantages in practice. For example, against the benefits to be derived from data integration, there are the potential costs associated with variations in priorities between departments. Similarly, the advantages of improved access to information may be offset by differences between departments in terms of their abilities to exploit the potential offered by GIS.

TABLE 8.1

The main advantages and disadvantages of corporate and departmental approaches to implementing GIS

Corporate approach		Departmental approach	
Advantages	**Disadvantages**	**Advantages**	**Disadvantages**
Integration of data sets	Variations in priorities between departments	Independence	Departmental isolation
Increased data sharing Improved access to information	Differences between departments in their ability to exploit GIS facilities Differences in the level of awareness and spatial data handling skills between departments	Control over: – priorities – the form and accessibility of information	Lack of support in terms of: – finance – technical specialists – training
More informed decision-making	Inter-departmental disagreements over: – access to information	– technical specialists – equipment	Absence of authority-wide including lack of system and data compatibility
Increased efficiency due to reduced duplication leading to time, staff and cost savings	– leadership – data standards – equipment – training	Clear lines of responsibility Continuity	Inertia

Furthermore, the benefits to be derived from corporate approaches in terms of more informed decision-making may be outweighed by differences between departments in terms of the level of awareness and the spatial and data handling skills at their disposal. In addition, the potential savings to be derived from increased efficiency in terms of the reduction of duplicated efforts may be offset by inter-departmental disagreements with respect to access to information, professional leadership, data standards, equipment and training needs.

On the other hand, the advantages of a departmental approach in terms of independence, control over resources, clear lines of responsibility and continuity which underlie the antithesis may be offset in practice by the corresponding disadvantages. For example, the benefits to be derived from independence may be outweighed by the costs incurred as a result of the department's isolation within the authority. Similarly, the advantages of retaining control over priorities, information content and information processing facilities may be offset by the lack of support in terms of finance and technical expertise. In addition, the benefits to be derived from clear lines of responsibility may not outweigh the lack of system and data compatibility which would enable authority-wide benefits to be derived from departmental activities. Under such circumstances, the advantages to be gained from continuity may be offset by inertia which seriously impedes the realization of the potential offered by the new technology.

It is also important to bear in mind that the advantages of corporate and departmental approaches are likely to change over time within any organization. To explore the dynamics of such processes two simple models of information-sharing are outlined in Figure 8.1. These describe information-sharing in terms of exchanges between different departments within an organization or different organizations and in terms of a functional model where one department or agency provides services for the other departments or organizations, respectively. Figure 8.1 also shows that the relative benefits and costs to be obtained from collaboration are likely to vary from participant to participant.

The dynamic impacts of these differential costs and benefits are illustrated in Figure 8.2 for both models. This shows that as soon as the costs outweigh the benefits for one of the partners involved the arrangements as a whole are likely to be threatened even though there may be net positive advantages for the other partners.

The most important conclusion to be drawn from Table 8.1 is that both approaches have their advantages and disadvantages. The extent to which the advantages overcome the disadvantages is likely to vary considerably in practice according to the organizational context within which GIS is being introduced. The extent of these variations is explored in the next section of the chapter with respect to preliminary findings of the study of GIS on British local government that is being carried out by the authors (Campbell and Masser, 1993).

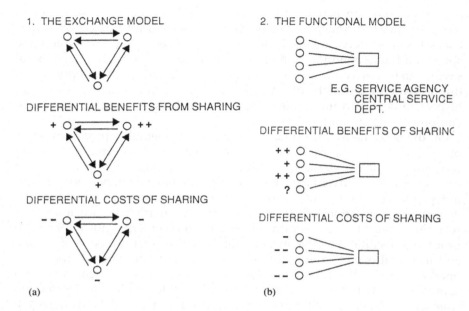

FIGURE 8.1
Two simple models of sharing dynamics: (a) the exchange model and (b) the functional model.

8.3.1 Research Strategy

The findings discussed below form part of a comprehensive survey of British local government, which is being undertaken by the authors. A combination of research methods has been adopted as the basis for this project, including a telephone survey of all 514 local authorities in Britain and detailed case study investigations. This chapter concentrates on the results of the telephone survey which was undertaken between February and June 1991. The 100 per cent response rate to the survey supports the adoption of this method, and also provided valuable subsidiary information during the interviews. This method also removes the ambiguity which exists in some surveys regarding the definition of GIS, as the capabilities of the software and the precise nature of its use in the authority can be related to operational definitions. The researchers adopted a broad interpretation of GIS including automated mapping/ facilities management and land information systems but excluding thematic mapping and computer-aided design (CAD) packages. A more detailed discussion of the survey findings with respect to the adoption and implementation of GIS in British local authorities can be found in Campbell and Masser (1993 and Chapter 7 of this book).

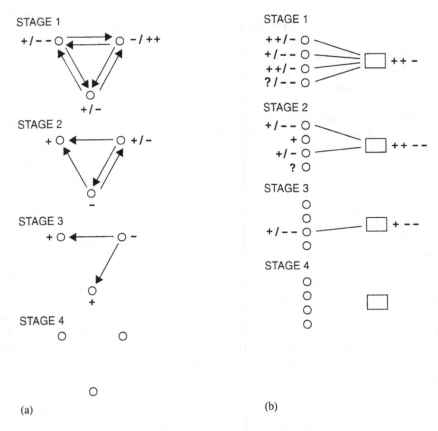

FIGURE 8.2
The dynamic processes: (a) the exchange model and (b) the functional model.

8.3.2 Some Findings from the Survey

The units of study in this analysis are systems rather than authorities. Systems are regarded as a distinct piece or combination of software which one or more departments within a local authority are implementing. The findings of the survey show that 98 systems have been purchased by 85 local authorities. Shire Counties are the most likely to have more than one system within an authority as there are 44 systems in 32 Counties. The only other authority to have adopted more than one system is a large Scottish District.

The presence of more than one system in many of the Shire Counties indicates the extent to which departmental-based approaches are still prevalent in some areas of British local government. In fact, there is a virtually even split between departmental and corporate approaches. Given this broad trend there are distinct variations between the different categories of authority. It is nearly twice as likely that a system in a County or Region will have been implemented departmentally rather than corporately. In contrast over three quarters of the systems in Metropolitan Districts and nearly 60 per cent of those in Shire and Scottish Districts were to some extent corporate in nature.

One important factor influencing the type of GIS adopted in many authorities is the division of responsibilities for funding. Although funding is more or less evenly divided between central and departmental sources, there are important differences in emphasis between the different types of authority. In the case of the counties and regions nearly 47 per cent of funding came from departmental sources. In contrast nearly 45 per cent of funding in Shire and Scottish Districts came from central sources. Nevertheless, it appears that the introduction of GIS into local authorities frequently requires a significant commitment to resources by departments even in cases where a corporate approach is adopted.

There are considerable differences between the number of departments involved in the various types of authority where some form of corporate approach has been adopted. Generally, system implementation involves no more than three or four departments and over 70 per cent of corporate systems are being developed by less than five departments. Only 12 per cent are being introduced in line with the classic corporate approach which would entail all departments in an authority participating in the implementation of GIS. Smaller authorities such as Shire Districts are most likely to favour this framework.

Planning is most likely to be the lead department followed by information technology/computer services where some form of corporate approach has been adopted. The findings suggest that, even in instances where a top down structure is favoured, the lead is taken by the central computing department rather than by the Chief Executive's department. However, only in the case of one in four systems is the key coordinating role within the authority taken by one of the central departments. If it is assumed that the lead department is the one with the strongest commitment to GIS, then this is particularly the case with the technical and property-related departments. Consequently, the introduction of GIS appears to reflect the bottom up demand for such facilities rather than a perceived corporate need by those at the centre. The contribution of planning is most pronounced in the case of the Shire/Scottish and Metropolitan Districts where 47 and 26.5 per cent of systems are led by planning departments, respectively. However, planning departments are far less significant in the case of systems being developed by County and Regional authorities with highways and information technology/computer services most likely to take the lead. There is also a high proportion of systems in these authorities for which responsibilities are shared by all participating departments.

These findings suggest that active interest in GIS has not permeated all local authority activities. Large community service departments such as housing, education and social services tend not to have become involved. This may be a reflection of the considerable administrative pressures faced by these departments and the limited resources that are available for new activities associated with data handling.

Departmental systems are associated with very similar types of activity to the GIS that are being implemented corporately. Again, technical services such as property management and planning are prominent with a particular tendency for departmental systems to be developed by the engineers and highways authorities.

Overall the findings of the survey indicate that there are very few examples of centrally led fully corporate arrangements for the implementation of GIS in Great Britain at the present time. In practice around half the systems are being developed by a single department while collaborative projects most often involve no more than 3 or 4 departments with a technical service department taking the lead. There is also evidence to suggest that within the counties in particular there may be several systems being developed within one authority.

8.4 Synthesis

The findings from the survey of GIS adoption in British local government must be interpreted from two different perspectives reflecting the thesis and antithesis that was set out in the previous section, respectively. In terms of the thesis the fact that half the systems currently involve more than one department points to the extent to which the benefits of information sharing have been recognized in British local government. In this respect it may be argued that the adoption of GIS is having a marked impact on existing administrative practices particularly in the Shire Districts/Scottish Districts and Metropolitan Districts.

At the same time the findings of the survey also indicate the diversity of current arrangements within local authorities where more than one department is involved in GIS implementation. The classically corporate approach led by the central Chief Executive or Computer Services Department and involving most of the departments in the local authority is only being implemented in a relatively small number of cases, primarily, small Shire Districts. More typical are the more pragmatic arrangements that had been made between three or four departments in the larger Shire Districts and Metropolitan Districts. These have usually been instigated by planning or property-related departments (Campbell, 1993).

The arguments behind the antithesis are evident in the number of departmental applications of GIS that are being implemented in British local government. This is particularly apparent in the Shire Counties and Scottish Regions. It is especially pronounced in respect of departments such as Highways which already possess in-house technical expertise. Under such circumstances the advantages to be derived from information sharing as a result of inter-departmental collaboration appear to be offset by the benefits of retaining control over GIS operations.

With these considerations in mind it can be argued that these two contrasting standpoints reflect the technological and organizational context within which GIS is being implemented. On the one hand the facilities and technical experience of the County Highway Departments makes it relatively easy for them to adopt and implement GIS without having to make major changes in administrative practices. Under such circumstances there are strong pressures towards a departmental rather than an inter-departmental or corporate approach. On the other hand, the position in many Shire Districts and Metropolitan Districts is entirely different. The more limited experience of the line departments together with their more limited resources tend to favour some form of collaborative venture. In cases of this kind the impetus to innovation often comes from the top initially, even though it may eventually be implemented by Computer Services or the Planning Department rather than the Chief Executive.

A critical factor in both cases is the nature of the geographical information that is being used. Apart from basic map information, it can be argued, for example, that a large proportion of the information required by Highways Engineers is operational information which is specific to the Department itself and the use of strategic information related, for example, to population and social trends is relatively limited in this case. Under such circumstances the case for information sharing is relatively weak by comparison with the advantages of retaining control. On the other hand, it can be argued that, especially in small authorities, property-based maps are a key component in the activities of several departments and that there are considerable advantages to be gained by combining efforts. This is often reflected in existing practices prior to the adoption of GIS whereby one department, for example, Planning, already provides a mapping service for other departments. In cases of this kind, it is not surprising that collaborative and/or corporate approaches to GIS will be implemented.

To explore these questions in greater depth, a number of detailed case studies are currently being carried out by the authors. In addition to providing valuable information about the technological and organizational context within which GIS are being implemented in different authorities, these case studies will also throw light on the dynamics of the GIS implementation process. Until the findings from the case studies are available, only interim conclusions can be drawn with respect to the effect of GIS on British local government. However, there is a great deal of evidence from the survey to suggest that the adoption and implementation of GIS has led to the emergence of a variety of corporate approaches which seek to exploit the advantages of information sharing especially in the Shire Districts/Scottish Districts and Metropolitan Districts. On the other hand, there is also evidence of the continuation of department-based approaches, especially in the Shire Counties, despite these apparent advantages. As a result, it must be tentatively concluded that the findings from British local government point to the extent to which organizational and political factors apparently offset in many instances, the theoretical benefits to be obtained from structures which seek to promote information sharing.

8.5 Information Sharing Research – Some Recent Developments

Information sharing is an issue that has had a great deal of attention over the 25 years since the above paper was published and this update can only highlight some of the more interesting developments during this period. In addition, many of the papers that were presented at the NCGIA workshop on 'Sharing geographical information' (Onsrud and Rushton, 1995) are still relevant for the discussion of information sharing. A case in point is the chapter in the book contributed by Azad and Wiggins (1995) entitled 'Dynamics of inter-organisational data sharing: a conceptual framework for research'. This sets out a research framework that covers some of the reasons that push public and private organizations to engage in inter-organizational relationships and also identifies some of the main levels and stages of inter-organizational data sharing within a research framework for future research. These themes recur in many of the contributions that are discussed below.

There are a number of interesting contributions to the information sharing literature before 2006 that have been summarized in chapter 8 on information sharing. These include Uwe Wehn Montalvo's (2003) book about data sharing in South Africa entitled *Mapping the Determinants of Spatial Data Sharing*, which draws upon the Theory of Planned Behaviour developed by the social psychologist Icek Ajzen (1991), as well as a number of studies of information sharing in the United States (see, for example, Harvey and Tulloch, 2006; Nedovic-Budic et al., 2004) and Australia (see, for example McDougall et al., 2005). It is worth noting that there has been a strong local government emphasis in much of this research that supports Nancy Tosta's (1998) dictum:

> Successful SDIs will be local in nature. This is as much a function of practical matters such as the challenges of coordinating large numbers of people over large areas, as it is recognising that all geography is local and issues, physical characteristics, and institutions vary significantly across nations and the world.

In Europe there has been a strong cross-cultural dimension to research on data sharing in recent research. For example, Omran et al. (2007) draw heavily on the work of Hofstede (1991) who argues that five dimensions can be used to classify societies according to their culture: power distance, uncertainty avoidance, individualism/collectivism, masculine/feminine and long-term/short-term orientation. Castelein et al. (2013) have also explored collaboration within two national case studies of the SDIs of the Netherlands and Spain. Their basic overall aim is to gain a better understanding of critical collaboration factors for the development and implementation of SDIs. The findings of their research suggest that the SDIs are getting more embedded in administrative organizational and legal structures, relying on the distributed competences and knowledge of the different stakeholders even though the attitude and level of engagement of stakeholders in data sharing remains a crucial factor for collaboration. Consequently, they argue that a balance has to be found between top-down formalized collaboration structures and informal, more spontaneous ways of collaboration.

More recent studies of data sharing have been focussed on technical rather than organizational matters. For example, Goodchild et al. (2006) note that, while the history of data sharing has evolved from an early disorganized phase through one centred on national governments to the current, somewhat chaotic network of producers and consumers, the US government's Geospatial One-Stop initiative represents a significant technological advance even though its potential to provide a general marketplace for geographic information beyond government data has not been realized. Similarly, the findings of Noucher et al.'s (2017) study of 45 French institutional infrastructures, and their analysis of interviews with spatial data infrastructure (SDI) coordinators and their answers to questionnaires, as well as their websites and their metadata catalogues, produced varied results concerning data accessibility, stakeholder networks, the interoperability of tools and informational equality in different regions. The finding of their research also show that interoperability remains restricted to specific types of actors and themes, and it only concerns the public sector.

In the last few years, a number of studies have been undertaken in the public administration and information science fields that raise questions about information sharing in the geographic information field. For example, Yang et al. (2014) have explored how information is shared across the vertical and horizontal boundaries of government agencies in their analysis of the flagship e-networking project of the Taiwanese national development plan. There are some interesting parallels between their discussion of the characteristic features of the vertical and horizontal approaches to information sharing in their case study and our own distinction between corporate and departmental approaches to GIS implementation in in British local government that is discussed in the first part of this chapter. Their findings also suggest that the Taiwanese government, 'while still strengthening information sharing in the vertical dimension, focussed primarily on information across horizontal boundaries' (p. 641). In practice, a competition-and-cooperation relationship exists among the different types of information sharing in both dimensions. Consequently, it is suggested that participants should aim to strike a balance between centralized and decentralized types of information sharing to obtain advantages and diminish disadvantages.

It is also worth noting that one of the authors of the previous papers, Tung-Mou Yang, also co-authored an extensive literature review of materials on information sharing in public organizations (Tang and Maxwell, 2011). This contains a large number of references to papers in the public administration and information science fields that discuss interpersonal, intra-organizational and inter-organizational success factors.

The starting point for another study by Welch et al. (2016), using data from a 2012 national survey of US municipal government managers, was the common observation that even though the rise of big data, open government and social media may imply greater data sharing, data exchange in the public sector involves a complex social process and critical organizational and managerial capacity. Based on prior research, they hypothesized that greater data sharing would occur in agencies that were subject to institutional coercion, inter-organizational persuasion and have higher levels of technical management and technical engagement capacities. However, the results of their statistical analysis show limited support for the coercion hypotheses. Consequently, the authors conclude that it may be possible that general influence or pressure from external constituencies or government, when they are not specifically tied to a task or activity, have little effect and that, in the local government context, departments are less likely to respond to such pressures as the risks of incurring negative consequences are low.

This section has reviewed some recent developments in research on information sharing in the geographic information field. It also highlights some ongoing developments in the public administration and information science fields with a view to stimulating more interchange between researchers in related fields.

Acknowledgment

The research described in this chapter is funded under the Economic and Social Research Council/Natural Environment Research Council Geographic Information Handling Project. An earlier version of the paper was presented at the NCGIA Workshop on information sharing in San Diego in February 1992.

References

Ajzen, I., 1991, The theory of planned behaviour. *Organisational Behaviour and Human Decision Processes, 50,* 179–211.

Audit Commission for local authorities in England and Wales, 1990, *Preparing an Information Technology Strategy: Making It Happen*, London: HMSO.

Azad, B. and L. Wiggins, 1995, *Dynamics of Inter-Organisational Geographic Data Sharing: A Conceptual Framework for Research*, in Onsrud and Rushton, pp. 22–43.

Bates, B. J., 1988, Information as an economic good: sources of individual and social value, in Mosko, V. and Wasko, J. (Eds). *The Political Economy of Information*, Madison, WI: University of Wisconsin Press.

Beath, C. M., 1991, Supporting the information technology champion. *Management Information Systems Quart*, 355–372.

Bryan, N.W., 1992, A review of pricing and distribution strategies: local government case studies. *Proc. URISA*, 92, 4, 13–25.

Campbell, H., 1992, Impact of geographic information systems on local government. *Computers Environment and Urban Systems*, 16, 531–541.

Campbell, H, 1993, GIS implementation in British local government, in Masser, I. and Onsrud, H. (Eds). *The Diffusion and Use of Geographic Information Technologies*, Dordrecht: Kluwer, 117-146

Campbell, H. and I. Masser, 1993, The impact of GIS on British local government: some findings from Great Britain. *International Journal of Geographical Information Science*, 6, 529–546.

Castelein, W.T., A. K. Bregt, and L. Grus, 2013, The role of collaboration in spatial data infrastructures. *URISA Journal*, 25, 2, 31–40.

Coulson, M. and R. Bromley, 1990, The assessment of user needs for corporate GIS – the example of Swansea City Council. *Proc. EGIS 90*, Utrecht, The Netherlands.

Croswell, P. L., 1991, Obstacles to GIS implementation and guidelines to increase the opportunities for success. *URISA Journal*, 3, 43–57.

Danziger, J. N., W. H. Dutton, R. Kling, and K. Kraemer, 1982, *Computers and Politics: High Technology in American Local Governments*, New York: Columbia University Press.

de Montalvo, U. W., 2003, *Mapping the Determinants of Spatial Data Sharing*, Aldershot: Ashgate.

Finkle, R.W., and F. M. Lockfield, 1990. Major cost savings in multi-participant AM/GIS: the Santa Clara case study, JMM Technical Publishing Service, JM Montgomery, Inc.

Gault, I. and D. Peutherer, 1989, Developing GIS for local government in the UK, paper presented at the European Regional Science Conference, Cambridge.

Goodchild, M. F., P. Fu, and P. Rich, 2006, Sharing geographic information: an assessment of the Geospatial One-Stop. *Annals of the Association of American Geographers*, 97, 2, 250–266.

Harvey, F. and D. Tulloch, 2006, Local-government data sharing: evaluating the foundations of spatial data infrastructures. *International Journal of Geographical Information Science*, 20, 7, 743–768.

Hepworth, M., 1990, Mobilising information capital in local government. *Local Government Policy Making*, 16, 42–48.

Hofstede, G., 1991, *Cultures and Organisations: Software of the Mind*, Maidenhead, UK: McGraw-Hill.

King, J. L. and K.L. Kraemer, 1985, *The Dynamics of Computing*, New York: Columbia UP.

Kraemer, K. L. and J. L. King, 1988, The role of information technology in managing cities. *Local Govt Studies*, 14, 23–47.

Levinsohn, A., 1990, Institutional issues in GIS implementation. *Proc. ISPRS Commission IV*. Tsukuba, Japan.

Masser, I. and H. Campbell, 1991, Conditions for the effective utilization of computers in urban planning in developing countries. *Computers Environment and Urban Systems, 15*, 55–67.

Masser, I. and H. Campbell, 1992, Geographic information systems in organizations: some conditions tor their effective utilization, in Lepper, M. de, Scholten, H.J. and Stern, R.M. (Eds). *The Added Value of Geographic Information Systems in Public and Environmental Health*, Dordrecht: Kluwer, 287-297

Masser, I. and H. Onsrud, (Eds). 1993, *The Diffusion and Use of Geographic Information Technologies*, Dordrecht: Kluwer.

McDougall, K., A. Rajabifard, and I. Williamson, 2005. Understanding the motivations and capacity for SDI development from the local level, *Proceedings GSDI 8*, Cairo, Egypt.

Mosco, V. and J. Wasko, (Eds). 1988, *The Political Economy of Information*, Madison: University of Wisconsin Press.

Nedovic-Budic, Z., J. K. Pinto, and L. Warnecke, 2004, GIS database development and exchange: interaction mechanisms and motivations. *URISA Journal, 16*, 1, 15–29.

Noucher, M., F. Gourmelon, P. Gautrrau, J. Georis-Creuseveau, A. Maulpoix, J. Pierson, N. Pinede, O. Pissoat, and M. Rouan, 2017, Spatial data sharing: a pilot study of French SDIs. *ISPRS International Journal of Geo-Information, 6*, 99–122.

Omran, E.-S.E., A. Bregt, and J, Crompvoets, 2007, Spatial data sharing: a cross cultural conceptual model, in H. Onsrud (Ed). *Research and Theory in Advancing Spatial Data Infrastructure Concepts*, Redlands, CA: ESRI Press, pp. 75–92.

Onsrud, H. and G. Rushton (Eds), 1992, Institutions sharing geographic information, *Tech Report 92-5*, National Centre for Geographic Information and Analysis, Santa Barbara University of California.

Onsrud, H. and G. Rushton, (Eds). 1995, *Sharing Geographic Information*, New Brunswick, NJ: Centre for Urban Policy Research Rutgers University.

Openshaw, S. and J.B. Goddard, 1987, Some implications or the commodification or information and the emerging information economy for applied geographic analysis in the United Kingdom. *Environment and Planning, A, 19*, 1428–1439.

Pickles, J., 1991, Geography, GIS and the surveillant society. *Proc. Applied Geog. Con*: Toledo, OH.

Rhind, D., 1992, Data access, charging and copyright and their implications for GIS. *International Journal of Geographical Information Science, 6*, 13–30.

Tang, T.-M., and T. A. Maxwell, 2011, Information sharing in public organisations: a literature review of interpersonal, intra-organisational and inter-organisational success factors. *Government Information Quarterly, 28*, 164–175.

Tosta, N., 1998. NSDI was supposed to be a verb: a personal perspective on progress in the evolution of the US National Spatial Data Infrastructure, in B. M. Gittings (Ed.), *Integrating information infrastructures with geographic information technology*, London: Taylor and Francis, 13–24.

Tveitdal, S. and O. Hesjedal, 1989, GIS in the Nordic countries market and technology, strategy for implementation – a Nordic approach. *Proc. GIS 89*, Vancouver, Canada.

Welch, E. W., M. K. Feeney, and C.-H. Park, 2016, Determinants of data sharing in US city governments. *Government Information Quarterly, 33*, 393–403.

Wellar, B., 1988, Institutional maxims and conditions for needs sensitive information systems and services in local governments. *Proceedings of URISA, 88,* 4, 371–378.

Yang, T.-M., T. Pardo, and Y.-J. Wu, 2014, How is information shared across the boundaries of government agencies: A e-government case study. *Government Information Quarterly, 31,* 637–652.

9

The Diffusion of GIS in Local Government in Europe

Ian Masser and Massimo Craglia

9.1 Introduction

Geographic information systems (GIS) are a multimillion dollar industry which is growing very rapidly at the present time. Estimates of its size vary according to the measures used. Payne (1993), for example, estimates that the European GIS market for hardware, software and services was worth some 4 billion ECU in 1993. This estimate does not include the costs to government associated with the collection and management of geographic information. A recent report for the European Community (Commission of the European Communities, 1995) estimates that government spending on geographic information alone accounts for around 0.1 per cent of gross national product or about 6 billion ECU for the European Union as a whole. Similarly, in the USA the Office of Management and Budget has estimated that Federal agencies alone spent $4 billion annually to collect and manage domestic geospatial data (Federal Geographic Data Committee, 1994, p. 2).

The growth of the core GIS market shows little sign of slowing down. Dataquest (Gartzen, 1995), for example, estimate the revenue from GIS software worldwide is likely to grow at an average annual rate of 13 per cent up to 1998. At the present time, North America and Europe dominate the world GIS market.

The scale of the GIS industry makes it necessary to study the diffusion of this technology and its impact on society as a whole in some depth. This is because the technology has the potential to fundamentally change the use of geographic information in a wide range of circumstances, and it must be recognised that not all of these changes will be beneficial (Wegener and Masser, 1996). The largest users of GIS technologies are central and local government agencies although the number of private sector applications has substantially increased in the last few years. In countries such as Germany and the UK, local government applications probably account for somewhere between a quarter and a third of the total market.

Given the size of the GIS industry and the extent to which it is dominated by a relatively small number of large users, it is surprising to find that relatively little systematic research has been carried out either on the development of the geographic information services industry itself or on the diffusion of geographic information technologies in key sectors such as local government (see, for example, Masser and Onsrud, 1993). Local government is a particularly interesting field for diffusion research because it covers a wide range of applications in a great diversity of settings. There is also a strong national dimension to local government, which is reflected in the expectations that people have with respect to local authorities, the tasks that they carry out and the professional cultures that have come into being to support them.

With these considerations in mind, this chapter summarises the findings of research on the diffusion of GIS in local government in Europe, which has been undertaken in the context of the European Science Foundation's GISDATA programme (Masser et al., 1996). This includes nine national case studies of GIS adoption in Denmark, France, Germany, the UK, Greece, Italy, the Netherlands, Poland and Portugal, as well as two more wide-ranging studies of organisational perspectives on GIS implementation and the development of contrasting scenarios which highlight some of the negative as well as the positive consequences of GIS adoption and implementation.

Of particular importance in this context is the extent to which the nine case studies, conducted by teams of national experts in each country, draw upon a common methodology based on the research on this topic that has been carried out at Sheffield over the last five years. This includes both survey and case study research on UK local government (see, for example, Campbell and Masser, 1992, 1995) as well as comparative research on GIS in the UK and other European countries (see, for example, Campbell and Craglia, 1992; Craglia, 1994). The methodology in the UK studies involved a comprehensive telephone survey of all 514 local authorities followed by detailed case studies, and was adapted in the other countries to allow for local circumstances. In some instances, like Denmark, it was possible to carry out a complete survey of all local authorities, while in some others like Germany and Italy the surveys focused on significant subsets: large urban areas in Germany, regions and provinces in Italy. The other national studies focused more on local case studies backed up by secondary data to provide an overall picture of current developments.

Despite these differences, the findings of these case studies make it possible to carry out a more systematic cross-national comparative evaluation of GIS diffusion in one of the key stakeholders in the European GIS community in this chapter than has hitherto been the case. The next section of the chapter summarises the main features of the findings. As a result of this analysis, a simple typology is developed in the third section of the chapter, while the last section sets out an agenda for future research on both GIS adoption and implementation.

9.2 A Comparative Evaluation of European Experience

The main features of GIS diffusion in local government in the nine countries investigated in the GISDATA project are summarised in Tables 9.1, 9.2 and 9.3. Table 9.1 compares the characteristics of these countries with respect to the institutional context within which GIS diffusion takes place in local government, the structure of local government and the extent of digital data availability.

9.2.1 Institutional Context

From Table 9.1 it can be seen that there are important differences between the nine countries with respect to their political and economic stability. Recent political upheavals in Italy, for example, have profoundly affected the political culture that underlies both central and local government in that country. Poland is in the middle of a major structural transformation from a command to a market economy. Even relatively stable countries such as Germany are also undergoing important structural changes as a result of the incorporation of the East German Länder into the former West German state. In contrast, Denmark, France and the Netherlands have relatively high degrees of political and social stability which is also reflected in local government.

In several of these relatively stable countries, however, there have been moves to decentralise powers from central to local government over the last two decades. The impact of decentralisation is most pronounced in Denmark following the reforms of the 1970s, but similar trends can be found in France and Italy during the 1980s and more recently in Greece and Portugal. In the UK, developments during the last decade and a half run counter to this general trend with more central government control over local government. Poland is also a country where central government control over local government is increasing once again after the decentralisation of power following the collapse of the former communist regime.

These changes have given rise to uncertainties about the responsibilities of the various layers of local government in several countries. In the UK, for example, this culminated in the abolition of metropolitan counties by the government in 1986. Similarly, the provinces in the Netherlands are struggling to find a role after changes in local government structure. In France, on the other hand, decentralisation has given new powers to the *départements* as it has to the provinces in Italy.

There are also important differences between countries with respect to the relationship between local government and the private sector. In

TABLE 9.1

Key contextual factors in nine European countries.

	UK	Germany	Italy	Portugal	Denmark
Institutional context	Increasing pressures towards privatisation of local government tasks has affected GIS diffusion. Uncertain impacts of ongoing review of local government in terms of the number and functions of authorities	Stable and well-established local government culture. Incorporation of former East German *Länder* presents new opportunities	The 1990 reform of local government which extends powers and degree of autonomy of lower tiers has both constrained and created opportunities for GIS diffusion. Uncertain impacts of recent political upheavals on local government. Limited use of IT in local government	Pro-active Government measures to modernise local government. Establishment of a National Centre for Geographic Information to coordinate the implementation of a national geographic information system. Availability of EC funding to support modernisation	Widespread decentralisation of powers to local government since 1970. Extensive use of IT in local government since the 1970s
Structure of local government	Two tiers: 47 shire counties and 9 Scottish regions with 333 shire districts and 53 Scottish districts. Single tier of 69 metropolitan districts. Few districts have populations of less than 100 000	Three tiers: 16 *Länder*, 543 counties and large cities and 14 809 municipalities. Most of municipalities have less than 10 000 population	Three tiers: 20 regions, 103 provinces and 8100 communes. 87 per cent of communes have less than 10 000 population, 12 new Metropolitan Authorities	Three tiers: 7 regions (including the Azores and Madeira), 29 districts and 305 municipalities. Average size of municipality is 34 000	Two tiers: 14 counties and 275 municipalities. Half the municipalities have less than 10 000 population.
Digital data availability	Comprehensive digital topographic data service provided by OS. Service Level Agreement reached with local authorities in March 1993 boosted GIS diffusion	Strong surveying traditions and collaboration between local authorities with respect to the development and maintenance of digital topographic databases (for example, ALK and MERKIS)	Slow progress made by National Mapping Agency and Cadastre, nearly half the regions developing their own digital topographic databases. Profusion of data sources	Diversity of sources. 65 per cent of country covered by 1:10 000 scale orthophoto maps from the National Mapping Agency. Some digital data at 1:25 000 scale in vector format from the Army Cartographic Services	Municipalities are the main provider of large-scale maps. They also co-manage the Cadastre with the Danish Survey and Cadastre and maintain a wide range of other register-based information systems

	Greece	France	Poland	The Netherlands
Institutional context	Limited financial and manpower resources at the disposal of local government leading to contracting out of IT functions to the private sector and University Laboratories. Role of EC funds in facilitating the diffusion of GIS	Long tradition of geographic information handling in large cities. Decentralisation measures since 1980 have given new powers to Départements and to new administrative structures for main urban areas	Very rapid change from centrally planned to market economy. Earlier attempts to decentralise power to local administration have recently been reversed, leaving some uncertainty over the responsibilities of different layers of government	Strong and stable central government with local authorities retaining a high degree of autonomy. Middle layer of provinces struggling to define its role
Structure of local government	Two tiers: 51 prefectures form upper tier, 361 cities and 5600 parishes form lower tier. Average population of parish is 1800	Three tiers: 22 regions, 96 Départements and 36 000 communes. Average population per commune is 1500. Variety of inter-communal structures especially in large urban agglomerations	Three tiers: 49 regions, 327 sub regions, and 2465 local authorities having an average population size of some 15 000	Two tiers: 12 provinces and local government divided between 650 municipalities and 100 polder boards in charge of water control. 50 per cent of municipalities have less than 12 000 people and 90 per cent less than 40 000
Digital data availability	Lack of digital data. 1: 5000 scale digital contour data available only for some cities in the Athens and Thessaloniki areas. Proliferation of data produced by private sector databases	Small-scale digital data available from Institut Geographique National. Large-scale data in hands of Cadastre. Ongoing digitisation programme underway in conjunction with local authorities	Almost half of the country is covered in good quality cadastral information. The process of conversion to digital format has been undertaken but slow progress and organisational difficulties have led a number of local authorities to acquire their own data, often in partnership with private sector	Strong awareness of geographic information in the Netherlands and tradition of automated registers (population, etc.). Slow progress of large-scale mapping project based on Cadastre has led to local authorities acquiring their own digital mapping individually or in partnership with utility companies and private sector

TABLE 9.2

Key features of GIS diffusion in the five countries where surveys were carried out using a similar methodology.

	UK	Germany	Italy	Portugal	Denmark
Coverage of findings	Comprehensive survey of all local authorities in England, Wales and Scotland in the second half of 1993	Survey of cities with 100 000 population in the first half of 1994	Comprehensive surveys of all regions and provinces in 1993/1994: some additional information for communes (1991) and Metropolitan areas (1993)	Survey of municipalities in late 1993/early 1994	Comprehensive survey of counties and municipalities in first half of 1993
Extent of diffusion	Almost universal in counties and Scottish regions. Half metropolitan districts. One in six shire/Scottish districts	Almost universal: 70/80 cities had GIS, 10 had firm plans	Two-thirds regions – one third provinces but another one in three with firm plans. Limited in medium/large cities	12 municipalities had GIS and a further 24 had AM/FM facilities	Over 80 per cent of all municipalities use register-based systems where georeferencing present. GIS/AM/FM almost universal in authorities with 50 000+. GIS in half counties
Geographical spread	North South divide: 32 per cent of Southern authorities had GIS. 24 per cent of Northern authorities had GIS	No significant geographic difference	Pronounced North/South divide among the regions (90 per cent N – 37 per cent S) and the provinces (50 per cent N – 20 per cent S)	Adoption levels highest in urban areas and in northern parts of Portugal	Urbanised regions generally have higher take up than less urbanised regions
Length of experience	70 per cent systems purchased since 1990	Most systems purchased since 1990	Half regional and 80 per cent provincial systems purchased since 1990	Some municipalities began GIS projects in 1990	Most GIS systems acquired since 1990

Main applications	Automated cartography and mapping for local planning and management	Surveying and topographic database management	Digital map production, strategic land-use planning. Environmental monitoring also strong in the provinces	Land use planning; automated mapping	Digital mapping. Also extensive use for utility management
Predominant software	Arc/Info in 25–30 per cent of county and metropolitan districts. Axis, Alper Records and G-GP in shire/Scottish/metropolitan districts	SICAD and ALK-GIAP (latter free to German local authorities)	Arc/Info for over half regional and provincial applications. Greater diversity at municipal level	Intergraph purchased by National Centre for Geographic Information facilities	Intergraph MGE and Autocad-based systems followed by Dangraf and GeoCAD
Perceived benefits	Improved information processing (60 per cent) especially improved data integration, better access to information and increased analytical and display facilities	Improved information processing: (65 per cent) faster information retrieval and increased analytical and display facilities	Improved information processing (51 per cent) especially automated map production and thematic mapping	Improved information processing, administrative reorganisation, data sharing with utilities	Improved information processing (66 per cent)
Perceived problems	Technical problems including lack of software and hardware compatibility. Organisational problems especially poor managerial structures and lack of skilled staff	Organisational problems especially lack of qualified staff and insufficient motivation of staff by management	Organisational problems especially lack of awareness, poor coordination and lack of skilled personnel	Bureaucratic inertia, lack of skilled personnel digital-data availability, lack of awareness, vendors, attitudes and limited follow-up support	Technical problems in small municipalities. Organisational problems in large municipalities

TABLE 9.3

Key features of GIS diffusion in the other four countries

	Greece	France	Poland	The Netherlands
Extent of diffusion	About 10 per cent of all cities have GIS facilities	Two-thirds of cities with more than 100 000 population have GIS. Widespread use in inter-communal agencies	30–40 mainly in small-medium sized towns (20 000–100 000 inhabitants). Few regional/sub-regional GIS	GIS is almost univ-eral among munici-palities above 50 000 inhabitants and in approx. 50 per cent of those above 20 000
Geographical spread	No marked regional variations	No discernible regional variations reported	No marked regional variations	No marked variations
Length of experience	Most systems purchased since 1990	Most systems pur-chased since 1990	Most systems pur-chased since 1990	Rapid take-up since late 1980s
Main applications	Surveying and topo-graphic data-base man-agement	Urban database management for surveying and planning	Topographic data, parcels data, cadas-tral information	Mainly topographic and thematic mapping
Predominant software	Arc/Info for over 80 per cent of local govern-ment applications	Arc/Info and Apic (French package)	Map Info, Autocad for small systems, Arc/Info for larger central government implementation	Intergraph MGE, Autocad and IGOS (Dutch package)

Greece, Italy and Poland, for example, contracting out of data collection and IT tasks to the private sector is commonplace because of the limited resources at the disposal of local government. In the UK, local authorities are also being increasingly obliged by central government to put key tasks out to competitive tender to reduce costs.

In Portugal, on the other hand, central government is taking an increas-ingly proactive role to modernise local government. Like Greece, the diffusion of GIS in this country has been substantially accelerated by the availability of funds from the European Community.

9.2.2 Structure of Local Government

Table 9.1 shows only the tip of the iceberg in that it refers essentially to the number of authorities at various levels of local government in the nine countries and gives some indication of the size of these units in terms of population. As such, it does not deal explicitly with the great diversity that also exists with respect to the ways in which the various functions are discharged by the different tiers of local government in the nine countries.

Nevertheless, Table 9.1 highlights some important differences between the countries. Germany is the only country surveyed which has a Federal system of government. France, Italy and Poland have developed three-tier local government structures, whereas all the other countries have two-tier structures with the partial exception of the UK which has single-tier authorities in the main centres of population since the abolition of the metropolitan counties.

The UK also stands out from all the other countries with respect to the size of its lower tier authorities which, with relatively few exceptions, contain populations of over 100,000. Elsewhere in Europe, small is generally beautiful. Over 80 per cent of Italian communes have less than 10,000 population. In France the average population per commune is 1500 and the comparable figure for Greece is 1800. Most German authorities have populations of less than 10, 000 and even in Denmark, where there was a major reorganisation of local government during the 1970s, which resulted in a substantial reduction in the number of authorities, half the municipalities have populations of less than 10,000.

9.2.3 Digital Data Availability

Once again, the UK stands out from all the other countries with respect to digital data availability given that its national mapping agency, the Ordnance Survey (OS), provides a comprehensive mapping service for both large- and small-scale maps. Digitisation of the former is nearly complete and local government access to this information has been substantially facilitated by the service-level agreement for the purchase of digital data reached between the local authority associations and the OS in March 1993. Through this agreement, all authorities operate within the same framework and no longer need to enter into individual negotiations with the OS.

At the other end of the spectrum comes Denmark which also has a very high level of digital data availability but makes the municipalities themselves the providers of large-scale maps. These authorities also maintain the Cadastre together with the Land Registry as well as a large number of register-based systems.

In terms of digital-data availability, Greece, Italy and Portugal in particular currently suffer from a proliferation of digital-data sources as a result of the limited progress made by their central government agencies with respect to the provision and coordination of digital data.

Elsewhere in France, Germany, the Netherlands and to some extent Poland, large-scale digital-data provision is closely linked to the maintenance of the Cadastre. In these countries, the establishment of the data infrastructure required for municipal GIS is proceeding rather slowly despite the very strong surveying traditions in some of these countries.

It might be useful to point out that the focus of the research has been on the availability of digital mapping because the increasing availability and resolution of satellite data has not yet made a significant impact on local government in the countries analysed.

9.2.4 Survey Findings

Table 9.2 summarises the main findings of the five national surveys. However, before comparing the findings of these surveys, attention must be drawn to a number of differences between them in terms of timing and coverage.

Both the UK and Danish surveys involved comprehensive telephone surveys of all local authorities which obtained a 100 per cent response rate. However, the Danish survey was carried out in the first half of 1993 between six months and a year before the other four surveys. This discrepancy in timing is particularly important given the rapid take-up of GIS in the smaller authorities.

In Portugal, a postal questionnaire was sent to all municipalities in late 1993 but not the regions. It achieved a response rate of 55 per cent, which is good for a postal questionnaire, but nevertheless the findings must be treated with some caution given that no information is available for non-responding authorities.

Because of the very large number of authorities in Germany, the German survey which took place in the first half of 1994 was restricted to the 86 cities with populations of over 100,000. A response rate of nearly 90 per cent was achieved in this case which is very high under the circumstances. Nevertheless, no information is available for either the higher levels of local government or authorities with less than 100,000 population.

In Italy, two comprehensive surveys of the regions and provinces were undertaken in late 1993 and early 1994 respectively, both achieving a 100 per cent response. However, the information on GIS in municipalities is limited to the findings of research carried out in 1991.

The main findings of the other four studies are summarised in a similar format to those of the five surveys in Table 9.3. It should be noted, however, that these studies are less comprehensive in coverage than the five surveys. In the case of Greece and Poland, this presents relatively few problems given the low level of GIS diffusion in these countries. In the case of France and the Netherlands, it should be noted that much of the information is drawn from secondary sources which predate the findings of the five national surveys by several years.

9.2.5 Extent of GIS Diffusion

Tables 9.2 and 9.3 show that the extent of GIS diffusion was lowest in Greece, Portugal and Poland. In these countries, between 10 and 15 per cent of the cities had acquired GIS or AM/FM facilities. However, as the case of Portugal indicates, the situation is changing very rapidly as a result of proactive government measures, and the extent of GIS diffusion is likely to increase dramatically in the immediate future.

Urban applications in medium and large cities predominate in Denmark, France and Germany. In Germany the use of GIS is almost universal in cities with populations of over 100 000, as it is in Denmark with reference to cities with over 50 000 population. Throughout these countries, the extent of GIS diffusion is considerably lower in small towns and rural districts. In France and Denmark, it is also lower at the departement and county levels.

This relationship is completely reversed in the UK and Italy. In the UK the take-up of GIS in the counties is almost universal, whereas only half the metropolitan districts had acquired facilities. In Italy, levels of diffusion at the regional and provincial levels are also very much higher than those for the cities. In both cases, however, the take-up in urban areas is generally higher than in rural areas.

In the UK, the obvious explanation for this difference is the extent to which the surveying and mapping functions that are attached to local government in most European countries are carried out, in this case by the OS. In addition, there is no requirement in the UK to maintain a Cadastre as is the case in most other European countries. The case of Italy, however, is more difficult to explain and may reflect the new planning powers that have been given to the provinces.

9.2.6 Geographical Spread

As noted above, a general distinction can be made between urban, especially large urban, and rural areas with respect to the extent of GIS diffusion. In some countries, however, there is also a distinction between regions with respect to the diffusion of GIS. This is most pronounced in Italy with respect to the traditional divisions between the north, centre and south. At the regional and provincial level, GIS adoption has reached nearly 100 per cent and 50 per cent, respectively, in the north as against less than 40 and 20 per cent for the equivalent authorities in the centre/south.

There are also clear differences in the UK between the north and Scotland and the south and eastern parts of the country. In this case, however, the ratio is reversed in favour of the wealthier south and eastern regions where the level of GIS adoption is 32 per cent as against 24 per cent in the north and Scotland.

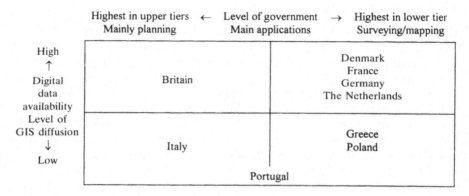

FIGURE 9.1
Typology of GIS diffusion in nine European countries

Elsewhere, no pronounced regional variations have been reported for Denmark, France, Germany and the Netherlands. In Greece, Poland and Portugal, the extent of diffusion so far is probably too low for any clear patterns to have emerged.

9.2.7 Length of Experience

Both Tables 9.2 and 9.3 show that the vast majority of GIS systems in local government in virtually all countries have been purchased since 1990. This highlights the extent to which GIS in local government is very much a recent development in Europe.

9.2.8 Main Applications

Tables 9.2 and 9.3 show that digital map production and digital mapping are the predominant local government applications in Denmark, France, Germany, the Netherlands and Poland. In most of them, these activities are closely linked to the maintenance of the Cadastre.

The main exceptions are Italy, Portugal and to some extent the UK. In Italy and Portugal, land-use planning in some form appears to be the main GIS application. This is particularly the case in Portugal where the dissemination of GIS is closely linked to the preparation and approval of municipal land-use plans as well as automated mapping. In the UK, the position is more complex, but the emphasis, nevertheless, is on automated cartography linked to planning activities as well as the other technical services provided by local authorities.

9.2.9 Predominant Software

Overall, the findings summarised in Tables 9.2 and 9.3 point to the remarkable dominance of North American software, especially Arc/Info, in European local government. Arc/Info itself accounts for over 80 per cent of all GIS applications in Greece, half the regional and provincial applications in Italy and between 25 and 30 per cent of applications in UK counties and metropolitan districts. It also has a strong presence in France. On the other hand, in Portugal, Intergraph is the GIS market leader as a result of its acquisition as part of the implementation of the National Council for Geographic Information's national strategy.

In contrast, the German local authorities have remained loyal to their local software developers, and the Danish authorities also make extensive use of local software packages. Many of these are custom-made for the specific tasks carried out in their local governments.

Despite the dominant market share of North American software in most countries, there are also a large number of local tailor-made packages in use especially at the municipality level. Examples of these from the UK include G-GP, Axis and Alper Records (now Sysdeco). Widely used local packages in France include Apic and in the Netherlands IGOS.

9.2.10 Benefits and Problems

Questions regarding the perceived benefits and problems to be derived from GIS were asked only in the five surveys. These revealed a high level of consensus of views throughout all the five countries involved. In the case of benefits, there was general agreement that improved information processing was by far the most important benefit to be derived from GIS. The main reasons for this view were improved data integration and faster information retrieval together with increased analytical and display facilities. Irrespective of their nationalities, most respondents also identified similar organisational problems associated with recent developments in GIS. These included poor management structures and bureaucratic inertia together with lack of awareness and the shortage of skilled personnel.

9.3 Towards a Typology of GIS Diffusion in European Local Government

In overall terms, the findings of the nine studies reveal a considerable measure of agreement regarding the perceived benefits and problems associated with GIS in local government in Europe. They also show that the length of local government experience in most cases is very similar between countries. It is useful at this stage, therefore, to concentrate on identifying some of the main differences between countries and exploring the extent to which a typology can be developed on the basis of these differences. From the evaluation above, it would appear that two main dimensions can be identified which account for a large number of the differences observed between countries. This provides a starting point for further comparative evaluation.

The first dimension that emerges from the findings is related to the overall extent of diffusion and the level of digital-data availability in these countries. It measures essentially the links between data infrastructure and diffusion.

The second dimension is associated with the nature of the main GIS applications that are undertaken and the level of government at which they are carried out in the different countries. This dimension largely measures the professional cultures surrounding GIS applications.

When the nine countries are plotted in relation to these two dimensions in Fig. 9.1, it can be seen that all four quadrants of the diagram are occupied. The largest grouping consists of countries with high levels of diffusion, reasonable data availability where GIS is used essentially at the municipal level for surveying and mapping operations. Denmark is probably the best example of this category, but France, Germany and the Netherlands possess most of these characteristics.

The second largest group consists of countries with low levels of diffusion and restricted digital-data availability where GIS applications in local government are predominantly at the municipal level for surveying and mapping. Greece and Poland are the best examples of this category although Portugal shares some of these characteristics. However, there is an important difference between Portugal and the other two countries with respect to the emphasis given to planning in Portugal. This places it somewhere between Greece and Italy in the table.

Each of the other two quadrants contains only one country. The UK stands out alone as a country with high levels of GIS diffusion and digital-data availability where GIS diffusion is greatest at the upper tier of local government and there is a strong planning tradition in local government. Similarly, the impact of GIS in Italy has been greatest at the upper tiers (regions and provinces) and mainly applied to planning even though the availability of digital data and the levels of GIS diffusion are lower than in the UK.

Two other matters are worth noting from Fig. 9.1. First, there is a broad correlation between relative stability and relative change in terms of the institutional context between the upper and lower halves of the diagram, respectively. Given the changes that are already in progress in Portugal, for example, and to a lesser extent in Greece, Italy and Poland, the positions of these countries in the table may change considerably in absolute if not in relative terms over the next few years.

Second, there are also some interesting links between location in the diagram and the extent of the dependency on imported versus local software. The highest levels of dependency on North American software are generally to be found in the bottom half of the diagram in countries such as Greece and Poland where the overall level of GIS diffusion is still relatively low. In the top half of the diagram, on the other hand, levels of dependency are generally much lower and there are also a large number of local software products in use. This is particularly the case in Germany where these products are also the market leaders.

9.4 Conclusions

The findings of this comparative evaluation highlight some of the main similarities and differences between GIS diffusion in local government in nine European countries. They show that in all the countries surveyed GIS is a very recent phenomenon in local government and so far, only a limited amount of operational experience has been built up to substantiate the claim that GIS implementation will lead to improved information processing. At this stage, it is important to emphasise the distinction between adoption, the decision to acquire GIS facilities and implementation, the use that is made of these facilities in practice.

9.4.1 GIS Adoption

From the standpoint of GIS adoption, it would appear from the analysis that two key factors account for most of the differences between the nine countries. The first of these measures the links between digital-data availability and GIS diffusion while the second reflects the professional cultures surrounding GIS applications in local government.

The question of digital-data availability is not simply a matter of the information rich versus the information poor. It is much more a question of central and local government attitudes towards the management of geographical information. For this reason, those countries with relatively low levels of digital-data availability tend also to be countries where there has been a fragmentation of *ad hoc* data sources as a result of a lack of central government coordination. Conversely, those countries with relatively high levels of digital-data availability tend also to be countries where governments have created a framework in terms of responsibilities, resources and standards for the collection and management of geographic information.

The question of professional cultures is closely linked to the responsibilities that have been given to local government in different European countries. In countries where local governments play an important role in maintaining land registers and/or cadastral systems, a highly organised surveying profession has come into being to carry out these tasks. Consequently, land information systems rather than GIS tend to predominate in local government applications in these countries. Conversely, in countries such as the UK where land registration and digital topographic mapping are dealt with centrally, there is a greater emphasis on applications which are broadly linked to the various planning activities of local government. In this case the predominant professional cultures tend to be those of the land-use planner and, to a lesser extent, the transport engineer.

Within these two groups, however, it is also necessary to consider the nature of the responsibilities that have been given to local government and the resources placed at their disposal when comparing professional cultures. There are considerable differences, for example, as Craglia (1993) has shown elsewhere between the local planning functions and associated cultures that have developed in the UK and Italy. Similarly, there are also important differences between Germany and Greece with respect to the resources at the disposal of local government for land-management functions, which affect the development of the surveying profession in each country.

As a result of this analysis, it is felt that these two factors provide a useful starting point for the further analysis of the diffusion of GIS in local government in both Europe and elsewhere. What is now needed is further research which evaluates and refines the typology developed for the analysis in the context of local government in other countries. There is also a need for cross-national comparative studies of other key GIS application sectors. In this respect, special priority should be given to the utilities sector because of its central position as the other key investor, alongside local government, in GIS systems technology.

The final question concerns the significance of the findings of the analysis with respect to diffusion theory as set out by Rogers (1993) and others. As noted above, the analysis has been primarily concerned with the adoption rather than the implementation of GIS. On this count, it shows that there are considerable differences between countries with respect to their positions on the S-shaped curve. GIS diffusion in local government in the UK, Denmark and Germany in particular is already well past the critical point of take-off and is approaching saturation in relation to particular levels of government. Elsewhere, in Greece, Poland and Portugal in particular, GIS diffusion in local government is still at a relatively early stage on the curve, but, as the findings of the survey clearly indicate, the situation is changing very rapidly at the present time. Given the speed of change, it will be necessary therefore to closely monitor events to keep track of the adoption process in these countries.

9.4.2 GIS Implementation

It must be recognised that the primary focus of the above discussion has been on GIS adoption and that matters relating to GIS implementation and its consequences have only been discussed in general terms. Nevertheless, it is clear from this discussion that implementation is likely to become an area of growing significance in GIS diffusion research. GIS implementation research is concerned with the process whereby new technologies are adapted to meet the specific needs of organisations such as local authorities. The starting point for such research, as Campbell (1996) has demonstrated, is not classical diffusion theory but theories of organisational change. Given that the focus of this research is on the organisation itself rather than on the national circumstances within which that organisation operates, it may be expected that some striking similarities will be found between organisations in widely differing institutional contexts. In this way the findings of such research are likely to further highlight the diversity of responses to the adoption and implementation of new technologies such as GIS.

9.5 Continuing Research on GIS Take up and Adoption throughout Europe

The timing of these studies must be borne in mind when considering the findings of this chapter 20 years later. As the findings of surveys that are described in the main body of the chapter show, the introduction of GIS in local government was still at the early stages of the diffusion process in the early nineties. Consequently, most of the adopters can be regarded as the early adopters with regard to the Rogers diffusion model (Rogers, 1993, 2003). Since that time the number of adopters has moved steadily through the early majority stages to reach the late majority. In the process GIS has become to be regarded as a standard feature of local government in most European countries. The main differences between countries reflect variations in the fundamental structures of local government and the nature of the main professional cultures of those involved in the GIS implementation and management process. The extent to which these features may have changed since the early nineties has yet to be fully explored.

It must also be borne in mind that there have been massive changes in the GIS technologies and the technical facilities that are available to local government staff since the original studies were undertaken. At the same time, the whole context of GIS practice in local government in Europe has also been changed by new management practices in local government in general such as the emergence of eGovernment as well as emerging concepts of spatial data infrastructures in the geographic information field that have been stimulated by the activities associated with the formulation and the implementation of the INSPIRE Directive.

Over the last 15 years, the primary focus of research in this field has shifted towards studies of emerging national spatial data infrastructures in the context of the formulation and implementation of the INSPIRE Directive 2007/2/EC of the European Parliament and of the Council of 14 March 2007 establishing an Infrastructure for Spatial Information in the European Community (INSPIRE) for improving environmental data management in the European Community by 2020 (Commission of the European Communities, 2007). Because of this, the Commission has funded a number of important studies in this field. These include the Methods for Access to Data and Metadata in Europe (MADAME) project (Blakemore et al., 1999), and the Geographic Information Network in Europe (GINIE) project (Craglia et al., 2003).

Since then the Commisson has also commissioned a large number of studies of the State of Play of SDI activities in all the European countries from the Spatial Applications Division at the Catholic University of Leuven (SADL) in Belgium. These studies began in 2002, and the reports assess the status and development of 32 European NSDIs (27 Member States, 4 EFTA countries and Turkey) (http://inspire.jrc.ec.europa.eu/). They used a common methodology and the reports have been revised and repeated annually between 2003 and 2011 with a break in 2008 (Vandenbroucke et al., 2008). This makes it possible to make cross national comparisons between these 32 countries and also to assess progress in each country. In this way, the series of the State of Play studies have documented the changing national situation in Europe at the start of the implementation of the INSPIRE Directive.

The findings of the workshop on Advanced Regional Spatial Data Infrastructures that was organised by the European Commission Joint Research Centre in May 2008 (Craglia and Campagna, 2009) also throw some light on developments at the sub-national level in Europe. The workshop discussed eleven case studies of regional and/or sub-national SDIs in Europe, the national State of Play studies, and international experiences in the United States and Australia. The discussions reviewed the state of progress, and analysed the different organisational models established with local and national stakeholders, as well as assessing the social and economic impacts of the regional SDIs. The findings of this workshop drew attention to the efforts that the regional SDIs reviewed were making in involving local government agencies.

> The main aspect worth highlighting here is the crucial role of this 'regional' dimension of SDIs which is often neglected by professional and academic debates that tend to focus more on the national dimension, subsuming the regional in a hierarchical view of SDIs. The regional experiences are not just an intermediate level from global to local, subservient to the higher administrative authority.

> They are often leading the field, pre-dating national developments, or setting the example and framework, including technical specifications, for the national levels. In some instances, as in Italy, Spain, Belgium, and Germany they are the key building blocks of the national SDIs, with the national level providing a thin layer on the regional infrastructures.
>
> (p. 127)

The next European initiative was the series of national and regional workshops that was organised throughout Europe as part of the eSDI-Net+ project (www.esdinetplus.eu/). This was a Thematic Network co-funded by the eContent*plus* Programme of the European Commission and coordinated by the Technical University of Darmstadt, Germany, to promote cross-border dialogue and stimulate the exchange of best practices on Spatial Data Infrastructures in Europe (Rix et al., 2011). The project started in September 2007 and ended in August 2010. During its lifetime, the eSDI-Net+ project brought together a substantial number of SDI players in a Thematic Network which provided a platform for the communication and exchange of ideas and experiences between different stakeholders involved in the creation and use of SDIs throughout Europe. Its activities will be discussed in more detail in Chapter 18 of this book.

Finally, it should be noted that the formulation and implementation of the INSPIRE Directive has had a profound effect on the GIS adoption and take up in local governments throughput the whole of Europe. The importance of monitoring the progress of INSPIRE implementation is discussed in the last section of the INSPIRE Directive (Article 21(1)). From the outset it was clear that two sets of types of information would be required for this purpose. The first is based on a quantitative approach based on indicators derived from the list of spatial data sets and network services. Alongside the substantial body of statistical material that the national Member States have assembled each year using these quantitative indicators, it was also recognised that it would be necessary for the Member States to provide qualitative information on their progress in the form of written reports every three years using a common template provided by the Commission covering developments since the previous report on the following topics: coordination and quality assurance, functioning of the infrastructure, usage of the infrastructure, data sharing arrangements and cost–benefit aspects. From this it can be seen that these topics cover a much wider range of issues than those covered by the quantitative monitoring indicators listed above. Three rounds of qualitative country reports have been completed so far: in 2010, 2013 and 2016. These have created another resource for researchers about developments in GIS adoption and implementation in different European countries (see, for example, Chapter 19 of this book).

References

Blakemore, M., Craglia, M., Evmorfopolou, K., Fonseca, A., Gouveia, C., Lefevre, A., Masser, I., and Pekkinen, P., 1999. *Comparative Evaluation of National Spatial Data Infrastructures*, Methods for Access to Data and Metadata in Europe (MADAME) project report, Sheffield: University of Sheffield.

Campbell, H., 1996. Organisational cultures and the diffusion of GIS technologies, in Masser, I., Campbell, H. and Craglia, M. (Eds.) *GIS Diffusion: The Adoption and Use of GIS in Local Government in Europe*, pp. 23–45, London: Taylor & Francis.

Campbell, H., and Craglia, M., 1992. The diffusion and impact of GIS on local government in Europe: The need for a Europe wide research agenda, in *Proceedings of the 15th European Urban Data Management Symposium*, pp. 133–155, Delft: Urban Data Management Society.

Campbell, H., and Masser, I., 1992. GIS in local government: Some findings from Great Britain, *International Journal of Geographic Information Systems*, 6, 529–546.

Campbell, H. and Masser, I., 1995. *GIS and Organisations*, London: Taylor & Francis.

Commission of the European Communities, 1995. *GI 2000: Towards a European Geographic Information Infrastructure*, Luxembourg: DGXIII.

Commission of the European Communities, (CEC), 2007. Directive 2007/2/EC of the European Parliament and of the Council of 14 March 2007 establishing an Infrastructure for Spatial Information in the European Community (INSPIRE), *Official Journal of the European Union, L108*, 1–14.

Craglia, M., 1993. *Geographic Information Systems in Italian Municipalities: A Comparative Analysis*, Doctoral dissertation, Department of Town and Regional Planning, University of Sheffield.

Craglia, M., 1994. Geographic information systems in Italian municipalities, *Computers, Environment and Urban Systems, 18*, 381–475.

Craglia, M., Annoni, A., Klopfer, M., Corbin, C., Hecht, L., Pichler, G., and Smits, P. (Eds.), 2003. *Geographic Information in the Wider Europe*, Sheffield: University of Sheffield.

Craglia, M., and Campagna, M. (Eds.), 2009. *Advanced Regional Spatial Data Infrastructures in Europe*, JRC Scientific and Technical Reports, Luxembourg: Office for Official Publications of the European Communities.

Federal Geographic Data Committee, 1994. *The 1994 Plan for the National Spatial Data Infrastructure: Building the Foundations of an Information-Based Society*, Reston: Federal Geographic Data Committee.

Gartzen, P., 1995. GIS in an ever changing market environment, in *Proceedings of the Joint European Conference and Exhibition on Geographic Information*, pp. 433–436, Basel: AKM Messen AG.

Masser, I., Campbell, H., and Craglia, M. (Eds), 1996. *GIS Diffusion: The Adoption and Use of GIS in Local Government in Europe*, London: Taylor & Francis.

Masser, I., and Onsrud, H. J. (Eds.), 1993. *Diffusion and Use of Geographic Information Technologies*, Dordrecht: Kluwer.

Payne, D. W., 1993. GIS markets in Europe, *GIS Europe, 2*, 10, 20–22.

Rix, J., Fast, S., Masser, I., Salge, F., and Vico, F., 2011. Methodology to describe, analyse and assess sub-national SDIs, *International Journal of Spatial Data Infrastructures Research, 6*, 23–52.

Rogers, E. M., 1993. The diffusion of innovations model, in Masser, I. and Onsrud, H. J. (Eds.) *Diffusion and Use of Geographic Information Technologies*, 9–24, Dordrecht: Kluwer.

Rogers, E. M., 2003. *Diffusion of Innovations*, 5th edition, New York: The Free Press.

Vandenbroucke, D., Jansen, K., and Van Orshoven, J., 2008. INSPIRE State of play: Generic approach to assess the status of NSDIs, in J. Crompvoets, A. Rajabifard, B. van Loenen and T. Delgado Fernandez (Eds.) *A Multi-View Framework to Assess Spatial Data Infrastructures*, pp. 145-172, Melbourne: Centre for SDIs and Land Administration, University of Melbourne,

Wegener, M., and Masser, I., 1996. Brave new GIS worlds, in Masser, I., Campbell, H. and Craglia, M. (Eds.) Chapter 2, *GIS Diffusion: The Adoption and Use of GIS in Local Government in Europe*, pp. 9–22, London: Taylor & Francis.

Part II

Spatial Data Infrastructures
Introduction

10

Managing Our Urban Future

The Role of Remote Sensing and Geographic Information Systems

Ian Masser

10.1 The Challenge

From the forecasts that are currently available it seems that rapid urban growth over the next ten to 20 years must be regarded as inevitable (Sadik, 1999). The likely scale of this likely growth is nothing less than awesome. It has been estimated that more than five billion people will be living in urban areas by 2025 and that 80% of these will be residents of cities in less developed countries. It is anticipated that the impacts of this growth will be particularly marked in megacities with a population of at least 5 million. Currently 41 cities in the world fall into this category, and a further 23 are expected to join this group in the next 15 years. All but 11 of these 64 megacities are located in less developed countries. However, it should not be forgotten that there are also thousands of rapidly growing smaller cities throughout the less developed world. In many countries, the problems of these smaller cities are no less challenging than those of the megacities especially where the latter attract more than their share of the limited resources that are available because of their sheer size and prominent positions.

The challenges facing urban managers are numerous and include: demographic pressure; inadequate infrastructures; inadequate resources for service delivery and planning; conflicting interests between groups; and the contradicting priorities of economic development, ecological sustainability and community quality of life (Devas and Rakodi, 1993). The inability to effectively manage these related challenges is rapidly increasing the human risks associated with poor housing conditions, uncollected solid waste, overconsumption of limited freshwater supplies, untreated waste water and urban air pollution.

There are no precedents in history for urban growth on this scale. The closest parallel is the rapid urbanisation of Western Europe and North

America that took place during the nineteenth and early twentieth centuries. However, this was accompanied by rapid economic growth which mitigated to some extent many of the problems accompanying urbanisation on this scale. This is not likely to be the case in most of the cities in the less developed countries. Consequently, there are marked discrepancies between the scale of the planning problems associated with rapid urbanisation and the resources that are available to deal with them.

With these considerations in mind, this chapter considers some of the ways urban planners can respond to the challenge of managing our urban future. The discussion is divided into four sections. The first of these considers the nature of the tasks involved. The second examines the potential of remote sensing and geographic information systems (GIS) to assist in these tasks in general terms. The second examines the potential of remote sensing and GIS to assist in these tasks in general terms. The third section presents some of the findings of three case studies carried out in the Division of Urban Planning and Management at ITC which give some insights as to how these tools can be used to respond to this challenge and the final section sets out a vision of sustainable urban development and its implementation at the local level.

10.2 The Nature of the Task

The word 'managing' rather than 'management' has been chosen to emphasise actions rather than processes. This does not mean that processes and procedures are unimportant. What it stresses, however, is that processes should be seen as only a means to an end and not as an end in themselves. What really matters is what actions are taken to respond to the urban challenges that are likely to face us over the next 10 to 20 years.

It must also be emphasised that the word 'managing' includes both knowledge and action. In this context knowledge refers to the activities of monitoring, analysing and evaluating, which are needed to increase our understanding of what is happening in our cities. Action refers to the tasks of integrating, planning and executing, which are the main components of any management process.

It must also be recognised that urban planners and managers are part of both the problem and the solution. In today's world, it is unthinkable that urban managers should be given the powers to impose their own preferred technical solutions on others without first getting their consent (Healey, 1997). In practice, the implementation of their plans is also likely to be very much dependent on their ability to persuade the key stakeholders to join together to realise their objectives. Consequently, it is necessary to add another set of activities to those listed above. These include tasks such as networking, mobilising and consensus building to build platforms for particular lines of action. Under these circumstances, with many different interests involved, managers will have to use their coordinating, facilitating and enabling skills within the planning process.

The growing interdependency of urban and rural areas and global and local matters in the modern world must also be taken into account in the process. Many actions in urban areas impact on rural areas and vice versa. The same also applies to the distinction between developed and less developed countries within an overall global economy. What this means then is that it is not possible to consider the urban areas of the less developed countries in isolation and that it will be increasingly important to see them also in a broader perspective.

10.3 Remote Sensing and Geographic Information Systems

10.3.1 Remote Sensing

The advent of satellite imagery has created new sources of information for those involved in urban management. With sensors such as the SPOT panchromatic system (10 metres resolution) and the Indian IRS (5.8 resolution), mapping is already possible at the 1:50,000 to 1:25,000 scales. The IKONOS satellite provides 1 metre resolution images which make mapping at the 1:10,000 scale feasible. With tools such as these, it will be possible to monitor urban land-use changes throughout the world with increasing accuracy over the next 10 to 20 years. For this reason, they are likely to be of considerable importance for many rapidly growing urban areas in less developed countries where alternative sources of information are limited because of lack of resources (see, for example, Donnay, Barnsley and Longley, 2001).

Nevertheless, it must be borne in mind that, despite these developments, conventional aerial photography is likely to remain the primary source of remotely sensed information for the foreseeable future at the land parcel level (i.e. 1:2000 to 1:500 scales) which is the basic building block of the data bases used by those involved in urban planning and land administration. With this in mind it is also worth noting that recent developments in aerial photography have substantially increased its potential usefulness for these purposes (see, for example, Warner, Graham and Read, 1996, Chapter 15). These include improvements in film quality, better quality lenses and the use of GPS for positioning purposes. The forthcoming arrival of the digital survey camera will also provide direct images that are compatible with a digital geoinformation production line.

10.3.2 Geographic Information Systems

Using GIS, it is technically possible to integrate large quantities of data collected by different people for different purposes. This makes this tool especially appropriate for applications in the field of urban planning and management, given that this often involves integrating information from different sources. The GIS also provides unprecedented opportunities for urban planners to manipulate their information in an almost infinite number of ways. This facilitates new avenues of exploratory spatial data analysis that were previously not feasible and also enables the integration of data collected by different media thereby substantially increasing the communications capabilities of those involved in urban management (see, for example, Raper and Camara, 1999).

However, unlike remote sensing which can be regarded as essentially a technical activity in its own right, the effective exploitation of GIS is very much dependent on a number of institutional and organisational factors. Institutional factors refer to the willingness of data collectors to share their information with potential users. Consequently, one of most important challenges in the next 10 to 20 years is to enable potential users to obtain access to the information they require. For this purpose, spatial data infrastructures will need to be created at the local, national and even the global levels. Organisational factors include the extent to which organisations are able to adapt themselves to exploit the potential opportunities opened up by GIS. This also requires the reinvention of the technology itself to satisfy the various and changing requirements of each organisation (Campbell and Masser, 1995).

10.4 Three Case Studies

10.4.1 Monitoring Urban Development Processes

Rapid urban growth means that most cities are in a constant state of transition. Consequently, monitoring and evaluating urban change is a major issue in urban planning and management throughout the third world. Key research topics within this general theme are the use of remote sensing and related technologies to map changes both around the urban-rural fringe and also within the built-up area itself (Yeh and Li, 1997). Other research topics of some importance from the standpoint of current planning practice are the construction of various types of urban indicator from conventional and remotely sensed information to facilitate national and cross-national comparisons of urban regions (Flood, 1997).

A project carried out to evaluate the growth of the city of Villavicencio in Colombia demonstrates the value of monitoring urban development processes in less developed countries (Turkstra, 1998). Using conventional black and white vertical air photographs obtained from the Instituto Geografico Agustin Codazzi together with small format oblique colour air photographs he was able to reconstruct the sequence of urban growth and urban land use change in the city since 1939.

Villavicencio lies on the eastern edge of the Andes mountains about 120 kilometres from the capital Bogota. It is a relatively small provincial centre serving a primarily agricultural region on the great eastern plain of Colombia. Over the last 60 years, its population has grown by a factor of nearly 40 from 6,300 to more than 228,000. The air photographs provide a graphic record of urban development processes over this period (Fig. 10.1).

The nine sets of air photographs cover the period from 1939 to 1996. They show that the built-up area of the city has grown during this period

FIGURE 10.1
Changes in the urban area of Villavincencio Colombia 1949–1994 (from Turkstra, 1998)

by a factor of more than 60 from 34 to 2128 hectares. The first air photograph depicts a city that consisted essentially of 27 blocks around the central square. From the analysis of subsequent air photographs, it can be seen that the built-up area steadily expands outwards at rates which often exceed 10% per year. By 1970 scattered settlements surround the continuous built up area particularly on its southern and southeastern edges. Given the physical constraints on further outward expansion elsewhere, this axis becomes the main trajectory for subsequent development.

The air photographs also depict the emergence of squatter and illegal settlements within the urban area during this period. The first squatter settlements along the Maizaro river can be seen in an air photograph taken in 1955. After this there has been a steady growth in both their number and size. The photographs also show that many of these settlements are located on steep mountain slopes that are liable to landslides or in canal beds that are liable to flooding. The first 'pirate' settlements that are the product of the illegal but well-defined subdivision of land are evident on a photograph taken in 1970. By 1994 the findings of the analysis indicate that these two categories of residential development accounted for 28% of the total residential area of Villavicencio.

Further analysis of the changes that have taken place during this period shows that the city centre is still the main commercial centre of the city and that residential development has largely been a process of urban expansion. Consequently, many of the high-income neighbourhoods near to the city centre remain more or less the same as they were when they were first established 20 to 30 years ago.

10.4.2 Planning Urban Service Provision

The delivery of urban social services and the supply of basic utilities infrastructure present a wide range of management problems in rapidly growing urban areas with limited resources. In addition, there are marked differences in levels of access to basic social services such as education and health care between different groups among the population and between different localities within urban regions (Smith, 1994). Similarly, there are considerable discrepancies in the supply of urban water and electricity to different parts of the city and this often fluctuates during the day. Key research topics within this theme focus on the development of management and decision support tools to make the best use of the limited resources that are available (Kersten, Mikolajuk and Yeh, 1999).

A project carried out to identify priority areas for the construction of health clinics in Dar es Salaam in Tanzania illustrates the practical need for planning urban service provision in many of the urban centres in the developing world (Amer, 1998). Over the past decades, the city has experienced high rates of population growth. With a population of around 2.5 million, it is by far the largest urban centre of the country. Estimates indicate that over 70% of the city's inhabitants live in unplanned settlements. In most cases, these settlements are characterised by high rates of unemployment, substandard housing and limited access to safe drinking water, sanitation, electricity and basic social services. In short, the expanding population of Dar es Salaam is confronted with serious problems of urban poverty and unhealthy living conditions.

The Tanzanian health system is also in a state of transition as it tries to cope with increasing demand and the task to develop an efficient workforce in an environment of public budget constraints. It is important to recognise that the ongoing adjustments in health policy are being made in the broader context of adverse economic circumstances and a re-thinking of government roles in social policy. The main objective of the project is to derive a spatial planning approach that contributes to equitable and efficient urban health care delivery as set out in the 'Health for All' strategy of the World Health Organisation.

The approach is firmly rooted in empirical data to enable a thorough understanding of the variety of factors that influence health-seeking behaviour across different socio-economic strata. Conventional air photographs were used to capture the spatial dimension of health seeking behaviour. This was necessary because of the absence of an up to date map and the non-existence of street names and house numbers within unplanned settlements. Straightforward statistical techniques then reveal the regularities in the spatial choice behaviour of different socio-economic population groups. During this stage of analysis, particular attention is devoted to the more disadvantaged and vulnerable population groups as they suffer most from unhealthy living conditions. A number of non-standard analytical GIS techniques are subsequently applied to visualise existing spatial inequalities in access to health services and suggest priority areas for the construction of new health clinics. The ongoing development of this methodology and GIS based planning support tools will bring local health planners and decision makers in a better position to identify spatial inequality and re-direct the allocation of scarce health resources to those most in need.

10.4.3 Managing Urban Planning Information

There is a need for capacity building in urban areas in less developed countries to enable more effective strategic planning and policy formulation. A key component in more effective planning is the development and implementation of urban information management strategies that are based on a sound understanding of the needs of urban planning and the resources that are available to the agencies carrying out this task (Masser and Ottens, 1999). Key research topics within this theme are the integration of urban planning and land administration data and the analysis of the institutional issues that must be taken into account in the implementation of urban information management and planning strategies (Reeve and Petch, 1999).

A World Bank funded project carried out for the Lilongwe City Assembly in Malawi highlights some of the problems that have to be tackled in managing urban planning information in a rapidly growing African capital city (Sliuzas and Kawonga, 1997). The city itself has a population of about half a million, which is currently increasing at a rate of over 6% a year. Lack of both human and financial resources within the City Assembly has led to a situation where it finds itself increasingly unable to collect and process the considerable amount of information that it requires to carry out its administrative responsibilities.

The basic objective of this project is to strengthen the capabilities and facilities of the Lilongwe City Assembly through the provision of training and urban information technology. This entails the development of a GIS facility to support the Assembly's role in urban development. A GIS pilot project is being carried out in conjunction with the Department of City Planning and Estates Management Services. It is a first step towards the goal of developing a comprehensive spatial information system for the Assembly as a whole.

The pilot project has had to take account of a number of organisational and institutional problems associated with the working environment of the City Assembly. These include the lack of computer skills among the staff, the poor quality of existing file handling systems, the lack of consistent large and medium scale map coverage of the urban area and the continuing financial difficulties in which the Assembly finds itself.

Successful database applications have been developed to assist in the work of development control and traditional housing area management. This data can now be used with the digital mapping facilities to produce various large scale base maps of different subdivisions and a range of thematic maps. The necessary GIS hardware and software facilities are installed in a small local area network. The Assembly is now in a position to build applications to improve its revenue generation systems through local land taxation, to support other local services such as in engineering and public health and to use land related information in its strategic planning activities.

Capacity building is a critical component of this project. Short in-house training in computer skills, basic GIS techniques and database management is complemented by extended training at ITC at the technician and Masters levels. A combination of project funding and ITC Fellowships has been used to support the extended training.

10.5 Some Guiding Principles

The case studies described above indicate some responses to the pressures of the urban challenge. Two of them come from two of the world's poorest countries in Africa where any response is severely constrained by the lack of both human and financial resources. For this reason, both of these emphasise the need for capacity building to make the most effective use of the limited resources that are available at the local level. With these considerations in mind the final section of this paper sets out some guiding principles for an urban future based on the concept of sustainable urban development and its implementation at the local level.

There are always three distinct urban development processes underway at the local level – economic, community and ecological – each with its own imperatives. The development imperatives favour market expansion, externalisation of costs and sustained private profit. The community imperatives are to meet basic human needs, increase economic and social equity and create community self-reliance. Whereas the imperatives of ecological development can be supported by humans by limiting the consumption of natural resources to a rate that allows nature to regenerate resources and by reducing the production of wastes to levels that can be absorbed by natural processes.

The imperatives of these three development processes can often be contradictory. For instance, externalising costs in order to maintain private profit can contradict the ecological imperative of valuing and conserving natural resources. Another example is the global expansion of markets and the integration of national economies where the free-trade agreements can undermine the community development imperatives of local self-reliance and meeting basic human needs.

Sustainable urban development at the local level is a process of bringing all these three development processes into balance with one another, through negotiation among the stakeholders involved, for example, formal and informal sector, community-based organisations, and local government, to produce and implement an action plan for sustainable urban development (ICLEI, 1996). This action plan is based on the following guiding principles:

- Partnerships and accountability: Alliances are established between all stakeholders for collective responsibility, decision-making and planning on the basis that each is accountable for their actions.

- Participation and transparency: All major sectors of society are directly involved and all relevant information is easily available to all.

- Systemic approach: The solutions address the underlying causes of economic, ecological and community problems and whole systems.

- Concern for the future: The plans and actions address short and long-term trends and needs.

- Equity and justice: Economic development must be equitable, ecological and environmentally sound and socially just.

- Ecological limits: All communities must learn to live within the Earth's carrying capacity and the action plan must ensure the sustainable use of natural resources.

- Local and global: Local actions must reflect the global context and co-dependence of local sustainability on overall global sustainability.
- Local relevance: Local actions must be feasible, realistic and above all relevant to the needs and priorities of the local community.

10.6 Recent Developments

According to the World Cities Report prepared by the United Nations Habitat (2016), the overall urban challenge that was outlined in the first part of this chapter has, if anything, intensified during the last 20 years. The report argues that, in an increasingly urbanised world where urban populations and the challenges of urban areas continue to grow, longstanding models of urbanisation need to adapt to new social, economic and environmental realities. It explores persistent issues and emerging trends in urbanisation, highlighting the critical role of good urban governance to harness the positive potential of cities. Its New Urban Agenda will create a good relationship between urbanisation and development, as well as links to other global agreements and agendas.

Urban growth, the growing number of urban residents living in slums and informal settlements, and the challenge of providing urban services are persistent issues that will need to be addressed. Alongside these issues, emerging trends of climate change, social exclusion and rising inequality, insecurity and an upsurge in international migration present new challenges. Well-managed, urbanisation has the potential to foster social and economic advancement and improve quality of life for all through leveraging economies of scale and agglomeration.

The report's New Urban Agenda is based on five principles that reflect five broad shifts in strategic and policy thinking:

(1) ensuring that the new urbanization model includes mechanisms and procedures that protect and promote human rights and the rule of law;
(2) ensuring equitable urban development and inclusive growth;
(3) empowering civil society, expanding democratic participation and reinforcing collaboration;
(4) promoting environmental sustainability; and
(5) promoting innovations that facilitate learning and the sharing of knowledge.

To implement this agenda, a city-wide approach to development will need to include: the adoption and implementation of national urban policies; the strengthening of urban legislation and systems of governance; a re-invigoration of territorial planning and urban design; and ensuring the urban economy works in a way that creates decent employment opportunities.

The *2018 Revision of World Urbanization Prospects* produced by the Population Division of the UN Department of Economic and Social Affairs (UN DESA, 2018) notes that future increases in the size of the world's urban population are expected to be highly concentrated in just a few countries. Together, India, China and Nigeria will account for 35% of the projected growth of the world's urban population between 2018 and 2050. By 2050, it is projected that India will have added 416 million urban dwellers, China 255 million and Nigeria 189 million.

With respect to remote sensing and GIS, a new set of surveying and mapping tools based on digital technology has emerged over the last two decades since the original paper was published (Horn, 2018). These include high-resolution digital cameras, highly precise GPS positioning and guidance systems and sophisticated computer software for image processing and data extraction. Airborne laser scanning from survey aircraft is another modern innovative tool. By taking a distance measurement at every direction, the scanner rapidly captures the surface shape of objects, buildings and landscapes and enables three-dimensional terrain models to be quickly and accurately constructed.

The capabilities of earth-orbiting remote-sensing satellites now offer the possibility to map very large areas of the Earth's surface at very high resolution. Compared with the tens of meters resolution offered by the likes of the Landsat and SPOT satellites of a few decades ago, the modern sensors carried by, for example WorldView-4, can provide approximately 30 centimetre resolution.

Whilst sophisticated systems in high flying aircraft and satellites represent the state-of-the art in wide area coverage, at the lower end of the scale, inexpensive lightweight drones can be used by suitably trained individual operators to explore areas where other solutions are either too costly or too slow to respond (see, for example, Gevaert, 2018). Furthermore, drones can operate in hazardous environments or in areas where minimal infrastructure or support is available.

As a result of developments over the last 20 years, an untrained user can now sit at their own computer, or even smart phone, and view free imagery of almost any location on Earth using a simple tool such as Google Earth.

The contents of two recent inaugural lectures given by my successors give some interesting insights into the range of current and future research in the Faculty of Geo-information Science and Earth Observation at the at the University of Twente.

Karin Pfeffer's inaugural lecture is entitled 'Knowing the city' (Pfeffer, 2018). She argues that research on urban infrastructures has two methodological directions. The first direction literally unpacks infrastructures by exploring the mutual relationships between material infrastructure, geotechnologies and the wider social systems.

The second direction makes use of the analytical strength of geotechnologies, including GIS. Pfeffer (2018, 25) points out:

> This also involves the interactive generation and visualisation of spatio-temporal information, which will, among other purposes, be used as an input in co-creation processes as a mediator (e.g. displayed on a MapTable). Interactive generation emphasizes the collaboration between different experts to combine their disciplinary strengths, for instance between a simulation model maker, a social scientist and field workers.

Richard Sluizas's inaugural lecture is entitled 'Grappling with the city-disaster nexus' (Sluizas, 2018). In this lecture he argues:

> If both planning and managing urban development and enhancing resilience through risk reduction are recognized as moving targets, then their nexus is a complex setting indeed. Grappling evokes images of close combat or engagement, of wrestlers or martial arts experts battling for strategic advantage and victory.
>
> *(p. 18)*

There is no time limit to the city-disaster nexus which may take the form of strategic and systematically applied measures applied through formal planning and disaster reduction management frameworks. However, Sluizas (2018, p. 19) notes, 'it is likely to also encompass numerous individual or collective informal actions, some of which may even be very important ingredients in gaining a strategic advantage over disasters in the longer term.'

It should be noted that both these lectures emphasise the socio-technical dimensions of the issues explored in my original lecture while focussing on new issues for planning research and practice.

Acknowledgements

The original paper was based on a public address to mark the Opening of the ITC Academic Year on September 9th 1999. Although it appears under my name its preparation has been very much a team effort on the part of the staff of my division. Frans van den Bosch, Mark Brussel, Karen Buchanan and Sandra Dimmendaal took over the responsibility for the visual presentation that accompanied this text. Sherif Amer, Frans van den Bosch, John Horn, Richard Sliuzas and Jan Turkstra generously contributed material for the text and Karen Buchanan played a key role in drafting the last section as well as suggesting the title of the paper. Benno Masselink also spent a lot of time preparing the digital images for the presentation.

References

Amer, S., 1998. Planning public health care facilities in Dar es Salaam, Tanzania: Users, interaction patterns and underserviced areas, *Proceedings of the First International Health Geographics Conference*, Baltimore, October 16–18, pp. 112–118.

Campbell, H., and Masser, I., 1995. *GIS and organisations: How effective are GIS in practice?* London: Taylor and Francis.

Devas, N. and Rakodi, C., 1993. *Managing fast growing cities: New approaches to urban planning and management in the developing world*. Harlow: Longman.

Donnay, J. P., M. Barnsley and P. Longley, 2001. *Remote sensing and urban change*. London: Taylor and Francis.

Flood, J., 1997. Urban and housing indicators, *Urban Studies, 34,* 1635–1665.

Gevaert, C. M., 2018. *Unmanned aerial vehicle mapping for settlement upgrading*. University of Twente.

Healey, P., 1997. *Collaborative planning: Shaping places in fragmented societies*. Basingstoke: Macmillan.

Horn, J., 2018. Personal communication.

International Council for Local Environmental Initiatives, 1996. *The local agenda 21 guide: An introduction to sustainable development planning*. Toronto: International Council for Local Environmental Initiatives.

Kersten, G. E., Z. Mikolajuk and A. G. O. Yeh, 1999. *Decision support systems for sustainable development: A resource book on methods and applications*. Boston: Kluwer Academic.

Masser, I., and H. Ottens, 1999. Urban planning and GIS, in J. Stillwell, S. Geertman and S. Openshaw (Eds.), *Geographic information systems and planning*. Berlin: Springer, 25-42

Pfeffer, K., 2018. *Knowing the city, Inaugural lecture*, University of Twente.

Raper, J., and A. Camara, 1999. *Spatial multimedia and virtual reality*. London: Taylor and Francis.

Reeve, D. E., and J. R. Petch, 1999. *GIS, organisations and people: A socio-technical approach*. London: Taylor and Francis.

Sadik, N., 1999. Meeting the urban population challenge, *City Development Strategies, 1,* 16–23.

Sliuzas, R. V. and Kawonga, A. J. C., 1997. Lessons from the GIS pilot project at Lilongwe city council, *Proceedings of the Joint European Conference on Geographic Information*, Vienna, April 15–18, pp. 1340–1349.

Sluizas, R., 2018. *Grappling with the city-disaster nexus, Inaugural lecture*, University of Twente.

Smith, D. M., 1994. *Geography and social justice*. Oxford: Blackwell.

Turkstra, J., 1998. *Urban development and geographical information: Spatial and temporal patterns of urban development and land values using integrated geo-data, Villavicencio, Columbia*, ITC publication no. 60. Enschede, The Netherlands: International Institute for Aerospace Survey and Earth Sciences.

United Nation Habitat, 2016. *World cities report 2016: Urbanisation and development emerging futures*. Nairobi: UN Habitat.

United Nations Department of Economic and Social Affairs, 2018. *2018 revision of world urbanization prospects*. New York: United Nations.

Warner, W. S., R. W. Graham and R. E. Read, 1996. *Small format aerial photography*. Caithness: Whittles Publishing.

Yeh, A. G. O., and X. Li, 1997. An integrated remote sensing and GIS approach to monitoring and evaluation of rapid urban growth for sustainable urban development in the Pearl River Delta, China, *International Planning Studies*, 2, 195–222.

Theory

11

All Shapes and Sizes

The First Generation of National Spatial Data Infrastructures

Ian Masser

11.1 Introduction

The advent of GIS technology has transformed spatial data handling capabilities and made it necessary for governments to re-examine their roles with respect to the supply and availability of geographic information (Masser 1998). This is because government agencies are not only the main external providers of geographic information for most operational applications of GIS but also because they exert a profound influence on national developments as a result of what Rhind (1996, 8) has called 'a cocktail of laws, policies, conventions and precedents which determine the availability and price of spatial data'.

Given these circumstances many governments throughout the world are beginning to think more strategically about information needs, data collection and the resources needed to deliver information to a wider market. This trend can be traced back to the mid-60s when the potential of computer-based surveying and mapping systems for creating multi user multi-purpose databases for public administration was first recognised. However, as McLaughlin (1991, cited in NRC 1993, 14–15) has pointed out, this vision was largely lost in the 70s and early 80s as the emphasis shifted away from questions of 'why' to questions of 'how' as spatial information systems were implemented for a wide variety of purposes within traditional institutional frameworks. Consequently, it was not until the late 80s that the focus of discussion began to shift back from the technology itself to matters of geographic information and its use in society.

Table 11.1 shows the titles of 11 different national geographic information initiatives that are underway at the present time. Although the terms used vary from country to country these have three elements in common:

- they are explicitly national in nature;
- they refer either to geographic information, spatial data, geospatial data, or in one case, to land information; and
- they also refer to terms such as infrastructure, system or framework, which imply the existence of some form of coordinating mechanism for policy formulation and implementation purposes.

Given the common features, it can be argued that these initiatives form the first generation of national spatial data infrastructures (NSDIs). With these considerations in mind this chapter reviews the experiences of these initiatives and considers what lessons might be learnt for the next generation that will come into being over the next ten years.

The presentation is divided into four parts. Section 2 provides a short profile of each of the 11 initiatives. Section 3 evaluates them with respect to the following questions:

- what are the driving forces behind them?
- what are their main features in terms of status, scope, access, approach to implementation and resources?

The final section of the chapter considers what lessons might be learnt from this experience and discusses the changing context within which the next generation of NSDIs are likely to formulated. It concludes with some thoughts as to which countries are most likely to be candidates for inclusion in the next generation.

TABLE 11.1

The First Generation of National Spatial Data Infrastructures

Australia	Australian Spatial Data Infrastructure
Canada	Canadian Geospatial Data Infrastructure
Indonesia	National Geographic Information Systems
Japan	National Spatial Data Infrastructure
Korea	National Geographic Information System
Malaysia	National Infrastructure for Land Information Systems
Netherlands	National Geographical Information Infrastructure
Portugal	National System for Geographic Information
Qatar	National Geographic Information System
United Kingdom	National Geospatial Data Framework
United States	National Spatial Data Infrastructure

It should be noted that the term 'national spatial data infrastructure'/ NSDI is used throughout this chapter to refer generally to these initiatives. This term is used to avoid potential confusions in terminology of the kind shown in Table 11.1.

11.2 The First Generation of NSDIs: an Overview

The main features of the first generation of NSDIs in the 11 countries listed in Table 11.1 in alphabetical order are described below in chronological order.

11.2.1 Australia

The Australian Land Information Council was set up in 1986 by an agreement between the Australian Prime Minister and the Heads of State Governments to coordinate the collection and transfer of land related information between the different levels of government and to promote the use of that information in decision making (ANZLIC 1992, 1). In 1991 New Zealand became a full member of the Council which was renamed the Australia New Zealand Land Information Council (ANZLIC). ANZLIC is serviced by the Spatial Data Infrastructure Program of the Australian Surveying and Land Information Group. (AUSLIG).

During its lifetime the Council has produced a number of major reports on the status of land information in Australia as well as four versions of their national strategy for the management of land and geographic information in 1988, 1990, 1994 and 1997. In 1997 ANZLIC produced a discussion paper setting out its vision of an Australian spatial data infrastructure (ANZLIC 1997). It argued that most of the components of this infrastructure already exist in some form but there is nevertheless a need for the community to 'more clearly define and describe the infrastructure as a coherent national identity'. ANZLIC's role in this respect is 'to lead the community in defining the components of the NSDI, the characteristics of those components, and provide a vehicle for the determination of national priorities and custodianship' (p. 5).

11.2.2 The United States

An inter-agency Federal Geographic Data Committee was set up in 1990 as a result of a revised Circular A-16 issued by the Office of Management and Budget to coordinate 'the development, use, sharing and dissemination of surveying, mapping and related spatial data.' (OMB 1990). Four years after the establishment of the FGDC President

Clinton signed Executive Order 12,906 entitled 'Coordinating Geographic Data Acquisition and Access: The National Spatial Data Infrastructure' on the 11 April 1994 'to strengthen and enhance the general policies described in OMB Circular A-16' (Executive Office of the President 1994, Section 2A). The FGDC is based in the National Mapping Division of the US Geological Survey and is chaired by the Secretary of the Interior. Its members include representatives from all the major ministries who have an interest in geographic information together with a variety of other public agencies concerned with its collection and management.

The Executive Order also set out in some detail the main tasks to be carried out and defined time limits for each of the initial stages of the National Spatial Data Infrastructure. Apart from the core task of interagency coordination through the FGDC, these include the establishment of a National Geospatial Data Clearing House and the creation of a National Digital Geospatial Data Framework through a variety of partnerships between agencies at different levels of government and also between the public and private sectors.

11.2.3 Qatar

In 1988 the Minister of Municipal Affairs and Agriculture witnessed a demonstration of GIS technology in Canada and saw its potential to revolutionise the way information is managed in the small Gulf State of Qatar (Al Thani 1997). His vision led to a government wide user needs study which recommended

> that a Digital mapping database be implemented for the entire country; that a comprehensive fully integrated nationwide GIS be created; and that a high-level National GIS Steering Committee be established to set standards and oversee the implementation and development of GIS in Qatar.

As a result of these recommendations a National Steering Committee was set up in 1990 and a National Centre for GIS was created to implement GIS in Qatar in an organised and systematic way.

One of the first tasks of the Centre was to implement a high resolution digital topographic database. This is used by all 16 national agencies involved in GIS through a single high-speed fibre optic network. The public can also access GIS data through this network (Tosta 1997a).

11.2.4 Portugal

A National System for Geographic Information (SNIG) was created by the Portuguese government under the Decreto Lei No. 53/90 on the 13 February 1990. Under the same law the government set up a National

Centre for Geographic Information (CNIG) 'to coordinate the integration of data at different levels of public administration and thus develop a National System of Geographic Information' (Arnaud et al. 1996, 115). CNIG is a research centre of the Ministry of Planning and Territorial Administration (MEPAT) and obtains part of its funding from national and international agencies.

In addition to promoting the development of the GIS market in Portugal as a whole CNIG has supported the implementation of regional GIS nodes in the five regions of mainland Portugal and is coordinating two major projects funded by the European Commission to develop local nodes at the municipality level with particular reference to the needs of land use planning (Henriques 1996). The launch of the SNIG network on the Internet in May 1995 is regarded as a major step in the modernisation of Portuguese public administration. As a result, Portugal can be regarded as 'the first European country that has an operational national geographic information infrastructure, fully distributed, based on the most recent developments in information technology' (Gouveia et al. 1997, 3).

11.2.5 The Netherlands

The Dutch Council for Real Estate Information (Ravi) is an independent non-profit organisation set up originally to advise the Minister for Housing, Spatial Planning and the Environment on matters relating to the operations of the cadastre. In 1992 it was reconfigured as the National Council for Geographic Information. Its masterplan contained a vision that 'the proper development of the National Information Infrastructure requires a well thought out policy, an adequate administrative organisation, and the intensive coordination of all the involved parties' (Ravi 1992, 7, author's translation).

Ravi's (1995) view of the National Geographic Information Infrastructure makes a basic distinction between core and thematic data. With this in mind it has played a leading role in the creation of digital core data sets for the Netherlands as a whole at the 1:10,000 scale and also at the larger scales required for the municipal administration and public utility management purposes. It has also initiated a National Geographic Information Clearing House project which builds upon the experience of a number of metadata initiatives by various agencies in the Netherlands.

11.2.6 Indonesia

An inter-agency working group was established in 1993 to identify the main land data users and producers with the objective of establishing a National Geographic Information System for planning purposes. This project is coordinated by the National Coordinating Agency for Surveying and Mapping (Bakosurtanal).

High priority is being given to the creation of a national framework to ensure that the information produced by different agencies has the same geographic referencing frame. Given that only 62% of the land area of Indonesia is covered by topographic base maps it was decided in 1993 to complete the coverage of the whole country using digital mapping methods (Godfrey et al. 1997, 19). A national GIS arrangement law is also under preparation and a number of GIS projects have been included in Indonesia's Sixth five year plan (REPELITA VI). These include GIS training and awareness raising activities as well as technology transfer and digital database development (Suharto 1996).

11.2.7 Malaysia

Although the need for an effective land information system to assist planning and development in Malaysia has been felt since the early 1970s, the first steps towards setting up a national infrastructure were not taken until 1994 when the Ministry of Land and Cooperative Development appointed Renong Berhad to carry out a feasibility study. This study produced a comprehensive set of proposals setting out its vision of a National infrastructure for Land Information Systems which would 'make it possible to access the entire range of information required for the planning and maintenance of expensive infrastructure systems and support the sustainable development of natural resources such as oil, gas, forests, water and soil' (Berhad 1995, paragraph 1.5).

Following the publication of this report a task force was set up to make proposals for implementation at both the federal and state levels (NaLIS 1996). Work on a prototype has also begun in the Kuala Lumpur area (Tamin 1997) and in January 1997 the Prime Minister's Department of the federal government issued its guidelines for the establishment of the national infrastructure for land information system.

11.2.8 South Korea

A National Geographic Information System (NGIS) was set up in 1995 by the Korean government to stimulate the development of digital spatial databases and the standardisation of geographic information. The implementation of this programme is overseen by a steering committee of representatives from 11 ministries chaired by the Vice-Minister of the Ministry of Construction and Transportation. The budget allocated to NGIS is $360m over a five year period. It is expected that about 64% of these costs will be met by central and local government and that the remainder will come from the private sector (MOCT 1995, 2).

Phase 1 of this programme lasts from 1995 to 2002. It is primarily concerned with the creation of a digital topographic map base for the country as a whole at scales ranging from 1:1,000 in urban areas to 1:25,000 in mountain regions. Special attention is also being given to the digital mapping of underground facilities in this phase. It is assumed that, 'although the development of databases including geographic information is largely controlled by the Korean government, the application of GIS will be carried out by private sectors and research institutes' (MOCT 1995, 10).

11.2.9 Japan

The starting point for the Japanese National Spatial Data Infrastructure initiative was the government's reaction to the Kobe earthquake of January 1995. This led to a major review of emergency management services and their related data needs. As part of these developments a Liaison Committee of Ministries and Agencies concerned with GIS was set up in September 1995 under the supervision of the Cabinet. This includes representatives from 21 government agencies including the Ministry of International Trade and Industry (MITI). The Committee is serviced by the Cabinet Councillors Office of the Cabinet Secretariat with assistance from the National Mapping Agency and the National Land Agency.

In December 1996 the Liaison Committee published its plan of action up to the beginning of the 21st century. The first phase of this plan lasts until 1999 and includes the standardisation of metadata, clarifying the roles of government, local governments and the private sector and promoting the establishment of the NSDI (Godfrey et al. 1997, 25). A separate NSDI Promotion Association has been set up to support these activities. Its membership includes over 80 companies from the private sector (Yamaura 1996).

11.2.10 Canada

In 1995 the Canadian Council on Geomatics which represents the provincial geomatics agencies asked Geoplan to prepare proposals for an integrated spatial data model for Canada as a whole and to make recommendations on its implementation. Its report makes the case for 'a cooperative development which builds on Canadian strengths and recognises the current restrictions under which Canadian public sector geomatics agencies must operate' (Geoplan Consultants 1996, 43). As a result, the federal Inter-Agency Committee on Geomatics was asked in December 1996 by the Canadian Council on Geomatics to take a leading role in guiding federal and provincial governments and the private sector to create a Canadian Geospatial Data Infrastructure. The Inter-Agency Committee is chaired by the Assistant Deputy Minister of the Earth Science Sector in Natural Resources Canada.

Five basic themes have been identified for the Canadian Geospatial Data Infrastructure. These are to foster geospatial data access, to provide a foundation of framework data, to foster the harmonisation of geospatial standards, to encourage the establishment of data sharing partnerships and to create a supportive policy environment which facilitates the wider use of geospatial data (Corey 1998).

11.2.11 United Kingdom

The recommendation of the Chorley Committee on handling geographic information to set up an independent national centre for geographic information (Department of the Environment 1987) was rejected by the government of the day. As a result, the British National Geospatial Data Framework is the most recent of the first generation of NSDIs. It dates from late 1996 when the first meeting of its Board took place. This consists of data producers from both the public and private sectors and is chaired by the Director General and Chief Executive of Ordnance Survey Great Britain. An Advisory Council consisting mainly of data users has also been set up. This is serviced by the Association for Geographic Information. Both the Board and Advisory Council are independent bodies who work closely with government (Nanson and Rhind 1998).

The main objective of the National Geospatial Data Framework is to unlock geospatial information (NGDF 1998, 1). Its strategy is built around three 'pillars': collaboration, standards and best practice, and access to data (Hobman 1997, 1). Its work programme is managed by a taskforce which reports to the Board. This has set up working groups on various priority areas including metadata, accreditation and research.

11.3 Evaluation

From these brief national profiles, it can be seen that the first generation of NSDIs come in all shapes and sizes. They include some recent initiatives which have as yet little to show other than good intentions as well as some more established initiatives which have already achieved a great deal. They mix together some very small countries with some very large ones as well as countries with and without federal systems of government. The profiles also point to the extent to which some of these initiatives are confined largely to the public sector whereas others have a strong private sector and user involvement.

Two main questions must be addressed when evaluating these strategies:

- what are the driving forces behind them?
- what are their main features in terms of status, scope, access, approach to implementation and resources?

When addressing these questions, it is useful to bear in mind the differences between these countries in terms of geography, levels of economic development and systems of government. Table 11.2 summarises these differences with respect to some of the key indicators involved. From this it can be seen that there are massive differences between them both in terms of area and population. The United States, for example, covers an area which is nearly a thousand times that of Qatar and has nearly 500 times the population. Even if Qatar is discarded on the grounds that it is essentially a city state, the differences remain considerable with both the Netherlands and Portugal being smaller in both area and population than many American states. The same is the case in terms of levels of economic prosperity as measured by Gross Domestic Product per capita. In this respect the US is nearly 10 times as wealthy as the poorest country, Indonesia, and between two and three times as wealthy as countries such as Korea, Malaysia and Portugal.

It should also be noted that five of these countries have some form of federal system of government (Australia, Canada, Indonesia, Malaysia and the United States) although the extent to which powers are devolved to state and local government agencies varies considerably. At one end of the spectrum is the United States where a wide range of responsibilities relating to geographic information have been delegated to over 80,000

TABLE 11.2

Some Key Indicators for the 11 Countries

	Area 000 sq. km	Population Millions	GNP 000 $ per capita 1990
Australia	7686.8	18.1	14.5
Canada	9970.5	29.6	17.1
Indonesia	1904.6	189.9	2.0
Japan	377.7	124.8	14.3
Korea	98.5	44.6	6.7
Malaysia	329.7	20.1	5.1
Netherlands	40.8	15.4	13.0
Portugal	92.1	9.9	7.5
Qatar	11.0	0.5	16.6[*]
United Kingdom	244.1	58.4	13.2
United States	9809.1	259.7	18.1

Sources: Whitakers Almanack 1997
Note: * 1989.
National Bureau for Economic Research www.nber.org.

separate state and local government agencies (Tosta 1997a). At the other end come Indonesia and Malaysia which retain a considerable degree of federal control over land related matters. In contrast, in the six countries with non-federal systems of government, most of the responsibilities for geographic information are handled centrally (see, for example, Masser 1998, 92–95).

11.3.1 Driving Forces

Two basic themes underlie the NSDIs described above. These are the growing importance of geographic information in the coming age of digital technology and the need for some form of government intervention to coordinate data acquisition and availability. The significance of the former is encapsulated by Vice-President Gore (1998, 1) in the following remarks:

> A new wave of technological innovation is allowing us to capture, store, process and display an unprecedented amount of information about our planet and a wide variety of environmental and cultural phenomena. Much of this information will be 'geo-referenced', that is, it will refer to some specific place on the Earth's surface.

The case for the latter is set out at the beginning of President Clinton's Executive Order for the US National Spatial Data Infrastructure:

> Geographic information is critical to promote economic development, improve our stewardship of natural resources and to protect the environment. Modern technology now permits improved acquisition, distribution, and utilisation of geographic (or geospatial) data and mapping. The National Performance Review has recommended that the Executive Branch develop, in cooperation with State, local and tribal governments and the private sector, a coordinated National Spatial Data Infrastructure to support public and private sector applications of geospatial data in such areas as transportation, community development, agriculture, emergency response, environmental management and information technology.
>
> *(Executive Office of the President 1994)*

The need for government intervention to create the necessary infrastructure for exploiting the potential of digital geographic information technology is particularly important in the eyes of the Korean government:

> The National Geographic Information System (NGIS) is recognised as one of the most fundamental infrastructures required in promoting national competitiveness and productivity. This enormous task is a national project that is led by the government since a substantial funding is required, and based on the fact that the usage of GIS [is]

mainly for the public sectors. Furthermore, since the geographical factors as well as the attribute information are the basic assets of our country, construction or development of the relevant databases has been recognised as a national project. Accordingly, the Korean government is exerting significant efforts to develop and improve NGIS.

(MOCT 1995, 10)

The Australia New Zealand Land Information Council also highlights the parallels between geographic information and other types of infrastructure:

ANZLIC views land and geographic information as an infrastructure, with the same rationale and characteristics as roads, communications and other infrastructure. As the peak coordinating body for the management of land and geographic information, ANZLIC believes that Australia and New Zealand should have the spatial data infrastructure needed to support their economic growth and their social and environmental interests, backed by national standards, guidelines and policies on community access to the data.

(ANZLIC 1997, 1)

In essence, then, the objectives of most NSDIs can be summarised as follows: to promote economic development, to stimulate better government and to foster environmental sustainability. The notion of better government is interpreted in several different ways in the strategies. In many countries it means better planning and development. This is particularly the case in developing countries such as Indonesia and Malaysia. Planning, in the sense of a better state of readiness to deal with emergencies brought about by natural hazards was also an important driving force in the establishment of the Japanese National Spatial Data Infrastructure while in Portugal the National Geographic Information System is seen as an instrument for modernising central, regional and local administration.

On the other hand, better government can also be interpreted in terms of more open government as a result of better access to information in the statements quoted above by President Clinton and ANZLIC. The importance of access is particularly apparent in the mission statement of the British National Geospatial Data Framework which seeks:

To provide a framework to unlock geospatial information for the benefit of the citizen, business growth and good government through enabling viable, comprehensive, demand-led and easily accessed services.

(1998, 4)

These views are echoed in the final report of the study carried out for the Canadian Council of Geomatics. It formulated its mission in the following terms:

to: 1. Provide easy, consistent and effective access to geographic information maintained by public agencies throughout Canada;

2. Promote the use of geographic information in support of political, economic, social and personal development by all Canadians. (Geoplan 1995, 33).

11.3.2 Key Features

Status

The 11 NSDIs can be divided into two broad categories with respect to their status: those which are the result of a formal mandate from government and those which have largely grown out of existing geographic information coordination activities.

The first category includes Portugal where the National Geographic Information System was created by the Decreto Lei of 53/90 and the United States where the National Spatial Data Infrastructure was the subject of an Executive Order of the President in April 1994. There is also clear evidence of strong government involvement in the establishment of the Japanese NSDI and the Korean NGIS as well as the NGIS in Qatar. The Indonesian NGIS is built into the country's sixth Five Year Plan and a GIS arrangement law is currently under consideration. Similarly, the prime mover of the Malaysian National Land Information System is the Ministry of Land and Cooperative Development.

The second category consists of countries where NSDIs have largely grown out of existing coordination activities. This is clearly the case in Australia where current discussions regarding spatial data infrastructure are essentially an expansion of earlier discussions regarding national land information strategies. Similarly, the reconstitution of the Dutch Council for Real Estate Information in 1992 as the National Council for Geographic Information marked a significant step towards the development of a national geographic information infrastructure. Canada also falls into this category as the Federal Inter Agency Committee on Geomatics was asked by the Canadian Council on Geomatics in 1996 to take a leading role in creating a Canadian Geographic Information Infrastructure and it should also be noted that the US Federal Geographic Data Committee was in itself an outgrowth from the previous Federal Inter Agency Committee on Digital Cartography.

In many respects the British National Geospatial Data Framework falls into a category of its own in that it has no direct mandate from government, nor is it in any real sense a direct product of any existing governmental coordination activities although it is strongly supported by professional bodies such as the Association of Geographic Information and leading government agencies such as Ordnance Survey of Great Britain. As a result, its formal status vis-à-vis government remains unclear.

Scope

Scope can also be looked at from two different standpoints: the range of substantive geographic information interests which is represented in the different coordinating bodies and the extent to which the main stakeholders are directly involved in the process.

With respect to the former the membership of the US Federal Geographic Data Committee covers a very wide range of substantive interests. These include the Departments of Agriculture, Commerce, Defence, Energy, Housing and Urban Development, Interior, State and Transportation as well as the Federal Emergency Management Agency, the Environmental Protection Agency, the National Aeronautics and Space Administration, the Library of Congress and the National Archives and Records Administration. The Canadian Inter Agency Committee on Geomatics also includes representatives from a wide range of federal agencies, but, unlike the FGDC, it also has a representative from the Geomatics Industry Association of Canada which represents the private sector.

The Portuguese and Qatar National Geographic Information Systems also involve a wide range of central government agencies. In contrast the Indonesian and Malaysian National Geographic Information Systems tend to be focused mainly on surveying and mapping activities associated with land management. The initial stages of the Korean National Geographic Information System and the Japanese National Spatial Data Infrastructure are also primarily focused around central government surveying and mapping activities. However, in the case of the latter an NSDI Promotion Association has also been set up to complement these activities. This is chaired by a representative of Mitsubishi Corporation.

The Australian coordinating body is concerned primarily with the coordination of the collection and transfer of land related information between different levels of government. Each of the 10 members of ANZLIC represents a coordinating body within their jurisdiction (i.e. the Commonwealth Spatial Data Committee, the respective coordinating bodies at the eight state and territory levels and Land Information New Zealand. These members have the responsibility for both expressing that jurisdiction's views at the council and also for promoting ANZLIC's activities within their jurisdiction (Masser 1998, 64).

Unlike the other NSDIs, the Dutch and British initiatives are dependent on voluntary rather than mandatory participation. Nevertheless, the board of the Ravi consists of most of the data providers and users in the Netherlands. These include the Cadastre, the Topografische Dienst (national mapping agency), and Statistics Netherlands together with representatives from various groups within the Ministry of Housing, Spatial Planning and the Environment (VROM), the survey department of the Ministry of Transport, Public Works and Water Management (V&W), the National Institute for Public Health and Environment (RIVM) and the Centre for

Land Development and Soil Mapping (Staring Centre) as well as representatives from the Association of Provincial Agencies (IPO), the consultative group of the Public Utilities Companies, the Royal Association of Civil Law Notaries, and the Association of Water Boards. Another major stakeholder, the Association of Dutch Municipalities (VNG) also supports Ravi by contributing to the costs of some of its projects but is not a member of its Board (Masser 1998, 47).

The Board of the British National Geospatial Data Framework includes key data providers in both the public and private sectors such as the Office for National Statistics, Ordnance Survey and HM Land Registry in the former, and Landmark and Property Intelligence in the latter as well as representatives from the Information Management Advisory Group of the Local Government Management Board and the Natural Environment Research Council. Its Advisory Council has an even wider remit with members drawn from government departments such as the Environment Agency, the Public Records Office, as well as ESRI (UK) and MVA Systematica in the private sector, academia and bodies such as Friends of the Earth.

There are important differences between the NSDIs particularly in terms of the extent to which the main stakeholders are involved in the management of them. The vast majority of these initiatives are primarily public sector in scope and most are largely concerned with central or federal government activities. Although essentially public sector in scope ANZLIC is unusual in that it is centrally concerned with the interface between different levels of government.

The exceptions to this general rule are the British and Dutch initiatives whose coordinating bodies are not a formal part of government and also include private sector and user representation. This is particularly well developed in the case of the Ravi where a separate business platform consisting of representatives from the main geographic information service providers has been established to complement the activities of its council and efforts have also been made to ensure that the needs of academic research are also taken into account when formulating policy. By comparison, representation on the British National Geospatial Data Framework bodies is more of a hit and miss affair particularly with respect to the Advisory Council whose members are elected on an individual rather than an institutional basis.

Given these differences it is useful to note some of the recommendations made in the US National Academy for Public Administration report on 'Geographic Information in the 21st Century: Building a Strategy for the Nation' to the Bureau of Land Management, the US Geological Survey, the Forest Service and the National Ocean Service as a result of its assessment of the public management issues created by recent developments in geographic information. One of the principal recommendations of this report is that a National Spatial Data Council should be set up to provide a forum for all organisations involved in the development and maintenance of the US National Spatial Data Infrastructure:

> While the FGDC has been instrumental in much of the progress achieved over the past few years the Panel is convinced that an organisation is needed which provides full participation by all the major parties and interests engaged in developing and maintaining the NSDI.
>
> *(NAPA 1998, 6)*

It is envisaged that the National Spatial Data Council will be a private, non-profit organisation, preferably authorised by Congress but clearly located in the private sector and that its activities will complement those of the FGDC by concentrating on coordinating geographic information functions and activities outside federal government.

Access to Public Information

The 11 NSDIs also differ considerably in terms of their positions with respect to access to public information. One end of the spectrum is the United States where federal government agencies are required by law to make the information they collect available to the public free from any copyright restrictions at no more than the marginal costs of dissemination. This is the policy followed by the National Mapping Division of the US Geological Survey and the US Bureau of the Census with respect to the distribution of their products (Tosta 1997b, 1). In countries such as Australia, Britain and Canada the position is more complex as a result of conflicts between the desire of their governments to promote more open government and public accountability and the need to recover some or all of the costs of database creation and maintenance through the sale or the licensing of access to these databases.

At the other end of the spectrum come countries such as Malaysia where topographic survey maps are classified documents and access to NaLIS is restricted to public sector agencies. This position also creates a major barrier to implementation as the Renong Berhad (1995, 3.42) report points out:

> For as long as access to NaLIS is restricted to the public sector it will be difficult to offset anticipated increases in costs of data collection and conversion to digital format. However, if and when the system is opened to the private sector, these costs should be more than outweighed by the consequent increase in revenue.

It is also worth noting that restrictions on access to public information are not limited to developing countries such as Malaysia. For example,

Pollard's (1997) guide to digital national land information in Japan (Kokudo Sûchi Jôhô) contains the following statement:

This data is not available to the general public. It is intended for use by government administrators and university researchers. Some of the usage restrictions that may apply to the Digital National Land Information database are listed below.

1. If you find error data you are requested to inform the Japan Map Centre.

2. If you have an idea for use of the data you are requested to inform the JMC.

3. The words 'Kokudo Sûchi Jôhô' must appear somewhere on any product resulting from the use of the data.

4. After purchase of the data if you decide that you would like to use the data for a specific purpose you must file an application for permission to use the data. If and when you receive permission you may not use the data for any other purpose.

5. You may not give copies of the data to anyone.

6. Once permission is granted to use the data for the applied for purpose the permission is good for one year. If you wish to use the data the year after the original application you must reapply for permission.

Implementation

Only 2 out of the 11 countries have set up specialist centres to implement their NSDIs. These are the National Centre for Geographic Information Systems in Qatar and the National Centre for Geographic Information (CNIG) in Portugal. Although in both cases the work programme is overseen by government ministries, it should be noted that both these centres have a considerable degree of autonomy regarding the planning and implementation of particular projects. The Portuguese case is especially interesting in that its activities include more general geographic information R&D activities as well as national geographic information system implementation. Part of its funding is also derived from projects funded by the European Community and other agencies.

Feasibility studies have been commissioned in Malaysia and Canada to explore the options for a national geographic information strategy. In the case of Malaysia these included the controversial option of privatisation where the consultants recommended that 'in the long term, the implementation of NALIS should be privatised in order to transfer the burden of funding from the government to private sector and accelerate implementation' (Renong

Berhad 1995, paragraph 6.2). The Canadian study reviews some of the other NSDIs described in this chapter and recommends a cooperative approach towards implementation which takes account of circumstances which are specific to Canada as against the US National Spatial Data Infrastructure model which has 'tried to accommodate the prevailing fragmented conditions, arrangements and the underlying culture defining geographic information management in the US at this time' (Geoplan 1996, 43).

Only in the case of Australia has there been any attempt to quantify the benefits associated with the implementation of a National Land and Geographic Data Infrastructure. ANZLIC commissioned Price Waterhouse to carry out a study of the economic benefits arising from the acquisition and maintenance of land and geographic information at the national level in 1994. The findings of this study suggest a benefit cost ratio for data usage of the order of 4:1.

They also showed that:

> The existing infrastructure for supplying data had provided information to users at a cost far lower than alternative methods. If this infrastructure had not been in place, and users had been forced to meet their data requirements from other sources, their costs would have been approximately six times higher. Over the past five years alone, an established infrastructure has saved users 5bn Australian dollars much of which has been reinvested to generate additional economic activity.
>
> *(Price Waterhouse 1995, 1)*

Resources

It is very difficult both to obtain and to interpret the information that is available for the 11 NSDIs with respect to resources. The only thing that is clear is that the task of coordination is relatively inexpensive in relation to the overall expenditure on geographic information whereas the task of core digital database development is relatively expensive. With respect to the former it is worth noting that the US Office of Management and Budget has estimated that federal agencies alone spend $4bn annually to collect and manage domestic geospatial data (FGDC 1994, 2). This sum is of a very different order to the $25m that has been spent by the USGS to date to support the FGDC and its work (Tosta 1997b, 4). With respect to the latter it is worth noting that $288.5m out of the $360m budget for the Korean NGIS is allocated to digital topographic, thematic and underground facility mapping (MOCT 1995, 8). Similarly, in Malaysia the federal government has allocated nearly $50m to cover the costs of the development of the initial NaLIS prototype in Wilayah Persekutuan Kuala Lumpur (NaLIS 1996) and it even cost $5m to build the database sets for the tiny state of Qatar (Tosta 1997a).

11.4 Towards the Next Generation of National Spatial Data Infrastructures

In the light of the findings of this analysis it is possible to consider what lessons can be learnt for the next generation of NSDIs which will come into being over the next ten years. Although the findings themselves show that there is a great deal of diversity in the first generation of NSDIs, they also point to some useful lessons for the future. With this in mind it is useful to reconsider the three criteria used for selection purposes in the first section of the chapter in the light of the findings of the evaluation: i.e. national scope, focus on geographic information and the existence of some form of coordinating mechanism. With respect to national scope there are some important differences between countries with federal governmental systems and those with non-federal systems. In countries such as Australia and Canada the main emphasis in infrastructure development has been on the state rather than the federal level and this is strongly reflected in their NSDIs. In contrast, local initiatives have been less important in shaping national policy in non-federal countries, although they may be of some significance in situations where national policies are not clearly articulated as is the case in France (Smith 1998).

With respect to the focus on geographic information the analysis shows that there are both advantages and disadvantages in the different positions that have been taken with respect to the scope of NSDIs. The case for a comprehensive approach is a very strong one given the importance attached by users to integrating data from a wide variety of sources. On the other hand, it must be recognised that some players are more central than others and that comprehensive coverage does not necessarily equal similar levels of commitment. As Tosta (1997a) has pointed out, even though President Clinton signed the Executive Order for the NSDI and the Secretary of the Interior, Bruce Babbit chairs the FGDC, no other Secretaries sit on this committee and many agency representatives are not the 'highest' level. Consequently, a good case can also be made for approaches which are more limited in scope and concentrate only on key stakeholders who have a strong vested interest in the success of national information strategies as is the case with respect to Indonesia and Malaysia in the land information field.

The findings of the evaluation also show that there are clear advantages associated with a formal mandate for a national geographic information strategy from the standpoint of coordination, provided that this is accompanied by the necessary resources to enable its implementation. Lack of dedicated resources is obviously the weak point where initiatives are essentially outgrowths of existing coordination activities, yet this model also has considerable advantages in that it builds upon existing cooperative procedures.

In the medium to long term, however, the success of these strategies is likely to be closely coupled with the extent to which they meet the requirements of users. In this respect questions of public access to public information are likely to be a critical factor not only from the standpoint of the extent to which geographic information is utilised in practice but also with respect to its impact on the economics of spatial database creation and maintenance. This is potentially one of the great strengths of the British National Geospatial Data Framework where producers and user interests are taken into account through the dual Board and Advisory Council structure.

Where there is little existing GIS activity and/or there is a lack of basic skills and resources, the establishment of a national centre can play a vital role in creating an NSDI as was the case in Portugal and Qatar. Similarly, feasibility studies and pilot projects of the kind that are underway in Canada and Malaysia can be of considerable value both for building up operational experience and technical skills also for raising overall levels of awareness of the opportunities opened up by NSDIs.

Awareness, not only within governments but also within the public at large is likely to be the critical factor in the success of these strategies (see for example, FGDC 1997, goal 1). Closely linked to this is the need for those politicians and decision makers that are most closely involved to recognise that geographic information is a national asset which must be effectively coordinated and managed in the national interest (Barr and Masser 1997).

It is also likely that the next generation of NSDIs will come into being in rather different circumstances to those which gave rise to the first generation. Two factors are of particular importance in this respect. These are the changes that are currently taking place in the nature of government itself and the growing globalisation of geographic information activities respectively.

In many countries, governments are increasingly expected to operate in a more commercial way. 'Reinventing government' as the title of Osborne and Gaebler's (1992) classic book suggests, implies reassessing the operations carried out by governments with particular reference to privatisation, deregulation and market testing. The impacts of developments such as these on many of the key government stakeholders in geographic information is likely to be profound. In Britain Ordnance Survey is approaching 100% cost recovery and the option of its privatisation has not been excluded by the government. In the Netherlands the Cadastre has become an independent administrative organisation within the Ministry of Housing, Spatial Planning and the Environment and it also set up a separate company in April 1996 to develop a new postal address coordinate product for all 7m postal addresses in the Netherlands (Masser 1998, 41). In other countries surveying and mapping activities are being restructured in ways that substantially alter their roles within government. For example, the subdivision of the old Department of Surveying and Land Information in New Zealand in 1996 into two new organisations, a national interest surveying and mapping agency (Land Information New Zealand) and a state owned commercial enterprise (Terralink) has introduced a new dimension into the national spatial information strategy debate in that country (Robertson and Gartner 1997). Developments such as these raise fundamental questions about the future of many national mapping agencies. In the eyes of O'Donnell and Penton (1997, 214):

> What we can be certain of is that the traditional surveying and mapping organisations are threatened with extinction. Their survival will depend on our ability to understand and adapt to the trends which are forging our future directions.

Alongside these developments is the emergence of global – i.e. transnational and regional – initiatives in many parts of the world (Masser 1997). The establishment of the European Umbrella Organisation for Geographic Information (EUROGI) in 1993 with help from the European Union was the first of these ventures (Burrough et al. 1993). Since then its membership has gradually expanded to include many central and east European countries and created a very useful platform for transnational strategic debate and professional collaboration in this region. More recently the creation of the Permanent Committee on GIS Infrastructure for Asia and the Pacific in 1995, following a resolution of the 13th United Nations Regional Cartographic Conference for Asia and the Pacific in Beijing, can be seen as the first stage of a broader UN strategy on this front (Godfrey et al. 1997). This committee has representatives from 55 countries in the region including China, mainly at the national mapping agency level. It should also be noted that the United Nations regional agencies are in the process of establishing a parallel body for Africa and the Middle East. If this comes into being and similar arrangements are made for the Americas, it can be argued that some of the basic organisational building blocks for a global spatial data infrastructure are already in place. These developments bring a new dimension to the work that is already been done as a result of established global remote sensing and mapping initiatives such as the UNEP Global Resources Information Database Programme (GRID) which began in 1985 and the International Geosphere Biosphere Programme (IGBP) which was established in 1986 as well as the activities of the International Steering Committee for Global Mapping which was set up in 1994 and has its secretariat at the Geographical Survey Institute in Japan.

With these considerations in mind it is worth considering which countries are likely to be candidates for the second generation of NSDIs which will come into being over the next ten years. In practice probably three different sets of countries are most likely to fall into this category. The first of these are members of the first generation who will substantially restructure their current approaches as a result of the experience that has been built up over the last few years. This trend can already be seen in the proposals outlined by the US National Academy for Public Administration (1998) for the future development of the NSDI. Secondly, there are a number of developing countries, especially in the Asia and Pacific region who are likely to follow the examples of Indonesia and Malaysia to facilitate the planning and management of economic development and natural resources. Last but not least are the group of central and east European countries such as the Czech Republic, Hungary and Poland where there had been a massive investment recently in cadastral and digital mapping programmes as part of national restructuring activities in the post-Communist era (see, for example, Bogaerts 1997).

Given the number of countries potentially falling into each of these categories it seems likely that there will be a considerable increase in the number of NSDIs that are being implemented throughout the world in the next ten years. Because of the very different circumstances within which these will be implemented it also seems likely that these will be at least diverse in character as those of the first generation.

11.5 An SDI Phenomenon?

Twenty years ago, the author reviewed the experiences of 11 initiatives from different parts of the world that made up the first generation of National SDIs that are described above. The findings of this review highlighted some of the similarities and differences that existed in the composition of the first generation of National SDIs particularly with respect to their status, scope, access, and approach to implementation and resources and considered some of the lessons that might be drawn from this experience.

Since the publication of this chapter SDI initiatives have spread to all parts of the world. This diffusion process was particularly fostered by the efforts of the Global Spatial Data Association (GSDI) (Holland and Borrero 2003) and also by regional bodies such as the European Umbrella Organisation or Geographic Information and the Permanent Committee for Geographic Information in Asia and the Pacific (Masser et al. 2003), together with a number of books (see, for example, (Williamson et al. 2003; Masser 2005), and project reports (see, for example, the Geographic Information Network in Europe (GINIE) project (Craglia et al. 2003)) as well as scientific publications with a worldwide circulation such as Nature (Butler 2006) and the Proceedings of the US National Academy of Sciences (Goodchild et al. 2012). In 2007 the SDI movement as a whole was boosted by Directive 2007/2/EC of the European Parliament and of the Council of

14 March 2007 establishing an Infrastructure for Spatial Information in the European Community (INSPIRE) This Directive establishes a Spatial Data Infrastructure for improving environmental data management in the European Community by 2020 in all the European Union member states (Commission of the European Communities 2007).

A particularly valuable initiative in the SDI diffusion process has been the preparation of the GSDI cookbook as a collaborative effort by members of the GSDI Association which has gone through a number of updates since its first publication in 2000 (GSDI 2000). This has been translated into a number of languages and stimulated a number of initiatives which reflect regional circumstances (see, for example, Africa (UN Economic Commission for Africa 2004) and the Americas (Permanent Committee for Geospatial Data Infrastructure of the Americas 2013). Since its launch in 2003, the GSDI Association's Small Grants Program has also supported more than 100 projects around the world. Support for the programme comes from a partnership between the GSDI Association, the U.S. Federal Geographic Data Committee, Canada's GeoConnections Program, and the GISCorps of the US Urban and Regional Information Systems Association (URISA).

One outcome of the rapid expansion of SDI is the changes that have taken place in the organisations that reflect the field. The most significant of these is the establishment of the United Nations Committee of Experts on Global Geospatial Information Management (UN-GGIM) in July 2011 as the official United Nations consultative mechanism on global geospatial information management. This mechanism provides a forum for coordination and dialogue among Member States, and between Member States and relevant international organisations, and to propose work-plans with a view to promoting global frameworks, common principles, policies, guidelines and standards for the interoperability and inter-changeability of geospatial data and services.

One unfortunate consequence of this development has been the decision to wind up the Global Spatial Data Association in 2018 as it was felt that, due to the creation of the United Nations Committee of Experts on Global Geospatial Information Management UM GGIM), the UN now offers its member nations and GSDI professionals from across the public, academic and private sectors the opportunity to advance the very principles and practices that the GSDI developed and promoted since its founding in 2004.

The extent of this SDI phenomenon is all the more surprising, as there is still no clear consensus about what constitutes an SDI. In some cases, the term refers to projects rather than SDI activities more generally (see the last section of chapter 14). Many SDI's have a strong project dimension which focuses on concrete goals such as the completion of the national topographical database. Others are much more process oriented and focus mainly on strategic issues such as capacity building and the modernisation of government. This is partly due to the different interpretations that can be given to the notion of infrastructure. To some people infrastructure means tangible physical assets like roads and railway networks. These people tend to the ones who are most comfortable with the database creation/project model of an SDI. To others it is a strategic process of policy formulation and implementation carried out by governments to ensure that their geographic information assets are managed in the interests of the nation as a whole (Barr and Masser 1997). This includes not only the tangible assets but also the individuals and institutions that are needed to make it a functional reality.

One thing that is clear from this discussion is that

> the overriding objective of SDIs (including NSDIs) is to facilitate access to geographic information assets that are held by a wide range of stakeholders in both the public and the private sectors in a nation or a region with a view to maximizing overall usage. This objective requires coordinated action by governments.
>
> SDIs must also be user driven, as their primary purpose is to support decision making for many different purposes. SDI implementation involves a wide range of activities. These include not only technical matters such as data, technologies, standards, and delivery mechanisms but also institutional matters related to organisational responsibilities, overall national information policies, and availability of financial and human resources.
>
> *(Masser and Crompvoets 2015, p. 17)*

With this in mind the next chapter addresses some of the broader issues regarding the future development of SDIs throughout the world.

References

Al Thani, A.B.H., 1997. GIS in Qatar: An integral part of the infrastructure, paper presented at GIS/GPS '97 Conference, Qatar, 2–4 March 1997.

Arnaud, A.M., L.T. Vasconcelos, and J.D. Geirinhas, 1996. Portugal: GIS diffusion and the modernisation of local government, in Masser, I., H. Campbell, and M. Craglia (eds.), *GIS diffusion: The adoption and use of geographic information systems in local government in Europe*, London: Taylor and Francis, pp. 111–124.

Australia and New Zealand Land Information Council (ANZLIC), 1992. *Land Information Management in Australasia 1990–1992*, Canberra: Australia Government Publishing Service.

Australia and New Zealand Land Information Council (ANZLIC), 1997. *Spatial Data Infrastructure for Australia and New Zealand*, www.anzlic.org.au/anzdiscu.htm.

Barr, R. and I. Masser, 1997. Geographic information: A resource, a commodity, an asset or an infrastructure? in Kemp, Z. (ed.), *Innovations in GIS 4*, London: Taylor and Francis, pp. 234–248.

Berhad, R., 1995. Feasibility study for the National Land Information System: Final Report.

Bogaerts, T., 1997. A comparative overview of the evolution of land information systems in Central Europe, *Computers, Environment and Urban Systems 21*, 109–131.

Burrough, P., M., Brand, F. Salgé, and K. Schuller, 1993. The EUROGI vision, *GIS Europe 2*(3), 30–31.

Butler, D., 2006. Virtual globes: The web-wide world, *Nature 439*, 776–778.

Commission of the European Communities (CEC), 2007. Directive 2007/2/EC of the European Parliament and of the Council of 14 March 2007 establishing an Infrastructure for Spatial Information in the European Community (INSPIRE), *Official Journal of the European Union L108*, 1–14.

Corey, M., 1998. The Canadian geospatial data infrastructure, paper presented to ISO/TC 211 Conference, Victoria, BC, 4 March 1998.

Craglia, M., A. Annoni, M. Klopfer, C. Corbin, L. Hecht, G. Pinchler, and P. Smits, 2003. *Geographic Information in the Wider Europe*, IST-2000-29493. University of Sheffield

Department of the Environment, 1987. *Handling Geographic Information: Report of the Committee of Enquiry Chaired by Lord Chorley*, London: HMSO.

Executive Office of the President, 1994. Coordinating geographic data acquisition and access, the National Spatial Data Infrastructure, Executive Order 12906, *Federal Register 59*, 17671–17674.

Federal Geographic Data Committee, 1994. *The 1994 Plan for the National Spatial Data Infrastructure: Building the Foundations of an Information Based Society*, Reston, VA: Federal Geographic Data Committee, USGS.

Federal Geographic Data Committee, 1997. *A Strategy for the National Spatial Data Infrastructure*, Reston, VA: Federal Geographic Data Committee.

Geoplan Consultants, 1996. The development of an integrated Canadian spatial data model and implementation concept, Final Report prepared for the Canadian Council on Geomatics.

Global Spatial Data Association, 2000. *Developing Spatial Infrastructures: The SDI Cookbook*, Reston, VA: Federal Geographic Data Committee.

Godfrey, B., P. Holland, G. Baker, and B. Irwin, 1997. The contribution of the Permanent Committee on GIS infrastructure for Asia and the Pacific to a global spatial data infrastructure, paper presented at the 2nd Global Spatial Data Infrastructure Conference, Chapel Hill, NC, 19–21 October 1997.

Goodchild, M.F., H. Guo, A. Annoni, L. Bian, K. de Bie, F. Campbell, M. Craglia, M. Ehlers, J. van Genderen, D. Jackson, A.J. Lewis, M. Pesaresi, G. Remety-Fulopp, R. Simpson, A. Skidmore, C. Wang, and P. Woodgate, 2012. Next generation Digital Earth, *Proceedings of the National Academy of Science 109*, 11088–11094.

Gore, A. 1998. Digital earth: Understanding our planet in the 21st century, *speech at California Science Center, Los Angeles*, 31 January 1998.

Gouveia, C., J. Abreu, N. Neves, and R. Henriques, 1997. The Portuguese national digital earth infrastructure for geographic information: General description and challenges for the future, paper presented at the GISDATA Conference on Geographic Information Research at the Millennium, Le Bischenberg, France, 13–17 September 1997.

Henriques, R.G., 1996. The Portuguese national network of geographic information (SNIG network), in Rumor, M., McMillan, R., and Ottens, H.F.L. (eds.), *Geographic*

Information: From Research to Application to Cooperation, Amsterdam: IOS Press, pp. 112–116.

Hobman, L., 1997. The national geospatial data framework: From concept to reality, paper presented *at GIS '97*, Birmingham, 7–9th October 1997.

Holland, P. and S. Borrero, 2003. Global initiatives, in Williamson, I., Rajabifard, A., and Feeney, M.-E. (eds.), *Developing Spatial Data Infrastructures: From Concept to Reality*, New York: Taylor and Francis, 43-58.

Masser, I., 1997. Geographic information goes global, *Computers, Environment and Urban Systems 21*, 303–305.

Masser, I., 1998. *Governments and Geographic Information*, London: Taylor and Francis.

Masser, I., 2005. *GIS Worlds: Creating Spatial Data Infrastructures*, Redlands CA: ESRI Press.

Masser, I., S. Borrero, and P. Holland, 2003. Regional SDIs, in Williamson, I., Rajabifard, A., and Feeney, M.-E. (eds.), *Developing Spatial Data Infrastructures: From Concept to Reality*, New York: Taylor and Francis, 59-77.

Masser, I. and J. Crompvoets, 2015. *Building European Spatial Data Infrastructures*, Third edition, Redlands, CA: ESRI Press.

McLaughlin, J., 1991. Towards a national spatial data infrastructure, paper presented at the Canadian Conference on GIS, Ottawa, March 1991.

Ministry of Construction and Transportation (MOCT), 1995. *A Master Plan for National Geographic Information System (NGIS) in Korea*, Seoul: Ministry of Construction and Transportation.

Nanson, B. and D. Rhind, 1998. Establishing the UK National Geospatial Data Framework, paper presented at *SDI 98*, Ottawa, 9–11 June 1998.

National Academy for Public Administration (NAPA), 1998. *Geographic information for the twenty first century: building a strategy for the nation*. Washington: National A. cademy for Public Administration.

National Geospatial Data Framework (NGDF), 1998. *Strategy Plan*, Southampton: National Geospatial Data Framework Taskforce.

National Infrastructure for Land Information System (NaLIS), 1996. Report of the NaLIS task force on the establishment of the National Infrastructure for Land Information System (NaLIS).

National Research Council (NRC), 1993. *Toward a Coordinated Spatial Data Infrastructure for the Nation*, Mapping Science Committee, Washington, DC: National Academy Press.

O'Donnell, J.H. and C.R. Penton, 1997. Canadian perspectives on the future of national mapping, in Rhind, D. (ed.), *Framework for the World*, Cambridge: Geoinformation International, pp. 214–225.

Office of Management and Budget, 1990. *Coordination of Surveying, Mapping and Related Spatial Data Activities*, Circular A-16 revised, Office of Management and Budget, Washington, DC: Executive Office of the President.

Osborne, D. and E. Gaebler, 1992. *Reinventing Government: How the Entrepreneurial Spirit Is Transforming the Public Sector*, Reading, MA: Addison Wesley.

Pollard, S., 1997. Japan GIS/Mapping Sciences Resource guide, Second edition, www. cast.uark.edu/jpgis (last accessed 16 February 2019). University of Arkansas.

Price Waterhouse, 1995. *Australian Land and Geographic Information Infrastructure Benefits Study*, Canberra: Australia Government Publishing Service.

Ravi, 1992. *Structuurschets vastgoedinformatie voorziening, deel 1: Bestuurlijke notitie*, Apeldoorn: Netherlands Council for Geographic Information.

Ravi, 1995. *The National Geographic Information Infrastructure*, Amersfoort: Netherlands Council for Geographic Information.

Renong Berhad, 1995. *Feasibility study for the National Land Information System*, Final report, Kuala Lumpur: Renong Berhad.

Rhind, D., 1996. Economic, legal and public policy issues influencing the creation, accessibility and use of GIS databases, *Transactions in GIS 1*, 3–12.

Robertson, W.A. and C. Gartner, 1997. The reform of national mapping organisations: The New Zealand experience, in Rhind, D. (ed.), *Framework for the World*, Cambridge: Geoinformation International, pp. 247–264.

Suharto, L., 1995. Moves towards a national geographic information system, Paper presented at the 5th South East Asian Surveyors Conference, Singapore 16-20 July 1995.

Smith, I., 1998. Local information and national infrastructure: Geographic information in French local government, paper presented at the 1st AGILE Conference, Enschede, The Netherlands, 23–25 April 1998.

Suharto, L., 1995. Moves towards a national geographic information system, paper presented at the 5th South East Asian Surveyors Conference, Singapore, 16–20 July 1995.

Tamin, M.Y., 1997. The National Infrastructure for Land Information System (NaLIS): Applying information technology to improve the utilisation of land data in Malaysia, paper presented at Oracle Open World Malaysia, Kuala Lumpur, 23–24 January 1997.

Tosta, N., 1997a. Data revelations in Qatar: Why the same standards won't work in the United States, *GeoInfo Systems 7*, 5.

Tosta, N., 1997b. The US National Spatial Data Infrastructure experience, paper presented at the GISDATA Conference on Geographic Information Research at the Millennium, Le Bischenberg, France, 13–17 September 1997.

UN Economic Commission for Africa, 2004. *SDI Africa: An Implementation Guide*, Addis Ababa: UN Economic Commission for Africa.

UN Permanent Committee for Geospatial Data Infrastructure for the Americas (CP-IDEA), 2013, *Spatial Data Infrastructure (SDI) Manual for the Americas*, Rio de Janeiro: CP-IDEA.

Williamson, I., A. Rajabifard, and M-E. Feeney, 2003. *Developing Spatial Data Infrastructures: From Concept to Reality*, London: Taylor and Francis.

Yamaura, A., 1996. National spatial data infrastructure: An Asian viewpoint, paper presented at the Global Spatial Data Infrastructure Conference, Bonn, Germany, 4–6 September 1996.

12

The Future of Spatial Data Infrastructures

Ian Masser

12.1 Introduction

This chapter addresses two central questions relating to the future of spatial data infrastructures (SDIs)

- Where have we got to now?
- Where should go from here?

The answers to the first question are to be found in a review of the main milestones of SDI development over the last two decades and a comparative evaluation of SDI experiences in different parts of the world which constitutes the first main section of the chapter. The answers to the second question are inevitably more speculative in nature. With this in mind the second main section examines some emerging trends and explores some of the main strengths and weaknesses of current SDI practices in relation to the perceived opportunities and threats that are likely to emerge in the foreseeable future. The chapter concludes with a discussion of four key issues that are likely to play a vital role in determining the future success of SDIs.

> Before beginning the discussion, however, it is important to clarify was is meant by the term 'SDI'. A comprehensive definition which conveys some of the complexity of the issues involved can be found 'A ... Spatial Data Infrastructure supports *ready access to geographic information*. This is achieved through *the co-ordinated actions of nations and organisations* that promote awareness and implementation of complimentary policies, common standards and effective mechanisms for the development and availability of interoperable digital geographic data and technclogies *to support decision making at all scales for multiple purposes*. These actions *encompass the policies, organisational remits, data, technologies, standards, delivery mechanisms, and financial and human resources* necessary to ensure that those working

at the (national) and regional scale are not impeded in meeting their objectives'.

(Author's italics)

The italicised sections of this rather complex definition show that there are four key concepts underlying SDIs. The first of these states that their overriding objective is to promote ready access to the geographic information assets that are held by a wide range of stakeholders in both the public and the private sector with a view to maximising their overall usage. The second concerns the need for concerted action on the part of governments to ensure that this overriding objective is attainable. The next part of this sentence gives some examples of the kind of actions that are required from governments. The third key element stresses the extent to which SDIs must be user driven. Their primary purpose is to support decision making for many different purposes and it must be recognised that many potential users may be unaware of the original purposes for which the data was collected. Finally, the last sentence illustrates the wide range of activities that must be undertaken to ensure the effective implementation of an SDI. These include not only technical matters such as data, technologies, standards and delivery mechanisms but also institutional matters related to organisational responsibilities and overall national information policies as well as questions relating to the availability of the financial and human resources needed for this task.

12.2 Where Have We Got to Now?

12.2.1 SDI Milestones

The first SDI milestone dates back 20 years to the establishment of the Australian Land Information Council (ALIC) in January 1986 as a result of an agreement between the Australian Prime Minister and the heads of the state governments to coordinate the collection and transfer of land related information between the different levels of government and to promote the use of that information in decision making (ANZLIC 1992, p. 1).

The second milestone was the publication of the Report of the British Government Committee of Enquiry on Handling Geographic Information chaired by Lord Chorley in May 1987 (Department of Environment 1987). This set the scene for much of the subsequent discussion about SDIs in the UK and elsewhere. The report reflects the committee's enthusiasm for the new technology: 'the biggest step forward in the handling of geographic information since the invention of the map' (para 1.7), and also their concern that information technology in itself must be regarded as 'a necessary, though not sufficient condition for the take up of geographic information systems to increase rapidly' (para 1.22). To facilitate the rapid take up of GIS the committee argued that it will be necessary to overcome a number of important barriers to effective utilisation. Of particular importance in this respect are the need for greater user awareness and the availability of data in digital form suitable for use in particular applications.

The third milestone occurred in 1990 when the United States Office of Management and Budget (OMB) established an interagency Federal Geographic Data Committee (FGDC) to coordinate the 'development, use, sharing, and dissemination of surveying, mapping, and related spatial data'.

Up to this point the term 'National Spatial Data Infrastructure' was not in general use although a paper was presented by John McLaughlin at the 1991 Canadian Conference on Geographic Information Systems in Ottawa entitled 'Toward National Spatial Data Infrastructure'. Many of the ideas contained in this chapter were subsequently developed and extended by the United States National Research Council's Mapping Science Committee in their report on 'Toward a coordinated spatial data infrastructure for the nation' (National Research Council 1993). This recommended that effective national policies, strategies and organisational structures need to be established at the federal level for the integration of national spatial data collection, use and distribution. To realise this goal, it proposed that the powers of the FGDC should be strengthened to define common standards for spatial data management and to create incentives to foster data sharing particularly among federal agencies.

The next milestone is the outcome of an enquiry set up by DG XIII (now DG Information Society) of the European Commission which found that there was a strong European wide demand for an organisation that would further the interests of the European geographic information community. As a result, the first continental level SDI organisation in the world was set up in 1993. The vision of the European Umbrella Organisation for Geographic Information (EUROGI) was not to 'replace existing organisations but ... catalyse effective cooperation between existing national, international, and discipline oriented bodies to bring added value in the areas of Strategy, Coordination, and Services' (Burrough et al. 1993, 31).

The milestone that marks a turning point in the evolution of the SDI concept came in the following year with the publication of Executive Order 12,906 signed by President Bill Clinton entitled 'Coordinating Geographic Data Acquisition and Access: the National Spatial Data Infrastructure' (Executive Office of the President 1994). This set out the main tasks to be carried out and defined time limits for each of the initial stages of the National Spatial Data Infrastructure. These included the establishment of a National Geospatial Data Clearing House and the creation of a National

Digital Geospatial Data Framework. The Executive Order also gave the FGDC the task of coordinating the Federal government's development of the National Spatial Data Infrastructure and required that each member agency of that committee held a policy level position in their organisation. In this way the Executive Order significantly raised the political visibility of geospatial data collection, management and use not only among Federal agencies but also nationally and internationally.

One of the outcomes of this debate in Europe was the decision to hold the first of what subsequently became a regular series of Global Spatial Data Infrastructure conferences at Bonn in Germany in September 1996. This conference brought together representatives from the public and private sectors and academia for the first time to discuss matters relating to NSDIs at the global level.

After the second GSDI conference in Chapel Hill, North Carolina, in 1997, the author carried out a worldwide survey of the first generation of NSDIs (Masser 1999). This showed that at least eleven NSDIs were already in operation in various parts of the world by the end of 1996. What distinguished these from other GI policy initiatives was that they were all explicitly national in scope and their titles all referred to geographic information, geospatial data or land information and included the term 'infrastructure', 'system' or 'framework'. This first generation included relatively wealthy countries such as the United States and Australia as well as relatively poor countries such as Indonesia and Malaysia.

The rapid rate of NSDI diffusion after 1996 is highlighted by the findings of a survey carried for the GSDI (www.gsdi.org). These show that 49 countries responded positively to his questionnaire between 1998 and 2000: 21 of these came from the Americas, 14 from Europe, 13 from Asia and the Pacific and one from Africa. The number of positive responses to this survey is more than four times the number of first generation NSDI countries identified up to the end of 1996 Subsequent data collected by Crompvoets and Bregt (2003) suggests that as many as 120 countries may be considering projects of this kind. These figures must be treated with some caution as they do not necessarily imply that all these countries are actively engaged in SDI formulation or implementation. Furthermore, it is also likely that many of them may be engaged in some aspects of SDI development without necessarily committing themselves to a comprehensive SDI programme. Nevertheless, it is felt that the term 'SDI phenomenon' is a reasonable description of what has happened in this field over the last 10 to 15 years.

12.2.2 SDI Diffusion – A Global Overview

A comprehensive and consistent global evaluation of SDIs has yet to be carried out but there are encouraging signs of such activities at the regional level, particularly in Europe (see Masser 2005, chap 3). The following section summarises the material that is currently available. These findings must also be treated with some caution as the nature of the sources and their content still varies considerably from region to region.

Europe

The development of SDIs has been studied extensively in Europe over the last five years. This is partly due to the interest of the European Commission in such activities that was expressed initially in the GI 2000 initiative and more recently in the INSPIRE programme. In the process the Commission has also funded a number of important studies in this field such as the Geographic Information Network in Europe (GINIE) project (Craglia et al. 2003). More recently the European Commission commissioned a series of 32 country studies of the state of play of SDI activities in all the European countries from the Catholic University of Leuven (http://inspire.jrc.it/sta te_of_play.cfm). The findings of these studies constitute a major resource for SDI research not only in Europe but also for the rest of the world.

On the basis of this research, the authors of these studies have developed a useful typology of SDIs that is based on criteria based mainly on the coordination aspects of these initiatives. Matters of coordination are emphasised because 'it is obvious coordination is the major success factor for each SDI since coordination is tackled in different ways according to the political and administrative organisation of the country' (SAD 2003, 13). A basic distinction is also made between countries where a national data producer such as a mapping agency has an implicit mandate to set up an SDI and countries where SDI development has been driven by a council of Ministries, a GI association or a partnership of data users. A further distinction is then made between initiatives that do and do not involve users in the case of the former and between those that have a formal mandate and those that do not in the case of the latter.

According to the authors more than half the SDI initiatives in Europe are led by national data producers. This is particularly the case in the central and eastern European countries that have recently become members of the European Union (Craglia and Masser 2002) and the Nordic countries. All the Nordic countries explicitly include data users in the coordination process whereas only a minority of former accession countries make provision for user involvement. However, not all these SDI initiatives are operational. This is the case in Greece and Luxembourg as well as several of the former accession countries.

The remaining countries have made other arrangements for the coordination of their national SDI activities. In two countries (Germany and Portugal) a government interdepartmental body has been formally mandated to create a national SDI which is now operational. In the Netherlands a national GI association (RAVI) has been encouraged by the government to take lead and it has succeeded in developing an operational national SDI.

The Americas

The findings of a survey of 21 countries in the Americas carried out in 2000 give a useful overview of the state of SDI development (Hyman et al. 2003). The overall impression that is created by this survey is one of a growing awareness of SDI concepts and approaches in the Americas in 2000, together with recognition that the main obstacles to be overcome in these countries were institutional rather than technical in nature. There was also some concern about the question of the resources that would be required for effective SDI implementation.

The findings of the survey highlight the range of different kinds of SDI initiatives that existed at that time. Formal mandates for the development and implementation of SDIs existed in only six out of the 21 countries. In the majority of cases, a single institute, normally the national mapping agency, or in some countries such as Mexico, the national mapping and statistical agency, was the lead organisation in these initiatives. In some other countries, the Ministries of the Environment, Science and Technology, and Transportation and Public Works acted as focal points. Generally, these initiatives were restricted to central government although the utilities were involved in several countries together with the private sector. An interesting example of the latter is the Uruguay clearing house, which is managed by a private company under contract to the Ministry of Works. In most countries the basic data with reference to topography, transport, hydrology, land cover and administrative boundaries was available in digital form but there was often a lack of standardisation and harmonisation.

Asia and the Pacific

The Asia and the Pacific region is the largest and the most diverse region in the world. Its 55 countries contain 60% of the world's population. They include some of the largest countries in the world as well as many small island countries in the Pacific with tiny populations. They also include countries from the Middle East, and the Indian sub-continent, as well as south east and eastern Asia and Australia and New Zealand.

This diversity may also be the reason for the relative lack of regional studies of SDI diffusion of the kind described above for Europe and the Americas. Nevertheless, Rajabifard and Williamson (2003) estimate that somewhere between 20 and 30% of countries in the Asia and the Pacific region are developing or have plans to develop national SDIs. This broadly confirms the findings of their earlier survey of regional fundamental data sets (Rajabifard and Williamson 2000) when 17 out of the 55 members of the Permanent Committee for GI in Asia and the Pacific, or just under a third of the members, responded to their questionnaire. These were essentially national mapping agencies.

Within this region an obvious distinction can be made between developed and developing countries in terms of their needs and aspirations. Within the developing countries category, a further distinction can be made between countries in the process of transition from a less developed to a more developed state, countries at an early stage of economic development, and the Pacific island nations. It can also be argued that developing countries face different challenges from those of developed countries. 'The main limitations are a lack of appreciation of what SDI can and cannot do, lack of resources and trained personnel, inefficient bureaucratic processes, lack of data, and lack of infrastructure' (Rajabifard and Williamson 2003, p. 34).

Africa

The Johannesburg World Summit on Sustainable Development in September 2002 stimulated several Africa wide studies on SDI related topics. These included a report entitled 'Down to earth: geographic information for sustainable development in Africa' prepared by the Committee on the Geographic Foundation for Agenda 21 on the US National Research Council (2002).

These studies build upon earlier work in the environmental field in Africa. A good example of this is the Environmental Information System Programme for Sub Saharan Africa that has played an important role in harmonising standards for data capture and exchange, coordinating data collection and maintenance and promoting the use of common data sets by the different agencies involved (www.EIS-Africa.org).

Kate Lance's (2003) overview of the current state of the art in SDI development in Africa highlights the diversity of SDI initiatives that have come into being over the last 10 years and the role that has been played by international agencies of all kinds in facilitating the development of SDIs. This is particularly evident in the publication of an African version of the GSDI cookbook (2003) based on the efforts of GSDI, EIS Africa, the UN Economic Commission for Africa and the International Institute for Geoinformation Science and Earth Observation (ITC) in the Netherlands.

Lance also lists 21 national SDI initiatives that are currently under way in all parts of Africa. These include countries from both anglophone and francophone Africa. Her review also identifies some of the main problems facing SDI development on this continent. One of the most important of these is the question of political support as very few of these initiatives have a legal status or enabling legislation to support their efforts and there are only a few countries where SDIs have achieved the status of funded activities with a budget from central government. Another particularly African problem is that of leadership. While national survey and mapping agencies are an important contributor to SDI development it is quite common in Africa to find that other entities have the political influence (and funding) that drives the initiatives.

Comparative Evaluation

The SDI initiatives described above show the extent to which they come in all shapes and sizes with respect to population size, land area, level of economic development and distribution of administrative responsibilities.

There is also a basic difference within the group between Europe and the Americas on the one hand and Asia and the Pacific and Africa on the other. Most of the former are classified as either upper middle or high income by the World Bank whereas most of the latter are low income countries. These differences reflect the considerable gap that exists between these two parts of the world with respect to wealth and also, to a large extent, the resources that are likely to be available to implement SDI initiatives.

The driving forces behind these initiatives are generally similar: i.e. promoting economic development, stimulating better government and fostering environmental sustainability. This can be seen, for example, in India's national SDI, which sets out its objectives as follows:

> The NSDI must aim to promote and establish, at the national level for the availability of organised spatial (and non-spatial) data and multilevel networking to contribute to local, national and global needs of sustained economic growth, environmental quality and stability and social progress.
> *(DOST 2002, para 8.0)*

Other driving forces include the modernisation of government and environmental management. One of the main objectives of the Chile's SNIT is to modernise the way that territorial information is handled by government agencies and to create a collaborative scheme for its future management. Environmental concerns feature prominently in Africa in general and the starting point for the Ghana NAFGIM was a World Bank funded project carried out for the Environmental Protection Agency as part of the Ghana Resource Management Project.

eGovernment has also emerged as an important driving force in many recent cases. It features prominently in the Czech SDI that is linked closely to that country's overall national information infrastructure programme. Specific factors in certain regions may also act as a strong driving force in SDI development. This is particularly the case in the accession countries in central and eastern Europe. The initial development of the building blocks for SDIs in these countries was directly funded by the European Union through the PHARE programme that was set up specifically to help these countries meet the requirements for EU accession. International donors, such as the World Bank, have also played an important role in SDI development in many Asian and African countries

The distinction between SDIs led by national data producers and those that are led by Councils of Ministries or partnerships of data users proposed by the Leuven study is a useful indicator of the status of a SDI.

A formal mandate is particularly important where interagency bodies are involved as it defines their position and status with respect to government but, nevertheless, some advisory bodies enjoy de facto recognition without the need for a formal mandate. This is the case, for example, with respect to the Dutch national GI association (RAVI).

There are also marked differences between countries with respect to the range of substantive interests represented in the coordinating bodies and the extent to which stakeholders are directly involved. There is still a strong coordination dimension to the work of the US Federal Geographic Data Committee (FGDC). Although its composition is broad in scope, its formal membership is restricted to federal government agencies. The existing position in Australia is similar in some respects to that of the United States but more inclusive in terms of representation. The Australia New Zealand Land Information Council (ANZLIC) is essentially an umbrella organisation consisting of representatives from both the Commonwealth and State level government public sector coordination bodies. In contrast the lead Canadian agency, GeoConnections, has always been an inclusive organisation that seeks to bring together all levels of government, the private sector and academia. These interests are reflected in the composition of its Management Board and also in the membership of its committees. It sees itself as a catalyst for successful implementation. There is also a strong industry connection in the CGDI through the Geomatics Industry Association of Canada.

Somewhat surprisingly, given that are large number of low income countries are involved, questions of funding and resources do not feature very prominently in their discussion of SDI development. A notable exception is India which devotes a complete section of its National Spatial Data Infrastructure Strategy document to this matter. Where national data producers are involved as the lead agencies in SDI development it is likely that some of the costs will come from their own budgets. In Kenya, for example, the Survey of Kenya was able to insert the Kenyan SDI into the National Development Plan for 2002–2008. This means that the Ministry of Lands and Settlements has a mandate to invest staff time and resources into this initiative.

Other possibilities include international funding through World Bank and similar projects. Projects of this kind played an important role in setting up Ghana's National Framework for Geospatial Information Management and creating Nemoforum in the Czech Republic. Similarly, the Japan International Cooperation Agency was involved in the workshop that led to the creation of the Kenyan SDI. However, projects such as these generally have a limited life span whereas SDI development requires sustained efforts over a long period of time. This is one of the reasons why Kate Lance sees SDIs as a hard sell in regions such as Africa.

'SDI is a hard sell. It is a "beast" of an initiative since it requires inter-institutional, cross sector, long-term coordination – something that defies the administrative and budgetary structures in Africa, as well as the donor agencies' funding cycles' (Lance 2003, p. 36).

12.2.3 Achievements

From the above discussion it can be seen that a critical mass of SDI users in all parts of the world has come into being as a result of the diffusion of SDI concepts during the last 10 to 15 years. This provides the basic networks and channels for communication that are essential for the future development of the field. In the process regional bodies at the continental level and global bodies have come into being to facilitate SDI development and promote a wide range of capacity building initiatives throughout the world. Alongside these activities there is a growing body of SDI related literature and research.

12.3 Where Should We Go from Here?

12.3.1 Emerging Trends

Some emerging trends in thinking about SDIs have been explored by Rajabifard et al. (2003). Their findings show that first generation of SDIs gave way to a second generation from about 2000 onwards. The most distinctive change that has taken place is the shift from the product model that characterised most first generation SDIs to a process led model of a SDI.

Database creation was to a very large extent the key driver of the first generation and, as a result, most of these initiatives tended to be data producer led. The shift from the product to the process model is essentially a shift in emphasis from the concerns of data producers to those of data users. The main driving forces behind the process model are the desire to reuse data collected by a wide range of agencies for a great diversity of purposes at various times. Also associated with this change in emphasis is a shift from the centralised structures that characterised most of the first generation of SDIs to the decentralised and distributed networks that are a basic feature of the WWW.

There has been also a shift in emphasis from SDI formulation to implementation over time. This is associated with the nature of multi-level SDI implementation. Under these circumstances it is necessary to think in terms of more inclusive models of governance. In many cases these developments will also require new kinds of organisational structure to facilitate effective implementation.

The Multi-Level Nature of SDI Implementation

The impression given by many national SDI documents is that they abide by the principle of 'one size fits all'. In other words, they suggest that the outcome of SDI implementation will lead to a relatively uniform product. However, there is both a top down and a bottom up dimension to the relationships between the different levels involved in national SDI implementation. National SDI strategies drive state-wide SDI strategies and state-wide SDI strategies drive local level SDI strategies. As most of the detailed database maintenance and updating tasks are carried out at the local level the input of local government has also a considerable influence on the process of SDI implementation at the state and national levels. The outcomes of such processes from the standpoint of a national SDI such as that of the US are likely to be that the level of commitment to SDI implementation will vary considerably from state to state and from local government to local government. Consequently, the US NSDI that emerges from this process will be a collage or a patchwork quilt of similar but often quite distinctive components that reflect the commitments and aspirations of the different sub national governmental agencies.

This vision of a bottom up SDI differs markedly from the top down one that is implicit in much of the SDI literature. While the top down vision emphasises the need for standardisation and uniformity the bottom up vision stresses the importance of diversity and heterogeneity given the very different aspirations of the various stakeholders and the resources that are at their disposal. Consequently, the challenge to those involved in SDI implementation will be to find ways of ensuring some measure of standardisation and uniformity while recognising the diversity and the heterogeneity of the different stakeholders. This will require a sustained mutual learning process on the part of all those involved in SDI implementation.

A particularly interesting example of multi-level implementation in practice is the European Union's INSPIRE (Infrastructure for SPatial Information in Europe) initiative. This was launched in 2001 with the objective of making available relevant, harmonised and quality geographic information to support the formulation, implementation, monitoring and evaluation of Community policies with a territorial dimension or impact (http://inspire.jrc.it). INSPIRE is seen as the first step towards a broad multi-sectoral initiative which focuses on the spatial information that is required for environmental policies. It is a legal initiative that addresses 'technical standards and protocols, organisation and coordination issues, data policy issues including data access and the creation and maintenance of spatial information'. A draft Directive to 'establish an infrastructure for spatial information in the Community' was published in July 2004 (CEC 2004) and the European Environment Agency, together with Eurostat and the Joint Research Centre, are currently engaged in a process that should lead to its approval in late

2006 or early 2007. When approved, the governments of all 25 national member states will be required to modify existing legislation or introduce new legislation to implement its provisions within a specific time period.

More Inclusive Models of SDI Governance

Many countries are moving towards more inclusive models of SDI governance to meet the requirements of a multi-level multi-stakeholder SDI. Recent developments in the US and Australia, for example, show a marked shift in this direction. In the US the FGDC is considering the recommendations of its Future Directions Project regarding the creation of a new governance model that includes representatives of all stakeholder groups to guide the NSDI (FGDC 2005). This is supported by a joint FGDC/NSGIC (2005) initiative which aims to get all 50 states actively involved and contributing to the NSDI. Similar developments are already under way in Australia. ANZLIC's proposals for an action plan (ANZLIC 2004) involve a new governance model that takes account of the balance between public and private sectors, data sources and data users. These developments will bring both these countries into line with Canada where the lead Canadian agency, GeoConnections, has always been an inclusive organisation that seeks to bring together all levels of government, the private sector and academia.

A good example that highlights the need for more inclusive models of SDI governance at the outset of an SDI initiative is the formulation of a GI strategy for Northern Ireland in the UK. Ordnance Survey of Northern Ireland and its parent ministry, the Department of Culture, Arts and Leisure, decided that a new approach was required to the development and implementation of geographic information policy in Northern Ireland (Masser 2005, Chapter 4). It was decided to make use of the Future Search method to develop an initial GI policy agenda for the province. Weissbord and Janoff (2000) claim that Future Search is 'a unique planning meeting that meets two goals at the same time, (1) helping large diverse groups discover values, purposes and projects they hold in common, and (2) enabling people to create a desired future together and start working towards it right away'.

The Future Search process worked well at a special workshop involving all the main stakeholders at Lusty Beg, an island on a remote lough in Northern Ireland. The participants collectively created a mind map with 32 main trends and an even larger numbers of sub trends within these trends. In the process the following issues emerged as key elements in the common ground for a future strategy: the importance of creating an overall GI strategy for Northern Ireland, the need to facilitate access and promote awareness, and the importance of partnerships in realising these objectives. On the basis of this experience the participants set up a number of working groups to further develop Northern Ireland's GI strategy.

The Emergence of New Organisational Structures

In many cases it is likely that the multi-level nature of national SDI implementation will also require the creation of new kinds of organisation. These can take various forms. Masser (2005, Chapter 5) shows some of the different kinds of organisational structures that have already emerged in the US, Australia and Canada to facilitate national SDI implementation. This shows that at least five different types of partnerships are in operation. These range from the restructuring of existing government agencies to the establishment of joint ventures involving different combinations of the key stakeholders.

The simplest case is the merger of various government departments with responsibilities for various activities based on geographic information. The driving force for this kind of restructuring is typically the perceived administrative benefits to be derived from the creation of an integrated database for the agency as a whole. This can be seen in the creation of Land Victoria in 1996 in Australia which is the product of a merger of various state government entities with responsibilities for various aspects of land administration. The objective of this merger was to establish an integrated land administration agency with a shared geographic information resource for the State of Victoria.

An alternative strategy is to set up a special government agency outside the existing governmental structure with a specific remit to maintain and disseminate core datasets. Service New Brunswick in Canada is a good example of such a strategy. It is a Crown Corporation owned by the State of New Brunswick. It was originally set up to deal with matters relating to land transactions and topographic mapping for the Province as a whole. Since 1998 it has shifted its position to become the gateway for the delivery of a wide range of basic government services as well as national SDI implementation.

There are also some interesting examples of joint ventures between different groups of the stakeholders in SDI implementation. The simplest case is a data producer driven joint venture involving the Australian public sector mapping agencies that was originally set up in 1993 to create an integrated national digital base map for the 1996 Census of Population (www.psma.com.au). The driving force behind this partnership was the recognition the whole is worth more than the sum of the parts in that there are clear economic and social benefits for the nation to be derived through the assembly and delivery of national data sets from the data held and maintained by the consortium members.

The other two types of joint ventures involve more complex structures. Alberta's Spatial Data Warehouse is very much a user data driven initiative. It is a not for profit joint venture between key data users including the State itself, the local government associations and the utility groups to facilitate the continuing maintenance and distribution of four primary provincial data sets. From the outset the partners recognised that they did not either the expertise or the resources to maintain and disseminate the existing databases. Consequently, they negotiated a long-term Joint Venture Agreement with two private sector companies in 1999 to carry out these tasks. This covers the reengineering of the databases and also makes it possible to implement new pricing and licensing options.

Finally, initiatives such as the MetroGIS collaborative in the Minneapolis St Paul metropolitan region of the US bring together a large number of data producers and data users. Such initiatives are both more ambitious and more open ended in their potential for development than either of the other joint ventures. The distinctive feature of this initiative lies in its insistence on voluntary, open and flexible and adaptive collaborations which optimise the interdependencies between citizens and organisations.

12.3.2 SWOTs Analysis

Strengths

The most important strength of the SDI concept is the way in which it enables a diverse group of users to access a wide range of geo referenced data sets. The underlying rationale for SDIs is to maximise the use that is made of local, national and global geographic information assets and their success or failure is likely to be measured largely in these terms. In this way SDIs also make an important contribution to economic growth and job creation at the local, national and global levels as well as promoting more effective and transparent decision making in both the public and private sectors.

The second main strength is the degree to which the SDI concept straddles existing professional and administrative sectoral boundaries. It is a truly integrating concept that facilitates the use of local, national and global geographic information assets many times in many different applications. Recognition of the importance of integrating data from many diverse sources is already encouraging the merger of previously separate professional bodies in some countries. In Australia, for example, a Spatial Sciences Institute was set up in 2003 to bring together the professional disciplines of surveying, mapping, engineering and mining, surveying, remote sensing and photogrammetry. At the global level a Global Spatial Data Infrastructure Association was set up in 2004 'to promote international cooperation and collaboration in support of local, national and international spatial data infrastructure developments that will allow nations to better address social, economic and environmental issues of pressing importance' (www.gsdi.org). Its membership includes organisations of all kinds in both the public and private sectors as well as not for profit organisations and academia from all parts of the world.

The third main strength of the SDI concept is the way it has exploited recent developments in location-based services and the Internet and the World Wide Web. The importance of the latter was recognised by the US Mapping Sciences Committee in their report on Distributed Geolibraries (National Research Council 1999, 31). In their view, 'the WWW has added a new and radically different dimension to its earlier conception of the NSDI, one that is much more user oriented, much more effective in maximising the added value of the nation's geoinformation assets, and much more cost effective as a data dissemination mechanism.'

Weaknesses

Each of the strengths referred to above also brings with it its weaknesses. SDIs can only facilitate access to a wide range of users if radical changes take place in existing organisational cultures. To be effective SDIs require data sharing on an unprecedented scale. Some indication of the nature of the barriers that must be overcome is given in Uta Wehn de Montalvo's (2003) study of spatial data sharing perceptions and practices in South Africa from a social psychological perspective. This study utilises the theory of planned behaviour. This theory suggests that personal and organisational willingness to share data depends on attitudes to data sharing, social pressures to engage or not engage and perceived control over data sharing activities of key individuals within organisations. The findings of her analysis generally show that there was only a relatively limited commitment amongst those involved to promote data sharing in high profile initiatives such as the South African national SDI.

Similarly, the extent to which SDIs straddle professional and administrative sectoral boundaries may lead to problems in building up and maintaining a consensus among the stakeholders involved over time. The old adage that Rome wasn't built in a day is equally applicable to SDIs. The creation of SDIs is a long-term process that may take years or even decades in some cases before they will be fully operational. Such a process is also dependent on sustaining political support and commitment for such initiatives. This is likely to present particular problems in some less developed countries where financial and human resources are scarce and governments may be politically unstable.

Opportunities

The most important opportunity is the growing public awareness of the potential for SDI development in an Information Society. This can be seen from the agenda for the UN World Summit on the Information Society in Tunis in November 2005. Key factors underlying this Summit are the extent to which the Digital Revolution is changing the ways people think, communicate and earn their livelihood and the need to bridge the digital divide between rich and poor both between and within countries (www.itu.int/wsis/). Because of the degree to which a large proportion of all data is geo referenced SDIs are likely to play a major role in the achievement of the Millennium Development Goals that the UN has set itself for improving the living standards of millions of people throughout the world.

Another important opportunity for SDIs arises from the growing pressure to make public sector information more readily available for reuse by the private sector. The rationale behind these debates in Europe is the recognition that the public sector is the largest single producer of information in Europe and that the social and economic potential of this resource has yet to be tapped. Geographic information held by public sector organisations has considerable potential for the development of digital products and services. With this in mind EU Directive 2003/98/EC sets out a framework for the conditions governing the re-use of public sector information (CEC 2003). The adoption of this Directive has important implications for the future development of the geographic information field in

Europe because the measures that it contains are mandatory on all 25 national member states which had until July 2005 to incorporate them in their respective national legislation.

Threats

As was the case with respect to the strengths and weaknesses the opportunities created by the Information Society and public sector information debates also bring with them threats. There is already some concern that the GI/SDI sector will be swallowed up by these broader debates and lose its identity in the process. As a result, some of the special qualities of geographic information may not be adequately considered in future applications. These include the questions such as those associated with transforming 3D information relating to the globe into two dimensions for display and analysis, the need to be able to deal with multiple representations of the same data at scales varying from 1:100,000 to 1:500, and the voluminous sizes of geographic databases which can easily exceed one terabyte in size.

12.4 Key Issues for the Future

In the light of the preceding analysis four sets of issues can be identified that will play a vital role in determining the future success of SDIs. These are listed below in order of priority.

12.4.1 Creating Appropriate SDI Governance Structures

Top priority must be given to the creation of appropriate SDI govern-ance structures which are both understood and accepted by all the stakeholders. This is a daunting task given the number of organisations that are likely to be involved. In the US, for example, there are more than 100,000 organisations engaged in SDI related GIS activities. Under these circumstances it will not be possible in most cases to bring all the stakeholders together for decision making purposes but structures must be devised that keep all of them informed and give them an opportu-nity to have their opinions heard. The simplest solution to this problem is to create hierarchical structures at the national, state and local level for this purpose. As noted above, this kind of structure is already operational to some extent in Australia and is implicit in the proposals for a fifty states initiative in the US.

It also important in this respect that such governance structures should as inclusive as possible from the outset of an SDI initiative so that all those involved can develop a shared vision and feel a sense of common ownership of an SDI. Otherwise it may be difficult or even impossible to bring new participants into an SDI initiative at a later stage. This is likely to be a challenging task that may slow down the progress of the work in the short term but building up a base for future collaboration is an essential prerequisite for the long-term success of the SDI.

12.4.2 Facilitating Access

Facilitating access is the second highest priority after developing appro-priate governance structures because one of the biggest problems faced by users is the lack of information about information sources that might be relevant to their needs. Consequently, without appropriate metadata ser-vices which help them to find this information it is unlikely that an SDI will be able to achieve its overarching objective of promoting greater use of geographic information. There is also a very practical reason why the development of metadata services should be given a high priority in the implementation of an SDI. This is because they can be developed relatively quickly and at a relatively low cost. In this respect they can be regarded as a potential quick winner which demonstrates tangible benefits for those involved in SDI development.

The establishment of Web-based metadata services that provide infor-mation to users about the data that is available to meet their needs is also one of the most obvious SDI success stories. The US Federal Geographic Data Committee Clearinghouse Registry, for example, lists nearly 300 registered users from the all over the world (http://registry.gsdi.org/server/status). In recent years the development of spatial portals has opened up new possibilities for metadata and application services (Tang and Selwood 2005). As their name suggests, spatial portals can be seen as gateways to geographic information (GI) resources. As such they provide points of entry to SDIs and help users around the world to find and connect to many rich GI resources. These portals also allow GI users and providers to share content and create consensus.

12.4.3 Building Capacity

Capacity building is the next priority because SDIs are likely to be most successful in maximising the use that is made of local, national and global geographic information assets in situations where the capacity exists to exploit their potential. It must also be recognised that the creation and maintenance of a SDI is also a process of organisational change management. Consequently, there is a need for capacity building initiatives to be developed in parallel to the processes of SDI development. This is particularly important in less developed countries where the implementation of SDI initiatives is often dependent on a limited number of staff with the necessary geographic information management skills. However, although much of the recent SDI discussion justifiably focuses on the need to need to devote considerable resources to capacity building in less developed countries (Stevens et al. 2004), it must also be recognised that there is still a great deal to done to develop GIS capabilities, particularly at the local level, in many more developed countries if the potential of an SDI is to be exploited to the full.

12.4.4 Making Data Interoperable

It may come as something of a surprise to find that matters relating to data interoperability come last in terms of priority for future SDI development. This is because the development and implementation of SDIs involves much more than database creation. This is clearly evident from the preceding discussion. It should also be noted that the potential for making data interoperable is heavily dependent on the specific institutional context of each country.

In countries where large scale topographic data sets are incomplete the creation of a national digital topographic database is also likely to be an expensive task that takes place over a relatively long period of time. In the meantime, those involved in SDI development must exploit alternative information sources such as remotely sensed data in addition to conventional survey technology. A great deal can be done in this way without incurring the delays that are inevitably associated with conventional data base creation.

12.5 The Future of SDIs in 2018

Given its title it is interesting to reflect on what has happened since the original text was written in 2005. First of all, it is worth noting that the four key concepts underlying SDIs have not fundamentally changed since 2005: i.e. the overriding importance of promoting ready access to a wide range of data sets, the need for concerted action on the part of governments, the extent to which SDs must be user driven and the wide range of activities that must be undertaken during the implementation process.

Second, three more milestones must be added to the 2005 list. The first of these was the establishment of an intergovernmental Group on Earth Observations in February 2005 to implement a Global Earth Observation System of Systems (GEOSS) to better integrate observing systems and share data by connecting existing infrastructures using common standards (www.earthobservations.org). In 2018 there were more than 400 million open data resources in GEOSS from more than 150 national and regional providers such as NASA and ESA; international organisations such as WMO and the commercial sector such as Digital Globe.

The second was the launch of the first scholarly journal in the spatial data infrastructure field in 2006. The International Journal of SDI Research is a peer reviewed journal that is operated by the Joint Research Centre of the European Commission and published exclusively online. The aim of the Journal is to further the scientific endeavour that lies behind the development, implementation and use of spatial data infrastructures.

The third milestone is the Directive 2007/2/EC of the European Parliament and of the Council of 14 March 2007 establishing an Infrastructure for Spatial Information in the European Community (INSPIRE) This aims to establish a Spatial Data Infrastructure for improving environmental data management in the European Community by 2020 (Commission of the European Communities 2007). The basic objective of this ambitious initiative is to make harmonised high-quality geographic information readily available to support environmental policies along with policies or activities which may have an impact on the environment in Europe. It is a legally mandated programme managed by the EC's Environment Directorate General together with its Joint Research Centre and the European Environment Agency which brings the 28 Member States together to build an SDI based on 34 related data themes. INSPIRE is the first multinational SDI initiative. The issues involved in its implementation will be considered in chapter 19 of this volume.

The last milestone dates from 2011 when the United Nations Committee of Experts on Global Geospatial Information Management (UN-GGIM) was established in July 2011 (ECOSOC Resolution 2011/24) as the official United Nations consultative mechanism on global geospatial information management. Its primary objective is to provide a forum for coordination and dialogue among Member States, and between Member States and relevant international organisations, and to propose work-plans with a view to promoting global frameworks, common principles, policies, guidelines and standards for the interoperability and inter-changeability of geospatial data and services.

With respect to the section on GIS diffusion there are lots of examples that could be used to fill out the 2005 text. One of the most interesting of these is the Global Geospatial Industry Outlook report that was produced by Geospatial Media and Communications (2017). This presents the findings of the analysis in the form of a geospatial readiness index of the fifty nations that account for 75% of the world's population and 89% of the global GDP.

A novel feature of the report is the methodology that was developed to calculate the geospatial readiness index for the 50 selected countries. This is based on four separate policy areas or pillars: geospatial infrastructure and policy framework, institutional capacity, user adoption level and industrial capacity. Each of these four topics is also broken down into four or five sub categories. For example, the first topic is subdivided into data infrastructure, positioning infrastructure, platforms and portals, open and linked data and standards, and policy framework while the second considers business incubation, fundamental sciences, professional education, interdisciplinary application courses and vocational training. To calculate the scores, each parameter within the four pillars comprises a series of five sub-pillars that are each converted to a score of 0–5 and summed up to generate pillar scores of 25% each.

> Once the scores of the four pillars are derived, they are summed to map the overall readiness of the ranking. It is to be noted that the weights are based on our assessment of the relative importance of each dataset, parameters and pillars to determine the readiness index.
>
> *(pp. 17–18)*

The findings of the analysis give a good overview of the current state of the art throughout the world. Notwithstanding this, the detailed findings of the analysis are very interesting both in terms of the sub pillar scores and the aggregate scores. For example, the scores for the fifty nations with respect to the data infrastructure (topographic and earth observation) sub pillar show that that developed countries such as the US, the Netherlands, Germany, Switzerland and, somewhat surprisingly, Russia, are the leaders while most of the developing countries like India Thailand and South Africa are in the either the middle rank, or as is the case in Nigeria, Kenya and, even more surprisingly, Brazil, at an early stage of development.

The aggregate scores for all 22 pillars show that the US holds the top position in the global geospatial readiness index because the country 'has it all – an efficient geospatial infrastructure, an enabling policy framework, an excellent institutional capacity, the strongest industry capacity and an in-depth user adoption across all industry verticals' (p. 112) followed by the United Kingdom, the Netherlands and Canada. Some developing countries such as South Africa are ranked close to the top of the lower half of the table although it 'has been progressive in the use of geospatial technology by developing and adopts latest technologies and formulating South Africa's frameworks and policies to enhance the uptake of spatial information in the country' (p. 113), while countries such as Bangladesh are still at an early stage of development.

The topics discussed in the first part of section 3 of the original paper will be discussed in more detail in the next two chapters but it is interesting to reconsider the general results of the SWOTS analysis. Given that the SWOTS methodology is essentially intuitive in nature, it is useful to begin the discussion by comparing two SWOTS exercises that were carried out in very different circumstances. The Netherlands is described by its authors as 'a geodata rich country' which has 'commanded a strong position in the field of geo-information from time immemorial'. (Bregt et al. 2008, p. 364). It scored highly in the Readiness index described above. In contrast, Mongolia (Munaa et al. 2016, p. 4) only 'began discussing forming their SDI in 2004' and 'has begun pushing steadily to establish NSDI with the purpose of spreading spatial use to environment and the socio-economic branches of government'. Mongolia was not even included in the Readiness index described above.

The Dutch SWOTS, reflects circumstances in the Netherlands around 2003, and sees its strengths as being based on 'strong networks of parties who work collaboratively and exchange knowledge on geo-information' despite a weakness that 'the dissemination of geo-data is very supply oriented' and that many organisations are 'reticent about making their data available'. Although the Mongolian SWOTS states that SDI 'has the attention and focus of the government' under the heading of strengths, it also lists under the heading of weaknesses, 'the lack of knowledge on NSDI – officials and experts have poor understanding of NSDI and use of GIS within the limits of their job, as well as insufficient budget and lack of software standards and training facilities'.

It is generally accepted that the strengths and weaknesses section of a SWOTS analysis are frequently related to the internal circumstances of the organisation whereas the opportunities and threats section commonly focusses on the external circumstances that surround them. With this in mind, it is interesting to note that the Mongolian SWOTS is optimistic about the external opportunities provided by SDI development to create a new labour market in Mongolia, as well as to learn from other nations experiences while also taking advantage of the rapidly growing number of internet users in the country. However, the authors also recognise that these benefits may come at a high cost for data collection and may threaten information privacy. Similarly, the Dutch SWOTS welcomes the 'ambition of the Dutch government to make information available for innovative purposes' it notes that it lacks the ability to take advantage of international social issues. It is also felt that the old coordination approach is not enough and that more powerful steering is needed.

From this brief summary of some of the findings of these two studies there are a number of interesting similarities between despite the fact that the Dutch study was based on thinking in around 2003 rather than the present time. Given these reservations, it should be noted that the differences between the two countries in the terms of both strengths and weaknesses is considerable whereas the comments on opportunities and threats is much more similar. This suggests that the view of the authors of the two studies were less different when it came to external circumstances that were less familiar to them than those with respect to internal circumstances surrounding the organisation.

Turning to the general results of the SWOTS analysis, it should be noted that, with respect to the strengths section of the paper, the most recent UN GGIM report on Future trends *in geospatial information management: the five to ten year vision* (UN GGIM 2015, p.6) places great emphasis on the role of geographic information for monitoring the 2030 Agenda for Sustainable Development that is being established by the United Nations. It argues:

> Many of the sustainable development challenges are cross-cutting in nature and are characterised by complex inter-linkages which will benefit from using location as a common reference framework. To effectively measure, monitor and mitigate challenges we need to link demographic, statistical and environmental data together with the one thing they have in common – geospatial data.

It should also be noted that the three main strengths listed in the above paper been reinforced by developments over the last 10 years which have led to the emergence pf the geospatial industry worth nearly $500 billion according to the Global Geospatial Industry Outlook report (Geospatial Media and Communications 2017, p. 19). These developments are also supporting the moves that are taking place in all parts of the world towards the creation of a spatially enabled society which promotes the development of effective SDIs in such a way that they will support the vast majority of society, who are not necessarily spatially aware, in a transparent manner.

With respect to weaknesses, it should be noted again that many of these strengths bring with them associated weaknesses. Another weakness must be linked to the length of time that it takes to implement SDIs in practice.

This makes them particularly vulnerable to changes in administrative structures in government and changing financial circumstances. This is particularly a problem in countries that have relied on external project funding to get their SDIs off the ground.

As for opportunities, as noted under the heading of strengths, there are lots of emerging technologies and potential innovations that may be considered as opportunities for the further expansion of SDIs in the future. For example, the 2026 Australian Spatial Industry Transformation and Growth Agenda (2016) see its strengths as follows:

> Nowadays, location intelligence is becoming ubiquitous, which is redefining the sector and its players: the potential for spatial information technologies goes well beyond a location and a map. In order to sustain a leadership position and to increase the sector's growth both domestically and globally, those willing to explore location- based opportunities will have to become innovative, to be prepared for implement reform and to do so collaboratively in nearly every part of the economy and across most of the community.
>
> *(2016, p. 5)*

The main threats to SDIs are still similar to those listed in the 2015 but there also problems with the mass of data that is becoming available as a result of the increasing use of air borne drones and sensor technologies for data collection purposes as well as the growing volume of information that is emerging as a result of crowd sourced volunteered geographic information (see, for example, Coleman et al. 2009). The emergence of big data presents a new range of threats to personal privacy and there are also the associated risks with what has become known as fake data. At the same time, the increase in eGovernment programmes in government has resulted in the inclusion of a great deal of government held geographic information into more generally defined and not necessarily spatially referenced resources such as the data.gov initiative in the UK and the Dutch basic registers programme as well as the European Union digital agenda programme. These threats are likely to challenge the future development of geospatial SDIs.

References

2026 Spatial Industry Transformation and Growth Agenda, 2016. 2026 Agenda: Ideas, 2026 Agenda

Australia and New Zealand Land Information Council (ANZLIC), 1992. *Land information management in Australasia 1990–1992*, Canberra: Australia Government Publishing Service.

Australia and New Zealand Land Information Council (ANZLIC), 2004. *Position paper on engagement with the spatial information industry*, Canberra: ANZLIC, www.anzlic.org.au (last accessed 16 February 2019).

Bregt, A., L. Grus, J. Crompvoets, W.T. Castelein and J. Meerkerk, 2008. Changing demands for Spatial Data Infrastructure assessment: Experience from the Netherlands, in J. Crompvoets, A. Rajabifard, B. van Loenen and T. Delgado Fernandez (eds.), *A multi view framework to assess spatial data infrastructures*, Wageningen University: Centre for Space Information (RGI), 357-369

Burrough, P., M. Brand, F. Salge and K. Schueller, 1993. The EUROGI vision, *GIS Europe*, 2, 3, 30–31.

Coleman, D. J., Y. Georgiadou and J. Labonte, 2009. Volunteered geographic information: The mature and motivation of producers, *International Journal of Spatial Data Infrastructures Research*, 4, 332–356.

Commission of the European Communities (CEC), 2003. The reuse of public sector information, Directive 2003/98/EC of the European Parliament and of the Council, *Official Journal of the European Union*, L345, 90–96.

Commission of the European Communities (CEC), 2004. *Proposal for a Directive of the European Parliament and the Council establishing an infrastructure for spatial information in the Community (INSPIRE)*, COM (2004) 516 final, Brussels: Commission of the European Communities.

Commission of the European Communities (CEC), 2007. Directive 2007/2/EC of the European Parliament and of the Council of 14 March 2007 establishing an Infrastructure for Spatial Information in the European Community (INSPIRE), *Official Journal of the European Union*, L108, 1–14.

Craglia, M., A. Annoni, M. Klopfer, C. Corbin, L. Hecht, G. Pichler and P. Smits, (eds.), 2003. *Geographic information in the wider Europe*, Sheffield: University of Sheffield.

Craglia, M. and I. Masser, 2002. GI and the enlargement of the European Union, *URISA Journal*, 14, 2, 43–52.

Crompvoets, J. and A. Bregt, 2003. World status of national spatial data clearinghouses, *URISA Journal*, 15, 1, 43–50.

de Montalvo, U.W., 2003. *Mapping the determinants of spatial data sharing*, Aldershot: Ashgate.

Department of Science and Technology (DOST), 2002. *NSDI strategy and action plan December 2002*, New Delhi: Department of Science and Technology, also available at www.nsdiindia.org (last accessed 16 February 2019).

Department of the Environment, 1987. *Handling geographic information: Report of the Committee of Enquiry Chaired by Lord Chorley*, London: HMSO.

Executive Office of the President, 1994. Coordinating geographic data acquisition and access, the National Spatial Data Infrastructure, Executive Order 12906, *Federal Register*, 59, 17671–17674.

Federal Geographic Data Committee (FGDC), 2005. *Future directions – Governance of the national spatial data infrastructure: Final draft report of the NSDI Future Directions Governance Action Team*, Reston VA: Federal Geographic Data Committee, www. fgdc.gov (last accessed 16 February 2019).

Federal Geographic Data Committee (FGDC) and National States Geographic Information Council (NSGIC), 2005. *Fifty states and equivalent entities involved and contributing to the NSDI*, www.fgdc.gov (last accessed 16 February 2019).

Geospatial Media and Communications, 2017. *Global Geospatial Industry Outlook*, Noida, India: Geospatial Media and Communications.

Hyman, G., C. Perea, D. I. Rey and K. Lance, 2003. *Survey of the development of national spatial data infrastructures in Latin, America and the Caribbean*, International Centre for Tropical Agriculture and Augustin Codazzi Geographic Institute

Lance, K., 2003. Spatial data infrastructure in Africa: Spotting the elephant behind trees, *GIS Development*, 7, 7, 35–41.

Masser, I., 1999. All shapes and sizes: The first generation of national spatial data infrastructures, *International Journal of Geographical Information Science*, 13, 67–84.

Masser, I., 2005. *GIS worlds: Creating spatial data infrastructures*, Redlands: ESRI Press.

Munaa, T., B. Chogsom and B. Tumurshukh, 2016. SWOT, PEST and 5C analysis of the Mongolian national SDI, Paper presented at the 15th Global Spatial Data Infrastructure Conference, Taipei.

National Research Council, 1993. *Toward a coordinated spatial data infrastructure for the nation*, Mapping Science Committee, Washington, DC: National Academy Press.

National Research Council, 1999. *Distributed geolibraries: Spatial information resources*, Mapping Science Committee, National Research Council, Washington, DC: National Academy Press.

National Research Council, 2002. *Down to earth: Geographic information for sustainable development in Africa*, Committee on the Geographic Foundation for Agenda 21, National Research Council, Washington, DC: National Academy Press.

Rajabifard, A., M. E. Feeney, I. Williamson and I. Masser, 2003. National spatial data infrastructures, in I. Williamson, A. Rajabifard and M. E. Feeney (eds.), *Development of spatial data infrastructures: From concept to reality*, London: Taylor and Francis, 95-109

Rajabifard, A. and I. Williamson, 2000. *Report on analysis of regional fundamental data sets questionnaire*, Melbourne: Department of Geomatics University of Melbourne.

Rajabifard, A. and I. Williamson, 2003. Asia-Pacific region and SDI activities, *GISdevelopment*, 7, 7, 30–34.

Spatial Applications Division, Catholic University of Leuven (SAD), 2003. *Spatial data infrastructures in Europe: State of play during 2003*, Catholic: University of Leuven. http://inspire.ec.europa.eu (last accessed 16 February 2019).

Stevens, A., K. Thackrey and K. Lance, 2004. Global Spatial Data Infrastructure (GSDI): Finding and providing tools to facilitate capacity building, *Proceedings 7th GSDI Conference*, Bangalore, India, 2–6 February 2004

Tang, W. and J. Selwood, 2005. *Spatial portals: Gateways to geographic information*, Redlands: ESRI Press.

United Nations Initiative on Global Geospatial Information Management (UN-GGIM), 2015. *Future trends in geospatial information management: The five to ten year vision*, Second edition, Southampton: Ordnance Survey, http://ggim.un.org/docs/Future-trends.pdf (last accessed 16 February 2019).

Weissbord, M. and S. Janoff, 2000. *Future search: An action guide to finding common ground in organisations and communities*, Second edition, San Francisco: Berrett-Koehler.

13

What's Special about SDI-Related Research?

Ian Masser

13.1 Introduction

From my own perspective as a social scientist with a longstanding interest in the political and institutional aspects of spatial data infrastructures, I can easily think of four very good reasons why SDI-related research needs special attention. First, there is a need to study the processes of SDI diffusion and the social networks that facilitate them both between and within countries. Such studies not only map the spread of the SDI concept but also indicate the ways in which it has to be adapted to meet the demands of the national or local political and institutional context. For this reason, there is a strong comparative dimension to SDI diffusion research. Second, given that the development of SDIs is a process rather than a product it will also be necessary to closely monitor the evolution of SDIs over time. The outcomes of such research will highlight changes in the political and institutional environment within which SDIs are developed and indicate the long-term viability of the SDI concept in different environments. Third, given the importance attached to data sharing for successful SDI development, it will be necessary to explore the motivations of those involved and also explore ways of facilitating the exchange and integration of geospatial information from a variety of sources. Finally, it must be recognised that SDI implementation takes place in a multi-level environment and that it will be necessary to study hierarchies of SDIs with very large numbers of stakeholders. This research theme also raises important questions about the kind of models that are required for effective SDI governance.

SDI-related research in each of these four areas should take advantage of related work in the social sciences. Research on SDI diffusion, for example, can draw upon the conceptual frameworks that have been developed in sociology and communications research. SDI evolution studies make use of the findings of research on management information systems development while the concepts such as actor network theory and of the theory of planned behaviour from social psychology provide useful insights into the circumstances that facilitate or inhibit data sharing. Similarly, hierarchical reasoning is a useful tool for understanding the forces underlying the nature of SDI implementation. What's special about this kind of SDI-related research, then, is the insights and understanding of the processes involved in SDI development that can be gained from examining critically with the help of conceptual frameworks and models that have been rigorously tested in other fields.

13.2 SDI Diffusion

It has been argued that more than half the world's countries claim that they are involved in some form of SDI development (Crompvoets et al. 2004), but these claims need to be treated with some caution until they have been backed up by factual evidence as it is likely that there is an element of wishful thinking in some of them. The findings of the state of play studies carried out the Spatial Applications Division of the University of Leuven for the European Commission (SAD 2005), for example, suggest that only a handful of European countries have anything like a full blown SDI and most of these initiatives can better be described as 'SDI like or SDI supporting initiatives'. Furthermore, the fact that some countries have reported that they are engaged in some aspect of SDI development does not necessarily mean that this will translate into a fully operational SDI over time.

The most obvious SDI success story is in the establishment of clearinghouses and portals to disseminate metadata. The US Federal Geographic Data Committee's Clearinghouse registry, for example, lists nearly 300 registered nodes from the US and other countries. Similarly, David Maguire and Paul Longley (2005) claim that there are on average 5622 user visits per week to the US Geospatial One-Stop portal site in April 2004. However, it is not enough to report that clearinghouses have been established without including some information on their usage and the arrangements that have been made for their continuing upgrading and maintenance. For example, the findings of a number of surveys carried out at different points in time by Joep Crompvoets and his colleagues (2004) suggest that the use that is being made of some of these metadata services may be declining over time rather than increasing and that this is due to unsatisfactory arrangements for continuing site management.

While considering the extent to which the diffusion of innovations model is an appropriate one for the study of SDI diffusion it is worth noting that SDIs generally fit the definition of an innovation as 'an idea, practice, or object that is perceived as new by an individual or unit of adoption' (Rogers 1995, 11). However, while the characteristics of the innovators, early adopters and early majority of SDIs show most of the features described by Everett Rogers, it will be interesting to see whether this is the case for the late majority and laggards.

It should also be noted that the diffusion of innovations model has been criticised for its pro innovation bias (Rogers 1993). This can be seen in the statements that are made in connection with SDI development which constantly stress its positive impacts in terms of promoting economic growth, better government and improved environmental sustainability. These and other similar claims need to be rigorously examined in further research.

In the process more attention should be given identifying possible negative impacts arising out of SDI development. A useful example of this kind of work can be found in the four brave new GIS worlds scenarios that were developed by Michael Wegener and Ian Masser (1996). Their trend, market, big brother and beyond GIS scenarios are easily translatable into the SDI field as can be seen from the Mapping Science Committee's Future of spatial data and society project (National Research Council 1997).

It is not always easy to define with any precision the moment in time when the idea was adopted. In some cases, there is a gradual transition from existing practices into an SDI. This happened, for example, in Australia and Finland where there was a tradition of SDI like thinking before the SDI itself formally came into being. In other cases, the position is complicated by differences in the terminology that is used to describe SDI like activities. When, for example, does a national GI system become an SDI?

Some of these problems can be resolved by developing more systematic ways for describing and classifying SDIs. The typology developed by the Leuven group (SAD 2005) as a result of their EU wide state of play study is a step in the right direction, even though, in its current version, it gives rise to some ambiguities and overlaps in practice. However, this typology only takes account of the approach that has been adopted towards coordination and it may be worthwhile extending it to incorporate other variables.

Another matter that needs consideration in future SDI diffusion research is the extent to which cultural factors are likely to influence SDI adoption. An interesting example of this kind of research is Erik de Man's (2006) analysis of the role of culture in SDI development. This draws upon the four dimensional model developed by Geert Hofstede (1997). As a result of extensive empirical research Hofstede found that national cultures varied with respect to four main variables: power distance (from small to large), uncertainty avoidance (from weak to strong), masculinity versus femininity and collectivism versus individualism. In an SDI environment, De Man argues that cultures where there are large power distances are likely to use SDI to reinforce the hand of management whereas those with small power distances will welcome their data sharing and accountability properties. Similarly, masculine cultures will be interested in SDIs because of their capacity to contribute to the visible achievements whereas feminine cultures will welcome their networking and relationship building properties.

13.3 SDI Evolution

The old adage that Rome wasn't built in a day is equally applicable to SDIs. The creation of SDIs is a long-term task that may take years or even decades in some cases before they are fully operational. This process is likely to be an evolving one that will also reflect the extent to which the organisations that are involved reinvent themselves over time. Everett Rogers (1995, 16–17) defines reinvention as 'the degree to which an innovation is changed or modified by a user, in the process of its adoption and implementation'. He also notes that, while some innovations are difficult or impossible to reinvent, others are 'more flexible in nature, and they are reinvented by many adopters who implement them in different ways'. The degree of reinvention involved in GIS implementation in British local government led Heather Campbell and Ian Masser (1995, 109–110) to conclude:

> The meaning of a technology such as GIS was constantly being reinvented at both the organisational and individual scales. This has important implications for studies of diffusion as it would appear that innovations such as GIS embrace a wide range of perceptions. These differences in emphasis are likely to lead to tensions and problems which will complicate the implementation process. It is also likely that such systems will be used to undertake activities not originally anticipated by their inventors.

There are clear parallels between these findings and SDI evolution. Given that SDI implementation is likely to take place over a long period of time when the technologies are also changing, together with the external political and institutional circumstances that surround an SDI, it may be necessary to distinguish between two levels of reinvention in this case. The first of these concerns the processes needed to initially adapt the notion of an SDI to the local or national context to take account, for example, of the impacts that the allocation of administrative responsibilities and the style of government will have on the form of SDI development in each case. The second relates to processes that are involved in its evolution over time in response to changing political, institutional and technological circumstances.

Given the extent to which SDIs can be expected to change over time, it will be necessary to set up research procedures to ensure that their progress is systematically monitored. Longitudinal studies will form an important part of this research strategy. To facilitate research of this kind, it will also be necessary to ensure that key documents are not lost when they become out of date. This is particularly a problem in SDI research which relies to a great extent on grey literature in the form of unpublished reports and memoranda. As a result, it is heavily dependent on materials obtained from web sites that are also changing constantly over time. This is already a matter of concern in some countries, and the author's analysis of the UK NGDF experience (Masser 2005, Chapter 4) was made more difficult by the fact that many key documents are no longer readily available following the closure of the NGDF website in 2001.

Yola Georgiadou and her colleagues (2005) also argue that more attention needs to be given to the infrastructure dimension of SDIs in future research. Their work on the Indian SDI makes use of three key concepts identified in previous work on information infrastructures: the installed base, reflexive standardisation and a cultivation approach to design. The concept of the installed base and its associated lock in effects describes the extent to which the existing structures influence the design of SDIs. Reflexive standardisation refers to the self-reinforcing mechanisms involved in the standardisation process whereby the adoption of standards raises the need for further standards as more users and technologies are incorporated into the network. The cultivation approach to SDI design emphasises the improvisational dimensions of SDI design. In this respect, SDI design is not seen as a well-defined process with clearly identifiable start and end states but rather as a process of ecological change reflecting the designer's inability to fully anticipate future events.

It is also worth noting that several studies have recently been undertaken to assess SDI readiness and maturity. For example, Tatiana Delgado and her colleagues (2005) proposed an SDI readiness index combines organisational factors, information awareness and access networks together with human and financial resources. The model used to assess SDI readiness in Cuba is based on fuzzy logic given the qualitative nature of most of the variables. Similarly, the SDI maturity matrices devised by Bastiaan Van Loenen (2006) identify four stages through which organisations develop from a standalone solution into a networked SDI structure.

13.4 Data Sharing in SDIs

Data sharing featured prominently in the initial discussions about SDIs. The US Mapping Science Committee's landmark report, 'Towards a coordinated spatial data infrastructure for the nation', devoted a whole chapter to the sharing of spatial data. The rationale for a spatial data sharing programme is

> to increase benefits to society arising from the availability of spatial data. The benefits will accrue through the reduction of duplication of effort in collecting and maintaining of spatial data as well as through the increased use of this potentially valuable information.
>
> *(National Research Council 1993, 89)*

It also argued that this programme 'must do more than just disseminate spatial data collected by federal agencies. The richness and utility of the program is substantially enhanced by having participation of donors from state and local governments, academic, and the private sector' (p. 104). With this in mind the Committee recommended that the 'FGDC should establish a data sharing committee with the objective of providing the policy making and leadership to launch, maintain, and operate the proposed program' (p. 104).

These proposals do not fully take account of the complexity of data sharing in practice. The intricate nature of the relationships involved in organisational and inter organisational data sharing and the legal, economic, cultural and personal-privacy-related issues associated with these activities were highlighted in the report of an expert meeting convened by the US National Centre for Geographic Information and Analysis. The editors of this report, Harlan Onsrud and Gerard Rushton (1995), define the issues involved in the following terms: 'Sharing of geographic information involves more than a simple data exchange. To facilitate sharing, the GIS research and user communities must deal with both the technical and institutional aspects of collecting, structuring, analysing, presenting, disseminating, integrating and maintaining spatial data.'

Uta Wehn de Montalvo (2003) has subsequently explored spatial data sharing perceptions and practices in South Africa in some depth from a social psychological perspective. This study utilises the theory of planned behaviour. This theory suggests that personal and organisational willingness to share data depends on attitudes to data sharing, social pressures to engage or not engage and perceived control over data sharing activities of key individuals within organisations. The findings of her quantitative analysis generally bear out the relationships postulated in this theory and give valuable insights into the factors that determine the willingness to share spatial data. They also show that there was only a relatively limited commitment amongst those involved to promote data sharing in high profile initiatives such as the South African national SDI whose successful implementation is dependent on a high level of spatial data sharing.

For this reason, there is a pressing need for more research on the nature of data sharing in a multi-level SDI environment. The studies that have been carried out by Zorica Nedovic-Budic and Jeffery Pinto (1999) and

Nedovic-Budic et al. (2004) in the US provide a useful starting point for work in other parts of the world. The earlier study focuses mainly on the motivations for data sharing, the coordination process and the costs of coordination whereas the more recent analysis of the responses of 245 respondents to a survey questionnaire provides some interesting quantitative indicators of the interaction mechanisms involved and the motivations of the respondents. Similarly, the findings of Francis Harvey and David Tulloch's (2004) survey of local governments in the US suggest that many local authorities remain largely unaware of SDI concepts and assess the benefits of sharing information in a limited time frame with an emphasis on supporting existing administrative and political relationships. In an earlier study, Harvey (2001) also makes use of actor network theory to explore the socio-technical context of data sharing. Kevin McDougall and his colleagues' (2005) study of the experiences of local government in three Australian states describes the extent to which the technical and physical capacity of smaller jurisdictions can impact on their ability to participate in with larger and better resourced authorities. In a different vein, William Craig (2006) highlights the extent that the motivation of key individuals can influence data sharing. The findings of his research suggest that their efforts are motivated largely by idealism, enlightened self-interest and involvement in particular professional cultures.

13.5 The Hierarchy of SDIs

Some of the most challenging research questions are posed by the scale for multi-level stakeholder participation in SDI implementation. The numbers of stakeholders in large countries such as the US are massive given that more than 100,000 public bodies alone are involved in some way. This task is made even more difficult by a governance model that is based largely on consensus building and the extent to which coordination bodies such as the Federal Geographic Data Committee in the US and the Australia New Zealand Land Information Council in Australia lack the powers to enforce their strategies or to impose sanctions on unwilling participants.

It must also be recognised that the vision of a bottom up SDI associated with multi-level stakeholder participation differs markedly from the top down one that is implicit in much of the SDI literature. While the top down vision emphasises the need for standardisation and uniformity, the bottom up vision stresses the importance of diversity and heterogeneity given the very different aspirations of the various stakeholders and the resources that are at their disposal. Consequently, the challenge to those involved in SDI implementation is to find ways of ensuring some measure of standardisation and uniformity while recognising the diversity and the heterogeneity of the different stakeholders. This will involve a sustained mutual learning process on the part of all those involved in SDI implementation.

For this reason, it may be worthwhile exploring hierarchy theory in greater depth in the context of SDI development. Abbas Rajabifard (2002), for example, has made use of hierarchical reasoning in his work on SDI structures. He has also identified three properties of hierarchies that give some useful insights into these structures:

- the part—whole property which describes the degree to which higher level entities can be subdivided into lower level parts;
- the Janus effect which relates to the relationships that an element has with the levels above and below it; and
- the near decomposability property which describes the nesting of systems within larger systems and the extent to which the interactions between the different systems decrease in strength with the distance between them.

It is also important to bear in mind that different levels of the SDI hierarchy perform different tasks. The role of bodies at the continental and global levels is primarily to foster SDI development by disseminating information about current developments and best practices to the levels below them, whereas local SDIs are primarily concerned with the operational needs of day-to-day decision-making. Despite these differences, all levels of the hierarchy are involved to some extent in the dissemination of information between the various levels. National-level bodies perform a similar task with respect down to sub-national and upwards, while regional bodies and state-level bodies do the same with respect to local ones.

13.6 Conclusions

Four challenging areas for SDI-related research have been discussed in this chapter. These are SDI diffusion, SDI evolution, data sharing in SDIs and the hierarchy of SDIs. The most important conclusion to be drawn from this analysis is that SDIs must be viewed as social phenomena. Consequently, there is a continuing need for interaction between those involved in the critical study of SDIs and scholars who are familiar with mainstream social science research.

13.7 Some Recent Developments in SDI-Related Research

As might be expected in view of my own research interests, the four themes that were identified in this chapter: i.e. SDI diffusion, SDI evolution, data sharing in SDIs and the hierarchy of SDIs, are discussed at some length elsewhere in this book. Chapter 9 deals with empirical research of the diffusion of GIS in local government in Europe and its final section considers some recent developments in the field. Matters relating to SDI diffusion and evolution also feature in Chapter 7, which presents the findings of a comprehensive survey of GIS in local government in Great Britain at a critical stage of its development. Chapter 8 is exclusively devoted to data sharing and GIS implementation, while Chapter 14 presents an alternative approach to the traditional hierarchical model as a result of the shift from single level to multi-level participation, within the context of an administrative hierarchy of SDIs. As a result of these developments the coordination models that had emerged as result of research on single-level SDIs need to be substantially modified as new complex models of multi-level governance have emerged.

It should also be noted that some other developments in SDI research are discussed in other chapters of this book. The most interesting of these revolve around three main issues: the concept of SDIs as information infrastructures that was introduced in the last section of Chapter 5, the emergence of many types of SDIs around at the present time as a result of the evolution of SDIs in response to the growth of volunteered geographic information and ubiquitous mobile technologies that was raised in the final section of Chapter 14, and the growing need to consider a number of matters related to open data policies and the governance of SDIs that has emerged out of the experience that has been built up during the implementation of large-scale SDIs such as the multi-national INSPIRE Directive in the European Union that is discussed in Chapter 19.

Matters such as these bring the discussion back to the question raised by the title of this chapter, 'What's special about SDI related research?' The four answers to the questions that were given more than 10 years ago still seem to make sense, given the contents of this volume, but it is also worth considering the matter from a broader perspective of 'Is spatial still special?' Specifically, geographical statistical problems such as spatial autocorrelation and the modifiable areal unit problem, as well as the choice of map projection still generate debates among scholars and location is an important integrating factor in connection with the analysis of big data derived from many different sources. From a government perspective, Vancauwenberghe and Van Loenen (2018) also note that spatial data has become more relevant as a result of the development of spatially enabled e-government services to citizens, businesses and non-governmental organisations in the context of moves towards an open data spatial data infrastructure in some European countries. As a result of these and other related developments, Schade et al. (forthcoming) conclude that while geographic information infrastructures appear to have become much more integrated into the much wider fields of computer science and data science in recent years, the mainstreaming of GI also provides immense opportunities, such as, the increasing market for companies specialised in GI and many new job opportunities for GI experts.

A similar question can be asked about spatial data infrastructures. This has been explored at some length by de Man (2007) with reference to the literature on information infrastructures. His general conclusion is that spatial data infrastructures are not fundamentally different from other information infrastructures. Nevertheless,

> the concept of space does not differentiate between SDIs and other information infrastructures but rather between development, adoption, and application of any information infrastructure at different spatial levels. Information infrastructures are different in content, role, and degree of complexity at these different spatial levels.
>
> *(p. 47)*

References

Campbell, H. and I. Masser, 1995. *GIS and organisations: how effective are GIS in practice?* London: Taylor & Francis.

Craig, W., 2006. White knights of spatial data infrastructures: the role and motivation of key individuals, *URISA Journal*, 16(2): 5–12.

Crompvoets, J., A. Rajabifard, A. Bregt and I. Williamson, 2004. Assessing the worldwide developments of national spatial data clearinghouses, *International Journal of Geographical Information Sciences*, 18: 1–25.

de Man, E., 2006. Understanding SDI: complexity and institutionalisation, *International Journal of Geographical Information Sciences*, 20: 329–340.

de Man, E., 2007. Are spatial data infrastructures special? In H. Onsrud (Ed.) *Research and theory in advancing spatial data infrastructure concepts*, Redlands CA, ESRI Press, pp. 33–54.

Delgado, T., K. Lance, M. Buck and H. J. Onsrud, 2005. Assessing an SDI readiness index. In *Proceedings GSDI 8,* Cairo, Egypt.

Georgiadou, Y., S. Puri and S. Sahay, 2005. Towards a research agenda to guide the implementation of spatial data infrastructures: a case study from India, *International Journal of Geographical Information Sciences*, 19: 1113–1130.

Harvey, F., 2001. Constructing GIS: actor networks of collaboration, *URISA Journal*, 13(1): 29–37.

Harvey, F., and D. Tulloch, 2004. How do governments share and coordinate geographic information issues in the United States, *Proceedings 10th EC & GIS workshop*, Warsaw, Poland.

Hofstede, G., 1997. *Cultures and organisations: software of the mind*, Beverley Hills: Sage Publications.

Maguire, D. and P. A. Longley, 2005. The emergence of geoportals and their role in spatial data infrastructures, *Computers Environment and Urban Systems*, 29: 3–14.

Masser, I., 2005. *GIS worlds: creating spatial data infrastructures*, Redlands: ESRI Press.

McDougall, K., A. Rajabifard and I. Williamson, 2005. Understanding the motivations and capacity for SDI development from the local level, *Proceedings GSDI 8*, Cairo, Egypt.

National Research Council, 1993. *Towards a spatial data infrastructure for the nation,* Council, Washington, DC: National Academy Press.

National Research Council, 1997. *The future of spatial data and society,* Council, Washington, DC: National Academy Press.

Nedovic-Budic, Z. and J. K. Pinto, 1999. Understanding interorganisational GIS activities: a conceptual framework, *URISA Journal,* 11(1): 53–64.

Nedovic-Budic, Z., J. K. Pinto and L. Warnecke, 2004. GIS database development and exchange: interaction mechanisms and motivations, *URISA Journal,* 16(1): 15–29.

Onsrud, H. J. and G. Rushton, (eds), 1995. Sharing Geographic Information, New Brunswick, NJ: Rutgers University.

Rajabifard, A., 2002. Diffusion for Spatial Data Infrastructures: particular reference to Asia and the Pacific, PhD dissertation, University of Melbourne.

Rogers, E., 1993. The diffusion of innovations model. In Masser I. and H. J. Onsrud (Eds.) *Diffusion and use of geographic information technologies,* Dordrecht, Kluwer, pp. 9–24.

Rogers, E., 1995. *Diffusion of innovations,* Fourth edition, New York: Free Press.

Schade, S., C. Granell, G. Vancauwenberghe, C. Kessler, D. Vandenbroucke, I. Masser, and M. Gould, forthcoming. Geospatial information infrastructures. In H., Guo, M., Goodchild and A., Annoni (Eds.) *Manual of digital earth,* Berlin: Springer.

Spatial Applications Division, Catholic University of Leuven, 2005. *Spatial data infrastructures in Europe: state of play spring 2005,* Ispra: EC Joint Research Centre.

Van Loenen, B., 2006. *Developing geographic information infrastructures: the role of information policies,* Delft: Delft University Press.

Vancauwenberghe, G., and B. Van Loenen, 2018. Exploring the emergence of open spatial data infrastructures: analysis of recent developments and trends in Europe. In S. Saeed, T. Ramayah, and Z. Mahmood (Eds.) *User centric e-Government: challenges and opportunities,* Berlin: Springer.

Wegener, M. and I. Masser, 1996. Brave new GIS worlds. In Masser I., Craglia M. and H. Campbell (Eds.) *GIS diffusion: the adoption and use of geographical information systems in local government in Europe,* London, Taylor and Francis, pp. 9–21.

Wehn de Montalvo, U., 2003. *Mapping the determinants of spatial data sharing,* Aldershot: Ashgate.

14

Changing Notions of a Spatial Data Infrastructure

Ian Masser

14.1 Introduction

The term 'National Spatial Data Infrastructure' was first used in a paper presented by John McLaughlin at the 1991 Canadian Conference on Geographic Information Systems entitled 'Towards a national spatial data infrastructure'. The ideas contained in this chapter were subsequently developed by the United States National Research Council's Mapping Science Committee in their report on 'Toward a coordinated spatial data infrastructure for the nation' (National Research Council 1993). This recommended that effective national policies, strategies and organisational structures should be established for the integration of national spatial data collection, use and distribution.

These concepts were expanded and developed during the following year in the Executive Order 12906 signed by President Clinton entitled 'Coordinating Geographic Data Acquisition and Access: The National Spatial Data Infrastructure' (Executive Office of the President 1994). The Executive Order significantly raised overall awareness of the need for governmental strategies that facilitate geospatial data collection, management and use not only among Federal agencies in the United States but also nationally and internationally (Masser 2005, Chapter 2).

Since then the number of SDI initiatives has increased dramatically in all parts of the world to the extent that Crompvoets et al.'s (2004) work on the development of clearinghouses suggests that as many as half the world's countries were considering SDI related projects. These figures must be treated with some caution as they do not necessarily imply that all these countries are actively engaged in SDI formulation or implementation. It is also likely that many of them may be engaged in some aspects of SDI development without necessarily committing themselves to a comprehensive SDI programme. Nevertheless, the term

'SDI phenomenon' seems to be a reasonable description of what has happened in this field over the last 15 years.

With these considerations in mind, this chapter examines some of the changes that have taken place in the notion of an SDI during this time. The discussion is divided into five parts beginning with technological developments and then moving on to institutional matters. The first of these considers the impacts of innovations in communications and information technology during this period on the nature of SDIs. The second examines the changes that have taken place in the conceptualisation of SDIs, while the third discusses the nature of SDI implementation with particular reference to the concepts of multi-level governance that have been developed by political scientists. Underlying a great deal of this discussion is the notion that SDI development and implementation is very much a social process of learning by doing. Some of the main features of this process are examined in the fourth section of the chapter with reference to the experience of the State of Victoria in Australia. The concluding section of the chapter considers the challenges facing SDI implementation and identifies a number of dilemmas that have yet to be resolved.

14.2 The Impact of Innovations in Information and Communications Technologies

New technologies have played an important role in the evolution of the SDI concept. The earliest SDIs were conceived before the Internet and the World Wide Web (WWW) came into being and the opportunities opened up by their development have dramatically transformed the way that way that data is delivered to users. This was recognised by the US Mapping Sciences Committee in their report on Distributed Geolibraries (National Research Council 1999, 31). In their view, 'the WWW has added a new and radically different dimension to its earlier conception of the NSDI, one that is much more user oriented, much more effective in maximizing the added value of the nation's geoinformation assets, and much more cost effective as a data dissemination mechanism'.

The WWW has developed very rapidly over the last few years and the term 'Web 2.0' was introduced around 2005 to highlight the changes that had taken place since the emergence of Web 1.0 in the 90s (O'Reilly 2005). The most important differences between the two can be seen from some contrasting examples which illustrate the interactive and participatory nature of Web 2.0. The Web 1.0 consisted largely of static sites such as the Encyclopaedia Britannica online whereas the Web 2.0 hosts dynamic sites such as Wikipedia that are constantly being revised and enlarged by the contributions from users. Similarly, the personal websites that characterised the Web 1.0 have been replaced by the interactive blogs that are an important feature of the Web 2.0. One of the standard bearers for Web 1.0 was the Netscape server while Google can be seen as the standard bearer for Web 2.0. Unlike Netscape, Google began life as a web application that was delivered as a service with customers paying directly or indirectly to use that service.

These differences are reflected in the development of the GeoWeb that underpins the emergence of SDIs. The most important of these from a user perspective have been summarised in Table 14.1. From this it can be seen that the GeoWeb 2.0 is essentially dynamic, participatory, user centric, distributed, loosely coupled and rich in content in contrast to the static, producer driven and producer centric, centralised and closely coupled basic content of the GeoWeb 1.0.

The launch of Google Earth in June 2005 brought many of the elements of the GeoWeb 2.0 within reach of millions of users. Google Earth combined the powerful search engines developed by Google with the ability to zoom rapidly in or out from space to the neighbourhood street level. It also created new opportunities for these users to overlay their own spatial data on the top of Google Earth's background imagery. As Butler (2006, 776) pointed out in an article in the science journal *Nature*: 'By offering researchers an easy way into GIS software, Google Earth and other virtual globes are set to go beyond representing the world, and start changing it.' For this reason, they must be regarded as 'disruptive technologies' that are transforming the GIS industry in ways that the market does not expect.

A position paper from the Vespucci initiative (Craglia et al. 2008) highlights some of the impacts of the developments in information technology, SDIs and earth observation that have taken place since the launch of Vice President Gore's (1998) vision of Digital Earth. It points out that many elements of his vision are now regularly being used by large numbers of people throughout the world and that geography has become an important

TABLE 14.1

Differences between GeoWeb 1.0 and 2.0 (Maguire 2005)

GeoWeb 1.0	GeoWeb 2.0
Static	Dynamic
Publishing	Participation
Producer centric	User centric
Centralised	Decentralised
Close coupling	Loose coupling (eg mash ups)
Basic	Rich

way of organising many different kinds of digital spatial information that are now regularly collected by sensors that provide multi-level spectral information about the earth's surface in large-scale intergovernmental initiatives such as the Global Earth Observation System of Systems (GEOSS). The paper also sets out its own vision for the next 5 to 10 years. Elements of this vision include the development of multiple connected infrastructures addressing the needs of different audiences, and the possibility of searches through time and space to find analogous situations with real-time data from both sensors and individuals.

14.3 From Producers to Users – the Generation Analogy

There are interesting parallels between the shift from producers to users that has occurred as a result of emergence of the WWW and the changes that have taken place in the governance of SDIs over this time. A good example of the latter can be found in the typology of SDIs that has been developed in the course of the State of Play studies that have been carried out by the Spatial Application Division at the University of Leuven for the European Commission over the last five years (Vandenbroucke et al. 2008). This typology is based on the coordination aspects of national SDI initiatives. Matters of coordination have been emphasised because 'it is obvious coordination is the major success factor for each SDI since coordination is tackled in different ways according to the political and administrative organisation of the country' (SADL 2003, 13). A basic distinction is made between countries where a national data producer such as a mapping agency has an implicit mandate to set up an SDI and countries where SDI development is being driven by a council of Ministries, a GI association or a partnership of data users. A further distinction is then made between initiatives that do and do not involve users in the case of the former and between those that have a formal mandate and those that do not in the case of the latter.

This distinction is also reflected in the generation analogy that has been used to highlight the main structural changes that have taken place in the notion of SDIs over the last 15 years. Some features of the first generation of 11 SDIs that had emerged during the first half of the 1990s were described by Masser (1999). What distinguished these from other GI policy initiatives was that they were all explicitly national in scope and their titles all referred to geographic information, geospatial data or land information and included the term 'infrastructure', 'system' or 'framework'.

The development of a second generation of SDIs began around 2000 (Rajabifard et al. 2003). The most distinctive feature of the second generation of SDIs was the shift that was taking place from the product model that characterised most of the first generation to a process model of an SDI

(Table 14.2). Database creation was to a large extent the key driver of the first generation and, as a result, most of these initiatives tended to be data producer, and often national mapping agency, led. The shift from the product to the process model is essentially a shift in emphasis from the concerns of data producers to those of data users.

This shift had profound implications for those involved in SDI development in that it has resulted in data users becoming actively involved in SDI development and implementation. The main driving forces behind the data process model are data sharing and reusing data collected by a wide range of agencies for a great diversity of purposes at various times. Also associated with this change in emphasis is a shift from the centralised structures that characterised most of the first generation of national SDIs to the decentralised and distributed networks that are a basic feature of the WWW.

There has also been a shift in emphasis from SDI formulation to implementation as those involved gained experience of SDI implementation and a shift from single-level to multi-level participation, often within the context of an administrative hierarchy of SDIs. As a result of these developments, the coordination models that had emerged for single-level SDIs have been substantially modified and more complex and inclusive models of governance have emerged. They may also require the creation of new kinds of organisational structure to facilitate effective SDI implementation.

In the last few years, there are also signs that a third generation of SDIs is emerging. The most important difference between the second and third generations is that the balance of power in the latter has shifted from the national to the sub-national level (Rajabifard et al. 2006). Most large-scale land-related data is collected at this level where it is used for collecting land taxes, land use planning, road and infrastructure development, and day-to-day decision-making. Alongside these developments, there has been a shift

TABLE 14.2

Current trends in SDI development (Masser 2005, 257)

From a product to a process model
From data producers to data users
From database creation to data sharing
From centralised to decentralised structures
From formulation to implementation
From coordination to governance
From single to multilevel participation
From existing to new organisational structures

from government-led approaches to whole of industry models where the private sector operates on the same terms as its government partners. One consequence is that national SDI activities are likely to be increasingly restricted to the strategic level while most of the operational level decisions are handled at the sub-national levels by local government agencies in conjunction with the private sector.

The concept of spatially enabled government that is emerging as a result of these trends presents important challenges for those involved. The initial development of SDIs was largely in the hands of small elite of spatially aware professionals from the fields of geography, planning, surveying, land administration and environmental science. This elite not only dominated the production of geographic information but were also its main users. In recent years, as a result of the development of location-based services and the expansion of eGovernment activities the position has substantially changed to the extent that the vast majority of the public are users, either knowingly or unknowingly, of spatial information (Masser et al. 2008). As a result, many traditional professional practices must be drastically altered to ensure that SDIs develop in such a way that they provide an enabling platform that will serve the wider needs of society in a transparent manner.

14.4 SDI Implementation in the Context of Multi-Level Governance

Many national SDI documents seem to abide by the principle of 'one size fits all'. They suggest that the outcome of SDI implementation will lead to a relatively uniform product at the sub-national level. However, there is both a top down and a bottom up dimension to national SDI implementation. National SDI strategies drive state wide SDI strategies and state-wide SDI strategies drive local level SDI strategies and the outcomes of these processes are likely to be that the level of commitment to SDI implementation will vary considerably from state to state and from local government to local government.

The top down vision of an SDI emphasises the need for standardisation and uniformity whereas the bottom up vision stresses the importance of diversity and heterogeneity given the different aspirations of the various stakeholders and the resources that are at their disposal. Consequently, the challenge to those involved in SDI initiatives is to find ways of ensuring some measure of standardisation and uniformity while recognising the diversity and the heterogeneity of the different stakeholders. This is likely to become increasingly important as sub-national agencies take over the operational activities associated with SDI implementation.

The SDI that emerges from this process will have many features in common with a patchwork quilt or a collage of similar but often quite distinctive components. The patchwork quilt analogy assumes that the SDI outcome will be like the product of similar pieces of cloth of various colours sewn together to form a bedcover. This is a particularly useful where the SDI participants are largely administrative regions with similar functions in the hierarchy. The collage analogy, on the other hand, is based on the notion of a picture that is built up from different pieces of paper and other materials. This is most useful where the participants such as transportation and environmental agencies cover overlapping administrative districts (Masser and Crompvoets, 2015).

These two analogies broadly correspond to the two types of multi-level governance identified by political scientists such as Hooghe and Marks (2003) whose key features are summarised in Table 14.3. Type 1 governance describes jurisdictions at a limited number of levels as in the patchwork quilt model. These jurisdictions are essentially general purpose in that they bundle together many different functions such a housing, education, roads and environmental affairs. Membership of such jurisdictions is usually territorial in terms of nation, region or community and they are characterised by non intersecting memberships between different jurisdictions at the same level. In other words, a citizen may belong to only one of these jurisdictions. A limited number of levels are involved in these jurisdictions which are intended to be stable for periods of several decades or more. In essence, every citizen is located in a Russian Doll of nested jurisdictions where there is only one relevant jurisdiction at each level of the administrative hierarchy.

Type 2 governance, on the other hand, is composed of specialised task specific jurisdictions such as school catchment areas, watershed management regions, and travel to work areas. Like a collage it is fragmented in nature with every piece fulfilling its own function. In type 2 governance there is no reason why smaller jurisdictions should be neatly contained in larger ones while others may define a small segment of a larger area as is the case with a site of special scientific interest within a National Park.

TABLE 14.3

Types of multi-level governance (Hooghe and Marks 2003, 236)

Type 1	Type 2
General purpose jurisdictions	Task specific jurisdictions
Non intersecting memberships	Intersecting memberships
Jurisdictions at a limited number of levels	No limit to the number of jurisdictional levels
System wide architecture	Flexible design

There is no limit to the number of jurisdictional levels that are designed to respond flexibly to new needs and circumstances.

Generally, type 2 governance activities are embedded in type 1 structures but the way that this works out varies considerably. Type 1 jurisdictions are rooted in community identities whereas type 2 structures are more pliable. The main benefit of multi-level governance lies in its scale flexibility. Its chief cost lies in the transaction costs of coordinating multiple jurisdictions. The coordination dilemma confronting multi-level governance is described by Hooghe and Marks (2003, 239) in the following terms: 'To the extent that policies of one jurisdiction have spillovers (i.e. negative or positive externalities) for other jurisdictions, so coordination is necessary to avoid socially perverse outcomes.'

One strategy for dealing with the coordination dilemma that underpins type 1 governance is to limit the number of autonomous actors who have to be coordinated by limiting the number of autonomous jurisdictions. An alternative approach is to limit coordination costs by constraining interaction across jurisdictions. Type 2 governance sets no ceiling on the number of jurisdictions but may spawn new ones along functionally differentiated lines to minimise externalities across jurisdictions. Both these strategies have important implications for those concerned with SDI implementation.

14.5 SDI Implementation as a Social Learning Process

The old adage that Rome wasn't built in a day is equally applicable to SDIs. The creation of SDIs is a long-term task that may take years or even decades in some cases before they are fully operational. This process is likely to be an evolving one that will also reflect the extent to which the organisations that are involved are changing themselves over time. As a result, major changes in the form and content of SDIs can be expected over time as they reinvent themselves. In some instances, this process may even lead to the closing down of an SDI as was the case with the British National Geospatial Data Framework in 2002 (Masser 2005, 100–103).

The experiences of the State of Victoria in Australia provide a good example of learning by doing during the implementation process at the sub-national level. It is worth considering because Victoria has been particularly innovative in recent years both in the applications field and also in terms of the steps that it has taken to promote spatially enabled government. It is also one of the few SDIs that have published regular reports during the implementation process and cross-referenced new developments in relation to previous work. This makes it possible to trace SDI evolution in more detail than is usually the case.

The main stages in the evolution of Victoria's Spatial Information Strategy are summarised in Table 14.4. From this it can be seen that they date back at least 20 years. Concerns about duplication in maintaining computerised databases led the Victorian government to set up LANDATA to coordinate the development of a common land information system for the state as far back as 1984. This body turned out to be both under resourced and capable of producing digital maps only in a format that was unsuitable for modern GIS applications. As a result, the state commissioned Tomlinson Associates to carry out a comprehensive GIS planning study in 1991. This study examined the work carried out by 40 state agencies and reviewed 270 data sets. It was a seminal work for the spatial information industry in the state which demonstrated both the strategic importance and the economic potential of land information (Thompson et al. 2003).

In 1997 the Land Titles, Surveyor General, Valuation and Crown Lands administrations were merged to form a single body, Land Victoria, which

TABLE 14.4

The evolution of Victoria's spatial information management framework (Department of Sustainability and Environment, 2008, 13)

1993

Core spatial information identified

Government wide planning methodology introduced

Victorian Geospatial Information Strategy 1997–2000

Core data improved

Spatial management framework put in place – policy, infrastructure, awareness, distribution, business systems

Core principles for managing spatial information introduced – metadata, quality management, privacy, liability, licensing, pricing, custodianship

Coordinating and cooperative arrangements between key stakeholders established

Victorian Geospatial Information Strategy 2000–2003

Spatial information management principles further codified

Introduction of the concepts of 'framework' and 'business' information

Role of custodians defined

Framework datasets identified and custodians assigned

Victorian Spatial Information Strategy 2004–2007

Best practice principles for spatial information management extended to custodians of all spatial datasets

Custodianship formally identified as the basis for spatial information management

Holistic spatial information management framework defined

Victorian Spatial Council established

---

is responsible for land administration in the state. In addition to these operational services, a Land Information Group was established to deal with the broader policy issues relating to spatial information. This group worked in conjunction with an advisory panel made up from participants in the state's spatial information industry in the development of Victoria's first two state-wide spatial information strategies. The first of these was the Victorian Geospatial Information Strategy (VGIS) for the period from 1997 to 2000. This focussed on the creation and maintenance of high-quality fundamental data sets for the state as a whole.

This was followed by another strategy for the period from 2000 to 2003. This further consolidated the creation of high-quality fundamental data sets for the state and set out best practice management principles for custodianship, metadata, access, pricing and licensing and spatial accuracy (Thompson et al. 2003). The concept of custodianship lies at the heart of the State's spatial information management strategy.

While the proposals for the next strategy were under consideration, the Land Information Group with its 70 staff took a new position in early 2004 within the Department of Sustainability and Environment as the Spatial Information Infrastructure component of its Strategic Policy and Projects group. This move made it possible for those involved to play a larger part in the development of spatial information policy for the Department as a whole.

The Victorian Spatial Information Strategy for 2004 to 2007 (VSIS) differed from its predecessor in several important ways. The VGIS strategies had focussed largely on issues of management and custodianship associated with the eight fundamental data sets that are the core of the State's SDI, but the VSIS was much broader in scope and presented a whole of industry approach rather than a governmental model. The implementation of the VSIS and the preparation of the latest strategy for 2008–2010 have been overseen by the whole of industry body, the Victorian Spatial Council, which was set up in 2004. Its membership includes representatives drawn from state government (3), local government (2), federal government (1), academia (2), the professions (2) and the private sector (2). An independent chairman of the Council has been appointed by the Secretary of the State Department of Sustainability and Environment.

The experiences of Victoria indicate that a combination of internal and external factors affects the evolution of SDIs over time. Internally, those involved participate in a process of learning by doing that takes account of the experiences of earlier stages of SDI implementation. Externally, important changes in the nature of the SDI may be a consequence of the restructuring of other activities within government as a whole. The interaction between these two strands will govern the trajectory of SDI development.

14.6 Some Challenges Facing SDI Implementation

SDIs have attracted a lot of attention from governments all over the world over the last 10 years. This raises the inevitable question as to how far they will be able to deliver the promised benefits over time. Bregt and Crompvoets (2004) have argued that some SDIs may have already raised unrealistic expectations and their benefits that are not proven. These are likely to attract few stakeholders and can be classified as 'hype'. In contrast, the successful 'hit' SDIs will be those that have developed in response to realistic expectations and can deliver proven benefits. Most or all the relevant stakeholders are likely to be involved in such SDIs.

In practice, as the discussion in Section 3 of this chapter shows, SDI outcomes are likely to be much more complex in practice because of the nature of the implementation processes. The discussion of multi-level governance in Section 4 of this chapter suggests that successful SDI implementation will be heavily dependent upon the extent to which sub-national agencies are actively involved. The challenges that arise at the sub-national level are highlighted in the findings of the Advanced Regional SDIs workshop that was organised by the Joint Research Centre last year (Craglia and Campagna 2009) and are the central focus of attention in the current ESDInet+ project that is funded by the European Commission (see Chapter 18).

The findings of these and other studies suggest that effective SDI implementation is often facilitated in countries such as Australia and Germany where many important administrative responsibilities are devolved to the state level and established institutions already exist at this level for policy making and implementation. However, it should also be noted that the information infrastructures that come into being at this level are often different in many respects from those at the national level (de Man 2007) and also that these fall essentially into the type 1 governance category. Consequently, further challenges may have to be overcome in these situations in order to respond to the needs of type 2 governance agencies.

The discussion in the fifth section of this chapter considers another of the dilemmas facing those involved in SDI implementation: i.e. the challenges presented by both internal and external organisational changes during the course of SDI implementation. To be successful, SDIs must be sustainable over long periods of time. The current INSPIRE road map, for example, covers the period up to May 2019 (www.inspire.ec.europa.eu). As a result, the SDIs that are most likely to succeed will be those that meet the three sets of conditions originally identified by Campbell and Masser (1995, 45–48) for effective GIS implementation. These are (1) a consistent strategy that identifies the evolving needs of users and takes account of the changing resources and values of the participants, (2) a commitment to and participation in the implementation of the SDI by groups and individuals from both types 1 and 2 of governance structures and (3) an ability to cope with all kinds of change.

14.7 One SDI or Many?

Around the time that I was writing this chapter I was spending a month each year in the Centre for Spatial Data Infrastructures and Land Administration at Melbourne University collaborating with Professor Ian Williamson and his colleagues on a number of SDI-related projects. During this time, I co-wrote two papers with them which were subsequently published in the International Journal for Geographic Information Science. The first of these was entitled 'The role of sub-regional government and the private sector in future SDIs' (Rajabifard et al. 2006). The second paper entitled 'Spatially enabling government through SDI implementation' (Masser et al. 2008). The discussion in both papers developed some of the issues discussed in the above paper.

The Australian experience also gave me the opportunity to critically evaluate some of the criticisms that have been made regarding the implementation of the SDI model. It can be argued, for example, that recent developments in technology provide alternative modes of SDI creation. For example, Jackson et al. (2010, 91) claim:

> In parallel with government programmes, technology advances have occurred that give most of the general public the ability to collect location-based data. Most significant in this context is the near ubiquitous ownership of mobile phones, many of which now incorporate GPS, high-quality cameras, sensors such as tri-axis accelerometers and digital compasses, and broad-band data communications. Such informal and ad-hoc data collection does not typically adhere to formal standards of geometric precision or metadata consistency, neither does it provide consistency in coverage or detail. It is, however, most prevalent where people are and where there are the exceptional circumstances that stimulate individuals to record data of the environment and happenings around themselves and these circumstances frequently apply in disaster and emergency scenarios.

The benefits of volunteered geographic information (VGI) are particularly evident in dealing with emergencies and disasters such as forest fires, as Goodchild and Glennon (2010, 240) have pointed out:

> Agencies are inevitably stretched thin during an emergency, especially one that threatens a large community with loss of life and property.

Agencies have limited staff, and limited ability to acquire and synthe-size the geographic information that is vital to effective response. On the other hand, the average citizen is equipped with powers of obser-vation, and is now empowered with the ability to georegister those observations, to transmit them through the Internet, and to synthesize them into readily understood maps and status reports. Thus, the fundamental question raised by this paper is: How can society employ the eyes and ears of the general public, their eagerness to help, and their recent digital empowerment, to provide effective assistance to responders and emergency managers?

It can be argued that the evolution of SDIs in response to crowdsourcing and ubiquitous mobile technologies presents an opportunity rather than a challenge to the large-scale public sector SDI that dominated the debate 20 years ago. It is also important to recognise there are not just one but many types of SDIs around at the present time (see, for example, Haklay 2013). Citizen involvement in data collection occurs in an increasing variety of applications ranging from tsunami watches to bird counts and crime hot spot identification. In these cases, data from official public-sector SDIs is not necessarily used and for each application there might be a small SDI which addresses its specific sharing and coordination needs (Harvey et al. 2013).

According to Coleman et al. (2009), the development of VGI also creates opportunities for national mapping and cadastral agencies to update their databases although it raises questions regarding people's motivation for parti-cipation. However, it is unlikely that this will be not without cost for organisa-tions themselves:

If a mapping organization wishes to capitalize on a distributed net-work of volunteer geospatial data producers, then it must start refocusing attention across what happens both inside that organiza-tion and also in the new social network of geo-information produc-tion. New rules and standards will be required to take into account the values of these volunteers – equity, security, community building, privacy – in evaluating the performance of this new production system.

(p. 352)

One consequence of these developments is that it will be necessary to reconsider the role of users of SDIs. With this in mind, Budhatkoki et al. (2008, 151) have pointed out that 'contemporary SDIs are created *for* expert users *by* expert organisations'. In circumstances such as this, 'the users of an SDI are often referred to as *end users* – a term which itself reflects their marginalised role as mere recipients of GI'. They argue that this misses the point because SDI and VGI are not separate but comple-mentary phenomena:

Indeed, these can be brought within a single framework when the role of the user is reconceptualised to *producer* and VGI is included in the SDI related processes ... this creates a hybrid model that draws on the synergy between the conceptual foundation of SDI and an extensive producer base of VGI.

(p. 153)

With respect to conventional SDIs, Dias et al. (2012, 382) return to the discussion about the characteristics of information infrastructures, in general, in Chapter 5 and point out that 'most SDIs are built upon an existing installed base'. This inevitably slows down acceptance and adoption. They also argue that SDIs will continue to evolve, notwithstanding recent criticisms. This is because:

SDIs are not only about technology. Agreeing on common policies, standards and organizational structures is essential for bringing these technologies into use, and thus, realizing their potential benefits. The classic definition of SDIs being a set of policies, technologies and institutional arrangements to assist user communities in collecting, sharing and exploiting geospatial information resources ...still proposes a valid set of requirements and a general means of achieving these goals. Its scope is broad enough to embrace even major shifts in its concepts and implementations.

(Dias et al. 2012, p. 398)

In a recent paper, Alvarez Leon (2018) traces the political economy of SDIs with respect to three key aspects: legal frameworks, politics and economic impact associated with the creating and maintaining SDIs at multiple scales. Drawing on the work of Harvey et al. (2013), he argues that:

SDIs are both technical and political projects that simultaneously respond to interscalar political dynamics characteristic to particular administrative arrangements. These dynamics are reflected in their respective legal frameworks governing SDIs in each jurisdiction, which are in flux due to the technological changes in geographic information and its role in society. The combination of these factors influences the role played in politics, government, and society by each SDI, while shaping its potential for economic impact.

(p. 151)

Consequently, it can be concluded that there are likely to be an increasing number of diverse SDI applications in practice. One of the most interesting builds upon the hybrid model put forward by Budhatkoki et al. (2008). This can make the transition from the second generation of SDIs to a third generation possible.

References

Alvarez Leon, L. F., 2018. The political economy of spatial data infrastructures, *International Journal of Cartography*, 4(2), 151–169.

Bregt, A. and J. Crompvoets, 2004. Spatial data infrastructures: hype or hit? *Proc GSDI 8*, Cairo.

Budhatkoki, N. J., B. Bruce and Z. Nedovic-Budic, 2008. Reconceptualising the role of the user of spatial data infrastructure, *Geojournal*, 72, 149–160.

Butler, D., 2006. The web-wide world, *Nature*, 439, 776–778.

Campbell, H. and I. Masser, 1995. *GIS and organisations: how effective are GIS in practice?* London: Taylor and Francis.

Coleman, D, Y. Georgiadou and J. Labonte, 2009. Volunteered geographic information: the nature and motivation of producers, *International Journal of SDI Research*, 4, 332–356.

Craglia, M. and M. Campagna, 2009. *Advanced regional spatial data infrastructures in Europe*, Ispra: Joint Research Centre Institute for Environment and Sustainability.

Craglia, M., M. F. Goodchild, A. Annoni, G. Camara, M. Gould, W. Kuhn, D. Mark, I. Masser, D. Maguire, S. Liang and E. Parsons, 2008. Next generation Digital Earth: a position paper from the Vespucci initiative for the advancement of geographic information science, *International Journal of SDI Research*, 3, 146–167.

Crompvoets, J., A. Rajabifard, A. Bregt and I. Williamson, 2004. Assessing the world wide developments of national spatial data clearinghouses, *International Journal of Geographical Information Science*, 18, 1–25.

de Man, E., 2007. Beyond spatial data infrastructures there are no SDIs – So what, *International Journal of SDI Research*, 2, 1–23.

Department of Sustainability and Environment, 2008. *Victorian Spatial Information Strategy 2008–2010*, Melbourne: Department of Sustainability and Environment.

Dias, L., A. Remke, T. Kauppinen, A. Degbelo, T. Foerster, C. Stasch, M. Rieke, B. Schaeffer, B. Baranski, A. Broring and A. Wytzisk, 2012. Future SDI – Impulses from geomatics research and IT trends, *International Journal of SDI Research*, 7, 378–410.

Executive Office of the President, 1994. Coordinating geographic data acquisition and access, the National Spatial Data Infrastructure, Executive Order 12906, *Federal Register*, 59, 17671–17674.

Goodchild, M. F. and J. A. Glennon, 2010. Crowd sourcing geographic information for disaster response, *International Journal of Digital Earth*, 3(3), 231–241.

Gore, A., 1998. The Digital Earth: understanding our planet in the 21st century, *Australian Surveyor*, 43(2), 89–91.

Haklay, M, 2013. Citizen science and volunteered geographic information: overview and typology of participation, in D. Sui, S. Elwood and M. F. Goodchild (eds.) *Crowd sourcing geographic knowledge: volunteered geographic information (VGI) in theory and practice*, Berlin: Springer, 105-122.

Harvey, F., A. Iwaniak, and S. Coetzee, 2013. SDIs past present and future: a review and status assessment, in A. Rajabifard and D. Coleman (eds.) *Spatially enabling government, industry and citizens*, Needham, MA: GSDI Association Press.

Hooghe, L. and G. Marks, 2003. Unravelling the central state, but how? Types of multi level governance, *American Political Science Review*, 97, 233–243.

Jackson, M. J., H. A. Rahemtulla and J. Morley, 2010. The synergistic use of authenticated and crowd sourced data for emergency response, in C. Corbane, D. Carrion, M. Broglia, and M. Pesarasi (eds.) *Second International Workshop on Geoinformation products on validation of for crisis management, EUR 24530,* Ispra, Italy: EC Joint Research Centre.

Maguire, D., 2005. GeoWeb 2.0: implications for ESDI, Proc 12th EC-GIS Workshop, Innsbruck, Austria.

Masser, I., 1999. All shapes and sizes: the first generation of National Spatial Data Infrastructures, *International Journal of Geographical Information Science, 13,* 67–84.

Masser, I., 2005. *GIS Worlds: creating spatial data infrastructures,* Redlands: ESRI Press.

Masser, I. and J. Crompvoets, 2015. *Building European spatial data infrastructures,* Third edition, Redlands: ESRI Press.

Masser, I., A. Rajabifard and I. Williamson, 2008. Spatially enabling governments through SDI implementation, *International Journal of Geographical Information Science, 22,* 5–20.

National Research Council, 1993. *Toward a coordinated spatial data infrastructure for the nation,* Mapping Science Committee, Washington, DC: National Academy Press.

National Research Council, 1999. *Distributed geolibraries: spatial information resources,* Council, Washington, DC: National Academy Press.

O'Reilly, T., 2005. What is Web 2.0: design patterns and business models for the next generation of software, www.oreillynet.com (last accessed February 18 2019)

Rajabifard, A., A. Binns, I. Masser and I. Williamson, 2006. The role of sub-national government and the private sector in future spatial data infrastructures, *International Jour Geographical Information Science, 20,* 727–741.

Rajabifard, A., M. E. Feeney, I. Williamson and I. Masser, 2003. National spatial data infrastructures, in I. Williamson, A. Rajabifard and M. E. Feeney (eds.) *Development of Spatial Data Infrastructures: from Concept to Reality,* London: Taylor and Francis, 95-109.

Spatial Applications Division, Catholic University of Leuven (SADL), 2003. *Spatial data infrastructures in Europe: state of play spring 2003,* Summary report. Spatial Applications Division, Catholic University of Leuven (SADL).

Thompson, B., M. Warnest and C. Chipchase, 2003. State SDI development, in I. Williamson, A. Rajabifard, M. E. Feeney (eds.) *Development of Spatial Data Infrastructures: from Concept to Reality,* London: Taylor and Francis, 147-164.

Vandenbroucke, D., K. Janssen and J. van Osterhoven, 2008. *INSPIRE State of play: development of the NSDI in 32 European countries between 2002 and 2007, Proc GSDI 10,* Trinidad.

Practice

15

A European Policy Framework for Geographic Information

Massimo Craglia and Ian Masser

15.1 Introduction

Michael Goodchild (1997) has discussed the impacts that technological and economic changes are likely to have on the distribution and operation of central facilities, using libraries as a case study. As he argues, we are likely to see in the near future a major shift from the traditional patterns of distribution analysed by classic location theory to new ones based on networked localities catering for the needs of their immediate constituency with 'pointers' to the whole universe of other facilities and knowledge out there in the hyperspace. These new forms of production and distribution will require new institutional arrangements, which, as he argues, are only likely to be smooth if there is a consensus on the need for change and a shared vision of how this may be achieved.

Underpinning this vision are two key assumptions. The first is the existence of some sense of direction among the key players in the field including government and the private sector, which may frame the development of these localized initiatives. The second is the existence of a technological community of practitioners, which creates and follows a tradition of practice associated with the evolution of a particular technology (Constant, 1987). In the United States, the National Spatial Data Infrastructure (Tosta and Domaratz, 1997) provides the broad framework, and social groups such as that of the librarians are also well established as a forum for discussion, innovation, and turf battles. What is the state of development in Europe, bearing in mind its linguistic, professional, and cultural variations? Is there an equivalent policy framework emerging on the future of society channelling policy initiatives and resources? Are European-wide social networks being established to disseminate innovation? To address these issues, this chapter explores recent developments at the European level focusing in particular on those initiatives which are supporting the creation of a European geographic information (GI) community and on the emergence of a European policy framework for GI.

The deliberate focus on the European Union (EU) level does not imply that important initiatives are not taking place at national or even local level in the member countries, or indeed outside the EU. It does however reflect the conviction that it is at this level that an important debate is taking place to which we must be able to contribute.

15.2 European Initiatives on Geographic Information

As an introduction to this section reviewing recent developments in the field of GI, it is worth noting that despite the increasing process of integration, Europe is still very much made of individual member states. Even within the EU itself, each of the 15 member countries (that were members at the time of writing) had different institutional and legal frameworks, cultural traditions, and languages. These differences are very clear in the context of GI as each member state has one or more agencies responsible for mapping, with different coordinate systems and projections, and military traditions, which by and large do not foster openness and sharing. Similar arrangements exist with respect to socioeconomic data which is collected by the national statistical offices with different time frames, methodologies, and definitions. Equally complex are the variations in basic administrative units within each country and in copyright legislation, to mention but two of the many issues affecting data availability at the European level.

Against this background, Masser (1991) has argued that the low level of GI awareness throughout Europe was largely due to the failure to develop adequate channels of dissemination of information, and opportunities to exchange ideas and experiences needed to overcome the obstacle referred to above. To deal with these obstacles, he suggested the following measures: European-level conferences to provide an international forum for the exchange of ideas and experience; European journals to stimulate the debate on issues of particular relevance to Europe as a whole; European Centre of Geographic Information to act as a voice for the European GI community (Masser, 1991, p. 10); and Masser also argued that there was a pressing need to develop a major European research programme and share resources and experiences in teaching.

The developments which have taken place in Europe over the last 7 years are best judged against these suggestions as they clearly show the significant progress made. In 1991, the European Science Foundation launched a 5-year research programme on GIS (GISDATA), which involved over 300 researchers from 20 European countries in the course of its activities. This programme has done much not only in furthering scientific research in the database design, data integration, and socioeconomic and environmental applications (Arnaud, Craglia, Masser, Salgé and Scholten, 1993), but also in fostering the establishment of a European research community and in developing the linkages with the United States through its collaboration with the National Science Foundation and National Center for Geographic Information and Analysis.

With respect to education, a special interest group was established in 1992 within the framework of the EGIS conference series. Academics from throughout Western and Central Europe have met annually within this framework to exchange ideas and teaching methods, and in September 1998 they organized the First European GIS Seminar. In addition, the GIS International Group (GISIG), established in 1992 in the framework of the COMETT programme of the European Commission, has actively promoted since its creation the exchange of students and staff between the academic and the private sector, and short training programmes on GIS, as well as developing a number of EU-funded projects linking western and central European research establishments.

With respect to conferences, the EGIS conference series developed successfully between 1990 and 1994 attracting hundreds of academics from across Europe. As other established conference series such as AM/FM and the Urban Data Management Symposium became also increasingly interested in the use of GIS technology for utility management and local government, respectively, they joined forces with EGIS to develop a Joint European Conference Series on Geographic Information. This series developed between 1995 and 1997, and as it came to an end, it soon became clear that an alternative was urgently required to keep the momentum built over the previous years. The new initiative is the Association of Geographic Information Laboratories in Europe (AGILE) the mission of which is 'to promote academic teaching and research on GIS at the European level and to ensure the continuation of the networking activities that have emerged as a result of the EGIS Conferences and the European Science Foundation GISDATA Scientific Programme.' AGILE was launched at its inaugural conference in April 1998 in Enschede, the Netherlands, with 50 founding laboratories from 20 European countries.

A major step forward in organizing the variety of experiences and interests in GI at the European level was the establishment in 1993 of EUROGI, the European Umbrella Organisation for Geographic Information, the aims of which are

> to promote, stimulate, encourage and support the development and use of Geographic Information and Technology at the European level and to represent the common interest of the Geographic Information community in Europe. Its 18 founding members included 13 national associations and 5 Pan European sectorial organisations.
>
> *(www.eurogi.org)*

Of particular significance are the activities of EUROGI in stimulating awareness of GI matters across Europe, which has encouraged several countries to set up new national associations acting as a link between the European and national debates. This is important, because in a context as diversified as Europe, raising awareness is one of the most crucial needs to continued progress in this field. To this end, the establishment of GIM in 1986 and GIS Europe in 1992 are also worth noting as they provide truly European-based journals channelling information within the emerging GI social and professional networks.

In addition to EUROGI, other important transnational organizations have been established including CERCO, the European Committee of the Heads of the National Mapping Agencies (NMAs). Given the context highlighted earlier of national fragmentation of the key providers of GI in Europe, the establishment of CERCO is a very important step. This organization is not only providing a forum for joint discussions and strategic thinking, but it has also set up an operational arm (MEGRIN), the objectives of which are: to make NMAs data more accessible to European users by successfully providing appropriate products and services which meet user needs; to build close relationships with customers and distributors to stimulate the European GI market; and to contribute to the European strategies of member NMAs.

To meet these objectives, Megrin has initiated a number of projects including the Geographic Data Description Directory, which is a web-based metadata service providing information in a standardized format about the digital map data available in 25 European countries, and SABE (Seamless Administrative Boundaries of Europe), which is a unique dataset containing the administrative boundaries for 25 countries in Europe down to the level of individual municipalities (Salgé, 1998).

Given the increasing convergence of remote sensing and GIS, it is also important to mention the important initiative of the EU in setting up the Centre for Earth Observation at its Joint Research Centre in Ispra, Italy. Through this centre, the EU has invested 100 million ECUs (1 ECU=approximately 1 US$) in developing a European-Wide Service Exchange, a web-based service to provide data and metadata, and put in touch data providers, brokers, users, and specialists.

This review of recent initiatives is of course far from exhaustive, but it already provides a sense of the significant progress that has taken place since 1991 in developing a complex web of transnational networks linking different disciplines, economic sectors, and activities. Most of these initiatives have started independently of the activities of the European Commission, which has however gradually become a major player in shaping further developments, particularly at policy level. With this in mind, the next section reviews the debate on European-wide policies for GI.

15.3 The European Geographic Information Infrastructure Debate

15.3.1 Introduction

This section reviews some recent developments which have led to the emergence of a European policy framework for GI. The first part discusses the political context of these developments, whereas the second summarizes the main proposals put forward by the European Communities (EC) in November 1996 (GI2000) to establish such a framework. To complete the picture, the last part of this section describes a number of related GI developments sponsored by the EC over the last few years.

To aid non-European readers, it may be worth noting that the Commission of the EC is not like a fully fledged national government. It is more akin to a relatively small administrative body, organized by sectoral competencies into a series of departments (the General Directorates or DGs), which is headed by 20 commissioners appointed by the national government. Although a powerful body, the EC shares decision making with two other bodies, the European Parliament, elected directly by the EU citizens, and the Council of (national) Ministers. The latter body is the most important of the three because it is where the ultimate decisions rest after both Commission and Parliament have expressed their positions.

Within the EC, several departments have an interest in GI, either because they are users or because it underpins other activities. For example, EUROSTAT, which is the statistical service of the EC collecting and harmonizing data coming from the national statistical institutes, has its own GIS service (GISCO) to support the activities of the EC. Similarly, DGXVI which is responsible for Regional Development is a major user of GI to monitor and evaluate the impacts of EC policies in this area. Each of these DGs may also commission research studies or subcontract certain activities. To give an indication of the wide range of GI-related activities within the Commission, a search on the ECHO database of the Commission found over 100 'hits' for projects dealing with or using GI(S). However, most of these projects are primarily concerned with the application of GIS technology in the fields of the environment, transport, and spatial planning rather than GI policy itself. Therefore, for the purpose of the remainder of this chapter, reference is only made to three DGs which play a particularly prominent role in shaping GI policy at the European level: DGXIII is the directorate responsible inter alia for developing the information market; DGIII is the directorate responsible for industry and information technology; and DGXII is the directorate responsible for research and development policies.

15.3.2 The Political Context

The starting point for much of the current discussion is the vision for Europe that was presented to the European Council in Brussels in December 1993 by the then President Jacques Delors in the White Paper entitled 'Growth, Competitiveness and Employment: the challenges and ways forward into the 21st century' (Commission of the European Communities, 1993). An important component of this vision is the development of the information society essentially within the triad of the EU, the United States, and Japan. One result of this initiative was the formation of a high-level group of senior representatives from the industries involved under the chairmanship of Commissioner Martin Bangemann. This group prepared an action plan for 'Europe and the Global Information Society' which was presented to the European Council at the Corfu Summit in June 1994 (Bangemann, 1994). In this plan, the group argued that recent developments in information and communications technology represent a new industrial revolution that is likely to have profound implications for European society. To take advantage of these developments, it will be necessary to complete the liberalization of the telecommunications sector and create the information superhighways that are needed for this purpose. With this in mind the group proposed 10 specific actions. These included far reaching proposals for the application of new technology in fields such as road traffic management, trans-European public administration networks, and city information highways. These proposals were subsequently largely incorporated into the Commission's own plan 'Europe's way to the information society' which was published in July 1994 (Commission of the European Communities, 1994). Issues relating to the information society also featured prominently on the agenda of the G7 Ministerial Conference in Brussels in February 1995 where the need for global cooperation was stressed.

15.3.3 GI2000

Parallel to these developments, a number of important steps were taken toward the creation of a European policy framework for GI. In April 1994, a meeting of the heads of national geographical institutes was held in Luxembourg, which concluded that 'it is clear from the strong interest at this meeting that the time is right to begin discussions on the creation and supply of harmonised topographic data across Europe' (DGXIII/E, 1994). This view was reinforced by a letter sent to President Delors by the French Minister M. Dosson, which urged the Commission to set up a coordinated approach to GI in Europe and the correspondence on this topic between the German and Spanish ministers and Commissioner Bangemann during summer 1994.

As a result of these developments, a meeting of key people representing GI interests in each of the Member States was held in Luxembourg in

February 1995. The basic objective of this meeting was to discuss a draft document entitled 'GI2000: towards the European geographic information infrastructure' (DGXIII/E, 1995) and identify what actions were needed in this respect. The main conclusion of this meeting was that 'it is clear from the debate that the Commission has succeeded in identifying and bringing together the necessary national representative departments that can play a role in developing a Community Action Plan in Geographic Information' (DGXIII/E, 1995, p. 12). As a result, it was agreed that DGXIII/E should initiate wide ranging consultations within the European GI community, with a view to the preparation of a policy document for the Council of Ministers (Masser and Salge, 1997).

Since 1995, the original document has been redrafted as least seven times to take account of the views expressed during different rounds of consultations. In the process, its status has changed from a 'discussion document' to a 'policy document' and back to a 'discussion document' again. In the most recent of these drafts, which is dated November 1996 (DGXIII/E, 1996a), its title has also been changed from 'GI2000: towards a European geographic information infrastructure' to 'GI2000: towards a European policy framework for geographic information' and the term 'infrastructure' has largely disappeared from the document to make it more palatable to policy makers who, it is argued, identify 'infrastructure' with physical artefacts such as pipes and roads and may get confused by the broader connotation of this term. Despite these changes in emphasis, the basic reasoning behind the argument that is presented remains essentially unchanged.

> The major impediments to the widespread and successful use of geographic information in Europe are not technical, but political and organisational. The lack of a European mandate on geographic information is retarding development of joint geographic information strategies which causes unnecessary costs, is stifling new goods and services and is reducing competitiveness.
>
> *(DGXIII/E, 1996a, p. 2)*

Consequently, GI2000 argued that what is required is a 'policy framework to set up and maintain a stable, European-wide set of agreed rules, standards, procedures, guidelines and incentives for creating, collecting, exchanging and using geographic information' (DGXIII/E, 1996a, p. 3). The seven main practical objectives for the policy framework were identified as follows:

1. To provide, at the European level, an open and flexible, framework for organizing the provision, distribution, and standardization of geographic information for the benefit of all suppliers and users, both public and private.

2. To achieve European-wide metadata dissemination, through appropriate information exchanges that conform to accepted world-wide practices.

3. To stimulate the convergence of national geographic information policies and to learn from experience at national level to ensure that EU-wide objectives can be met as well, at little additional cost and without further delay or waste of prior work already completed.

4. To lay the foundations for rapid growth in the marketplace by supporting the initiatives and structures needed to guarantee ready access to the wealth of geographic information that already exists in Europe, and to ensure that major tasks in data capture are cost effective, resulting in products and services usable at national and pan-European scales.

5. To develop policies which aid European businesses in effective and efficient development of their home markets in a wide range of sectors by encouraging informed and innovative use of geographic information in all its many forms, and promoting new and sophisticated analysis, visualization, and presentation tools (including the relevant datasets), which can be used by non-experts.

6. To help realize the business opportunities for the European geographic information industry in a global and competitive marketplace.

7. To position Europe in a global context (DGXIII/E, 1996a, p. 17).

Implementation of this framework was envisaged at two levels. The first and most important was the political level with actions aimed at achieving agreement among the Member States: to set up a common approach to create European base data, and to make this generally available at affordable rates. This must include the adoption of the newest coordinate and projection systems on a Europe-wide basis applicable to European GI; to set up and adopt general data creation and exchange standards and to use them; to improve the ways and means for both public and private agencies and organizations to conduct European-level actions, such as creation of seamless pan-European datasets to ensure that European solutions are globally compatible (DGXIII/E, 1996a, p. 3)

The second more detailed level involved the setting up of a High-Level Working Party representing all the key players in the field, which would develop an action plan to implement the framework. Specific actions identified in GI2000 aimed at improving European cooperation and coordination, fostering the creation of base data and metadata services, lowering legal barriers, raising awareness and training, and fostering the development of the market and research and development in the GI field. In respect to the latter, GI2000 argued:

> the existing efforts regarding geographic information and geographic information systems in the Communities R&D programmes should be

clustered under the 5th Framework Programme, ensuring better exploitation of synergy, concertation and reflection of users' and industry's needs.
This will be an improvement to the present situation where projects
pertaining to geographic information are scattered throughout the programmes with little or no interaction between them. _ The approach
developed under the 4th Framework Programme for research and industry task forces will be applied in the area of geographic information.

(DGXIII/E, 1996a, p. 21)

The GI2000 draft entered the internal consultation process of the Commission in November 1996. It was envisaged that after being discussed among
all the directorates with a stake in GI and being approved by the Council
of Ministers of Telecommunications or Industry, it would become
a Communication to the European Parliament, the Council, and the Committee of the Regions in 1997. In fact, the Rolling Action Plan for Information Society (Commission of the European Communities, 1997) of
28 January 1997 already included GI2000 as Action 117 of the calendar for
1997 with the adoption by the Commission 'to be followed by the development of an action plan' (p. 6).

At the outset, it was recognized that the involvement of many organizations and institutions within Europe will be required to create such policy
framework and that strong leadership and political support will be needed to
carry the process forward. For this reason, as no organization exists with the
political mandate to create GI policy at the European level, the European
Commission was seeking such a mandate, and it envisaged that a high-level
task force would be set up to implement the policy outlined in the document.

What was probably underestimated at the time was not just the lack of
awareness among national political decision makers on the strategic
importance of GI, but also the difficulties within the Commission itself in
building sufficient momentum and political support behind GI2000. In
January 1998, the unofficial changes to the November 1996 draft were
suggesting that:

Action required

1. The High-Level Working Group proposed by the draft communication will be the European focal point to exchange information and
   debate on GI policy strategies.

2. A specific debate on public data dissemination, data harmonization,
   and the role of public services, at local, national, and European level,
   must be launched. Based on the existing public data, it should be
   possible to prepare value-added products for sale 'off the shelf,'
   creating new opportunities for European SMEs.

3. The European Commission, due to its own pan-European information needs, will provide strong incentives to the market, creating
   economic activity and new jobs.

4. One of the central developments affecting GI2000 has been the decision to increase substantially the funding of the overall research and development programme of the EU, the Fifth Framework Programme, but to withdraw funding to any programme outside the Framework. Hence, the action plan originally envisaged to implement GI2000 has been cancelled. This is not to say that GI2000 is no longer important as a Communication to the European Parliament on the strategic importance of GI, as this will do much to raise the overall level of awareness, which in turn is a prerequisite to initiate the necessary political actions. It must also be emphasized that a number of other initiatives have been developed parallel to GI2000 that contribute in sustaining the momentum. These are discussed in the next section.

15.3.4 Related Research and Developments

A number of related research and development projects have been commissioned by DGXIII/E within the context of the European policy framework. Some projects were part of the IMPACT-2 programme (1992–1995), three projects (1995–1997) dealt with specific aspects of European GI policy arising out of the GI2000 debate, while new GI projects have been taking place within the INFO-2000 programme.

IMPACT-2: The call for tenders for this project to develop new information services using GI technology was published in November 1992 and work began on 28 definition phase projects in July 1993. At the end of this stage, eight of these projects were given further funding for an implementation phase which continued until 1995. From the standpoint of the creation of a European policy framework, the most important feature of these projects is the extent to which they have built up operational experience with respect to the institutional, cultural and legal barriers that need to be resolved with respect to the creation of transnational European databases within Europe (Longhorn, 1998).

The three GI studies: In August 1995, DGXIII/E issued calls for tenders for three studies specifically related to issues arising out of the discussions regarding the European policy framework. The basic objectives of these GI studies are reflected in their titles:

- Study on policy issues relating to geographic information in Europe (GI-POLICY).
- On demand and supply for geographic information in Europe, including base data (GI-BASE).
- Feasibility study for establishing European wide metadata services for geographic information in Europe (GI-META).

Work on these projects began at the start of 1996 and was completed in 1997. Of particular interest here are the findings of the GI-POLICY study which interviewed in-depth 20 senior personnel in key organizations throughout Europe who are responsible for the provision and management of GI. The study indicated the crucial importance given by the respondents to issues of copyright and the protection of intellectual property rights, data quality and data access policies at the European level to develop the market for GI-based services. It also confirmed that lack of awareness, education, and training were viewed as the main barriers. What is also interesting is to note the regional variations in the ranking given to these core issues. As an example, copyright and data security were perceived as more important in northern Europe than in the centre and south, where lack of a critical mass of digital data and users was seen as a major problem. Somewhat unexpectedly, issues of privacy were perceived as more serious in southern Europe than in the north where the market is more developed and one might expect a greater concern for data matching. However, this variation seems to be explained by the relatively poor protection perceived by the respondents in the south compared to those in the north, where privacy legislation has long been established (Burrough, Craglia, Masser and Rhind, 1997). As a whole though, this study confirmed the crucial importance of these policy issues in the development of the European GI market, and the need for a fresh set of initiatives aimed in particular at making the data held by public sector agencies across Europe easier to access, add value to, and disseminate in partnership with the private sector. These issues have been particularly addressed by the INFO-2000 programme.

INFO-2000: DGXIII/E launched this 4-year research programme in 1996 with a budget of 65 million ECUs. The basic objectives of this programme are to stimulate the multimedia content industry and to encourage the use of multimedia content in the emerging information society. The rationale behind it is that this is a strategic area for future job creation in Europe, and that whilst Europe is lagging behind the United States with respect to the current market size for electronic information, technology, and low tariffs, it can build on its cultural diversity to excel in applications and development of content. With this in mind, the programme involves three key actions on stimulating demand and raising awareness, exploiting Europe's public sector information, and triggering European multimedia potential. Nearly half the budget for this programme is allocated to Action Line 3 for projects that will accelerate the development of the European multimedia industry in four key areas: cultural heritage, business services, GI, and science, technology, and medical information. To receive support from the GI component of this programme, projects had to satisfy at least one of the following objectives:

- to demonstrate through innovative pilot applications the advances made in integrated base data and thematic information content;
- to provide pan-European information about GI, what is held, in what format, and how it is accessed (meta-data services and their linking);
- demonstrate integration or inter linking of base data of a pan-European, or transborder nature that may form the building block of future commercial applications, especially where such projects will be held to establish common specifications for pan-European GI datasets;
- demonstrate methodologies for collecting, exchanging and using pan- European or transborder GI, including provision for networked access to other services (DGXIII/E, 1996b, p. 13).

Whilst at the outset it appeared that support for GI developments were only to be found in this action line of INFO-2000, the November 1997 call of proposals on Exploiting Europe's Public Sector Information has also identified GI as a specific example domain on which proposals could focus. Of particular interest is also the shift in emphasis in this call from near-market products as in the IMPACT-2 proposals to projects that address substantive bottlenecks such as data access policies, pricing, IPR, and copyright through examples of best practice in public/private sector partnerships, studies, and demonstrations of implementation.

15.4 Towards the Fifth Framework for Research and Development

The overall structure of the Fifth Framework Programme has been set out in the proposal of 30 April 1997. This was revised in January 1998 in the light of the comments received by the European Parliament and is expected to have an overall budget of 16.3 billion ECUs, a 23% increase on the size of the previous Framework programme. (https://cordis.eu ropa.eu/programme/rcn/624_en.html).

In essence, the programme is structured around four vertical thematic programmes and three horizontal ones. The vertical programmes are improving the quality of life and the management of living resource (2.6 billion ECUs); creating a user-friendly information society (3.9 billion ECUs); promoting competitive and sustainable growth (3.1 billion ECUs); and preserving the ecosystem (2.1 billion ECUs).

The three horizontal programmes will share approximately 2.5 billion ECUs and include confirming the international role of European research; innovation and participation of the small and medium enterprises; and improving the human potential.

Although GI plays a key role in many of the key actions which constitute these programmes, such as the 'integrated development of rural and coastal areas' in the first thematic programme or 'the city of tomorrow' in Thematic Programme 3, its most explicit role is within the information society. This thematic programme is organized around four key actions: systems and services for citizens; new methods of work and electronic commerce; multimedia content and tools; and essential technologies and infrastructure. Again, GI can be found subsumed in many of the activities of each of these key actions, but it is only specifically targeted as part of the multimedia content, the aim of which is 'to confirm Europe as a leading force in the multimedia content field, by building on its creativity, competitiveness, and culture' (DGIII/DGXIII, 1997, p. 19). The research activities of this key action are structured around four main axes: interactive electronic publishing; knowledge acquisition, skills, and cultural heritage; human language technologies; and information access, handling, and filtering.

It is in the first of these activities that 'Geographical and Statistical Information. - providing complex information content to non-specialist users' (DGIII and DGXIII, 1997, p. 21) is identified as one of the application areas, together with the electronic publishing of scientific and technical content (Knowledge Publishing), and the publishing of news and leisure (Lifestyle Publishing). Hence, it is perceived as just one type of 'information content' to be stimulated within the developing multimedia industry. This is a very different vision from the one advocated by the GI2000 initiative discussed earlier in which the strategic importance of GI is fully recognized. This short-sighted view of GI has been criticized by the European Science Foundation, which argued in its response to the Fifth Framework Programme proposals that geographical information research needs to be given a higher profile within the plans for the Fifth Framework. The ESF therefore recommended:

- moving away from the current emphasis on socioeconomic studies into the impact of technology toward promoting research on societal and user needs in relation to new information resources and technologies; and

- adding data resources as a focal element of the Essential Technologies and Infrastructures for the Information Society, with geographic and spatially referenced data as a strategic core subset (European Science Foundation, 1997, p. 10).

Whilst at this stage of preparation of the Fifth Framework Programme it does not seem that these recommendations will be taken on board within the Information Society Programme, a positive development is that a stronger social science dimension to the research framework is being developed within one of the three horizontal programmes of the Fifth Framework: 'Improving Human Potential'. This programme, although smaller than the big four vertical ones, addresses very important issues including social trends and structural change, new models encouraging growth and employment, and governance and citizenship. Moreover, it is within this line of the programme that support is provided for 'Large Scale Infrastructures', an action aimed at jointly funding large and expensive research facilities that no single state could fund on its own and in which the research conducted transcends national boundaries.

Traditionally, these large-scale facilities have been addressing research problems in the physical and natural science domains. However, a very positive feature of current discussions is that the European Commission now recognizes that the social sciences also need large-scale facilities. These need not take the form of physical buildings but can be established by networking centres holding the core resource for social science research data. In many cases, datasets may be shared over the networks, while in others the particular conditions attached to the data collection and the specialized support services needed to access it and analyse it may require that researchers travel to a designated facility where these conditions are met. In any event, the principle that data is a core infrastructure for research and development is being established within this programme, thus addressing at least in part some of the concerns expressed by the European Science Foundation.

15.5 Discussion

At the outset of this chapter, a number of questions were posed with respect to the policy framework governing GI in Europe and the extent to which a European GI community has come into being to promote the diffusion of ideas and experience. The findings of this analysis show that considerable progress has been made at the European level in connection with the latter over the last decade, and that the EU has played a significant role in this respect. However, the slow progress made in connection with the establishment of a European policy framework via GI2000 highlighted the barriers that still have to be overcome before Europe develops a shared vision, which is comparable to that underlying the US National Spatial Data Infrastructure.

Looking to the future, there were many encouraging signs that the momentum that has been built up within the fledgling European GI and technological communities will be sustained. This was likely to be fostered by the EU's growing interest in transborder planning and broader regional issues within its boundaries. However, it should be noted that these boundaries were also in a state of constant flux following the accession of Austria, Finland, and Sweden to the Union in 1995, and the proposals that have been made for the accession of the first wave of former communist countries in the early years of the next century (see Chapter 17). These developments, together with the challenges posed by a single European currency, were likely to have far-reaching repercussions throughout Europe as a whole and created new challenges to the GI community and further delay the acceptance of a shared vision of an overall policy framework for Europe.

15.6 From GI2000 to INSPIRE

Two years after this paper was published, there were some important changes within the European Commission. In March 1999, the president and all the commissioners were forced to resign in the face of mounting criticism regarding their conduct of EU affairs and it was not until September 1999 that the new commissioners began to take up their positions under the Presidency of Romano Prodi.

By this time Ian Masser had been elected as the president of the European Umbrella Organisation (EUROGI) in March 1999. He had extensive experience of EUROGI affairs prior to taking over this position. He had participated in the two foundation meetings in 1992 and 1993 and became the British Association for Geographic Information member on the Executive Committee in March 1998. He had also played an active part in the discussions surrounding the GI2000 and the Global Spatial Data Infrastructure initiatives. In his election manifesto, he argued that a number of important changes were likely to take place in Europe over the next few years, which would have a big impact on the future development of the organization. To meet these challenges, he called for a refocusing of EUROGI's core activities and a greater involvement of its member bodies in them.

One of the first things that he did on taking over the presidency was to write to Mr. Prodi in August 1999 to express his concern about the continuing failure of the Commission to establish an appropriate policy framework for GI at the European level. Despite the assurances given by Mr. Prodi in his reply to this letter, it had become increasingly clear by the end of the year that the GI2000 initiative was destined to be put on the shelf together with many other projects launched by the Commission. These developments were largely due to the lack of funds to support these activities and the relatively low priority given to this initiative by senior officials.

Given these circumstances, EUROGI (2000) published a consultation paper in October 2000 entitled 'Towards a strategy for geographic information in Europe.' The starting point for this chapter was the belief that positive steps were needed to fill the void with respect to GI strategy at the European level following the demise of GI2000. This identified five strategic objectives for EUROGI:

- Encouraging greater use of GI in Europe: This is the overarching goal as it is vital to ensure that GI is used as widely as possible in both the public and private sectors as well as by individual citizens in the interests of open government.

- Raising awareness of GI and its associated technologies: There is a continuing need to raise awareness in the community as a whole regarding the importance of recent advances in both technology and their potential for an increasing range of applications.

- Promoting the development of strong national GI associations: An important element of EUROGI's strategy is to create the institutional capacity to take a lead in SDI formulation and implementation. This is particularly important given the need for national associations to maintain some measure of independence from government.

- Improving the European GI infrastructure: Although many of the main elements of a European infrastructure are already in place in different countries, there is a lack of effective mechanisms at the European level to promote greater harmonization and interoperability between countries in this respect.

- Representing European interests in the global spatial infrastructure debate: In an era of increasing globalization, it is essential that Europe does not evolve in isolation.

Following the publication of this paper, a proposal was submitted to the European Commission as an accompanying measure under its Fifth Framework for Research and Development. This involved EUROGI, together with the Joint Research Centre of the EC, the Open GIS Consortium Europe, and the University of Sheffield. The main objective of the resulting Geographic Information Network in Europe (GINIE) project was to develop a deeper understanding of the key issues and actors affecting the wider use of GI in Europe, and to articulate a strategy that was consistent with major policy and technological developments at the European and international levels. The GINIE project took place between November 2001 and January 2004 (see Chapter 19).

To achieve these objectives, the project coordinators organized a series of specialist workshops, commissioned analytical studies, collected numerous case studies of GI in action, and disseminated widely its findings across Europe and beyond in more than 10 different European languages. Through its activities GINIE involved more than 150 senior representatives from industry, research, and government in 32 countries, and contributed to building up the knowledge necessary for an evidence-based GI policy in Europe. The project consortium presented its findings to a high-level audience of senior decision makers in government, research, and industry at its final conference in Brussels in November 2003. A summary of the main findings of this project was also published in book form under the title 'Geographic Information in a wider Europe' (Craglia et al., 2003).

During the lifetime of the GINIE project, two important developments took place at the European Commission level. The first was the debates leading up to the adoption by the Council of Ministers and the European Parliament of a Directive on the reuse of public sector information (EC Commission of the European Communities (CEC), 2003). The second was the launch of a new initiative by the European Environmental Agency, Eurostat, and the Commission's Joint Research Centre to create a European Environmental Spatial Data Infrastructure

The public-sector information debates within the Information Society Directorate ran in parallel during the late 1990s to those concerning GI 2000. The rationale behind these debates was the recognition that the public sector is the largest single producer of information in Europe and that the social and economic potential of this resource has yet to be tapped. Among the many types of information produced by the public sector, GI stands out as having considerable potential for the development of digital products and services. The Reuse of Public Sector Information Directive was approved in November 2003 after extensive consultations with the key players in the field. The adoption of this directive is mandatory for the governments of all EU member states, and they were legally obliged by the EU to incorporate its provisions into their respective national legislation by June 2005. As a result of a review of the Directive the Commission has approved a revised Directive (CEC, 2013). This is one of the key actions in their Digital Agenda for Europe (CEC, 2010).

The main driver behind the European Spatial Data Infrastructure initiative was the adoption of the Sixth Environment Action Programme of the EC for the period 2002–2012, which was approved by the European Parliament and the Council of Ministers in July 2002 (CEC, 2002). The programme identified four areas for priority action: (1) climatic change, nature, and biodiversity, (2) environment and health, (3) sustainable use of natural resources, and (4) management of waste. It called for policy making based on participation and reliable and up-to-date information to support and monitor environmental policies.

Following an initial expert meeting in September 2001, the Commission was asked to prepare proposals for the establishment of an Environmental European Spatial Data Infrastructure (E-ESDI). Each EU nation was invited to nominate two experts – one with a background in environmental matters and another with experience in the GI field – to become the nucleus of the expert group that would develop and implement the

E-ESDI initiative. This group, together with other experts from different sections of the European Commission and the international GI community, met in Vienna in December 2001 to discuss the proposals put forward by the Commission.

After the Vienna meeting, the initiative was renamed INSPIRE. Its primary objective was to make 'available relevant, harmonised and quality geographic information to support formulation, implementation, monitoring and evaluation of Community policies with a territorial dimension or impact'. (http://inspire.jrc.it). INSPIRE was seen as the first step toward a broad multi-sectoral initiative which focuses initially on the spatial information that is required for environmental policies. It was a legal initiative of the EU that addresses 'technical standards and protocols, organisation and coordination issues, data policy issues including data access and the creation and maintenance of spatial information'.

To carry out the tasks that were needed to develop this initiative the three commissioners responsible for the Environment, Economic and Monetary Affairs (including Eurostat), and Research (including the Joint Research Centre) signed a memorandum of understanding (MOU) in April 2002, which set out in some detail the roles of each of these bodies in the first developmental phase of INSPIRE. This move broke new ground, as it was the first time that three commissioners had signed an MOU to jointly develop a legal framework. As a result of these decisions, the Commission began developing the INSPIRE initiative (CEC, 2007), an activity that has already taken nearly 20 years. The story of INSPIRE's development and implementation will be described in Chapter 19.

References

Arnaud, A., M. Craglia, I. Masser, F. Salgé, and H. Scholten. 1993. The research agenda of the European Science Foundation's GISDATA scientific programme. *International Journal of GIS, 7*, 463–470.

Bangemann, M. 1994. *Europe and the global information society: Recommendations to the European Council.* Brussels, Belgium: Commission of the European Communities.

Burrough, P., M. Craglia, I. Masser, and D. Rhind. 1997. Decision makers' perspectives on European geographic information policy issues. *Transactions in GIS, 2*, 61–71.

Commission of the European Communities (CEC). 2010. *Communication from the Commission to the European Parliament, the Council, the European Economic and Social Committee and the Committee of the Regions: A digital agenda for Europe, COM (2010) 245.* Brussels, Belgium: Commission of the European Communities.

Commission of the European Communities (CEC). 2013. *Directive 2013/37/EU of the European Parliament and of the Council of 26 June 2013 amending Directive 2003/98/EC on the re-use of public sector information.* Brussels, Belgium: Commission of the European Communities.

Commission of the European Communities (CEC). 1993. *Growth, competitiveness and employment: The challenges and ways forward into the 21st century*. Brussels, Belgium: Commission of the European Communities.

Commission of the European Communities (CEC). 1994. *Europe's way to the information society: An action plan, COM (94) 347 Final*. Brussels, Belgium: Commission of the European Communities.

Commission of the European Communities (CEC). 1997. *Rolling Action Plan for the Information Society*. Brussels, Belgium: Commission of the European Communities.

Commission of the European Communities (CEC). 2002. Decision 1600/2002/EC of the European Parliament and of the Council of 22 July 2002 laying down the Sixth Community Environment Action programme. *Official Journal of the European Union, L242*, 1-16.

Commission of the European Communities (CEC). 2003. The re-use of public sector information, Directive 2003/98/EC of the European Parliament and of the Council. *Official Journal of the European Union, L345*, 90–96.

Commission of the European Communities (CEC). 2007. Directive 2007/2/E of the European Parliament and of the Council of 14 March 2007 establishing an Infrastructure for Spatial Information in the European Community (INSPIRE). *Official Journal of the European Union, L108*, 1–14.

Constant, E. W. 1987. The social locus of technological practice: Community, system or organisation? In W. E. Bijker, T. P. Hughes, and T. J. Pinch (Eds.), *The social construction of technological systems* (pp. 223–242). Cambridge, MA: MIT Press.

Craglia, M., A. Annoni, M. Klopfer, C. Corbin, L. Hecht, G. Pichler, and P. Smits (Eds.). 2003. *Geographic information in the wider Europe*. Sheffield: University of Sheffield.

DGIII and DGXIII. 1997. *Creating a user-friendly Information Society. Working document on the specific programme draft 2 July Version 0.2*. Brussels, Belgium: Author.

DGXIII/E. 1994. *Heads of the National Geographic Institutes: Report of meeting held on 8 April 1994*. Luxembourg: Commission of the European Communities DGXIII/E.

DGXIII/E. 1995. *GI2000: Towards a European geographic information infrastructure*. Luxembourg: Commission of the European Communities DGXIII/E.

DGXIII/E. 1996a. *GI2000: Towards a European policy framework for geographic information: A discussion document*. Luxembourg: Commission of the European Communities DGXIII/E.

DGXIII/E. 1996b. *Info 2000: Stimulating the development and use of multimedia information content*. Luxembourg: Commission of the European Communities DGXIII/E.

European Science Foundation. 1997. *Further considerations on the EC's proposal for a Fifth Framework Programme*. Strasbourg, France: European Science Foundation.

European Umbrella Organisation for Geographic Information (EUROGI). 2000. *Towards a strategy for geographic information in Europe: A consultation paper*. Apeldoorn: EUROGI.

Goodchild, M. 1997. Towards a geography of geographic information in a digital world. *Computers Environment and Urban Systems, 21*, 377-391.

Longhorn, R. 1998. Data integration for commercial information products: Experiences from the EC's IMPACT-2 programme. In P. Burrough and I., Masser (Eds.), *European geographic information infrastructures: Opportunities and pitfalls*. London: Taylor & Francis, 101-111.

Masser, I. 1991. Promoting GIS awareness: The European dimension. *Mapping Awareness, 5,* 9–13.

Masser, I., and F. Salge. 1997. The European geographic information infrastructure debate. In M. Craglia and H. Couclelis (Eds.), *Geographic information research: Bridging the Atlantic.* London: Taylor & Francis, 28-36.

Salgé, F. 1998. From an understanding of European GI economic activity to the reality of a European dataset. In P. Burrough and I. Masser (Eds.), *European geographic information infrastructures: Opportunities and pitfalls.* London: Taylor & Francis, 17-29.

Tosta, N., and Domaratz, M. 1997. The US national spatial data infrastructure. In M. Craglia and H. Couclelis (Eds.), *Geographic information research: Bridging the Atlantic.* London: Taylor & Francis, 19-27

16

Reflections on the Indian National Spatial Data Infrastructure

Ian Masser

16.1 Introduction

The Indian National Geospatial Data Infrastructure was launched at a workshop held in New Delhi on 5 and 6 February 2001. The workshop was organised by the Centre for Spatial Database Management and Solutions and took place under the sponsorship of the Department of Science and Technology and the Department of Space. Cosponsors included the Ministries of Rural Development, Information Technology, Environment and Forests, Urban Development, Surface Transport, Mines and Minerals, and Agriculture and Cooperation. The workshop attracted a large audience from all sections of the Indian GI community and the organisers also invited a number of overseas experts to participate in the discussion.

The centrepiece of the workshop was the report prepared by the Task Force on National Spatial Data Infrastructure (NSDI) set up by the Department of Science and Technology under the chairmanship of the Surveyor General of India, Lt. Gen. A. K. Ahuja (Task Force on NSDI, 2001). This sets out an overall framework for a decentralised Indian NSDI base, which takes account of the need to maintain standard digital collections of spatial data, the importance of developing common solutions for the discovery, access and use of such data to meet the requirements of diverse user groups, and the need to build relationships among the organisations involved to support its continuing development. With these considerations in mind, the report pays particular attention to metadata standards and provision and the creation of an organisational framework which is inclusive of all the stakeholders. To achieve its objectives, it recommends that the Government should pass enabling legislation that lays down guidelines for the commitment of the key players. Its authors see the NSDI as a national endeavour towards greater transparency and e-governance and propose the creation of a high-level National Spatial Data Commission with a senior Cabinet minister as the chairperson to oversee its implementation. The implementation of such an infrastructure is likely to cost at the very least 1000 crore rupees (i.e. about $2 billion) and a mix of options ranging from Government funding, public private partnerships, and international loans will need to be considered to make this possible.

It was clear from the presentations at the workshop that the authors of the report had done their homework and were familiar with recent NSDI developments in other parts of the world. Nevertheless, it may be still useful to reconsider some of the lessons that might be drawn from this growing body of experience as the Indian NSDI moves from the proposal to the implementation stages.

16.2 Some Key Issues

My own evaluation of the experience of the eleven countries that make up the first generation of NSDIs (Masser, 1999) shows considerable variations in both their composition and the driving forces that are behind them. In practice, NSDIs come in all shapes and sizes. Not only are there massive differences between countries with respect to size and economic circumstances but there are also large variations between NSDIs in countries with federal as against centralised administrative structures.

Despite this diversity, those involved in the further development of the Indian NSDI can draw useful lessons from this experience. However, it is very important in this respect to adopt a critical stance when evaluating these experiences and to bear in mind that much of the material that is available from the responsible agencies involved does not adequately explain the national institutional context within which they have developed. The extent to which NSDI initiatives reflect such circumstances is highlighted in my comparative evaluation of the experience of Australia, Britain, the Netherlands, and the United States (Masser, 1998). Another useful antidote to official compilations can also be found in critical appraisals of national experiences such as the report prepared by the US National Academy of Public Administration (1998), which views the US NSDI from a very different perspective to that of the Federal Geographic Data Committee.

With this in mind, I would like to highlight four issues that are likely to need special consideration by those involved. In order of priority, these are the nature of the machinery for coordination, the need to develop metadata services, the importance of capacity building initiatives and the need to promote data integration.

16.3 The Machinery for Coordination

This is one of the most important factors in the development and implementation of NSDIs. To be effective NSDIs must be given clearly defined mandates by their respective governments. This can be done in various ways: through enabling legislation as is proposed in India, or alternatively through the modification and adaptation of well-established coordination mechanisms as was the case in the Netherlands. In either case the mandate from government should make clear the driving forces behind the NSDI and create the machinery that is needed for its coordination. This will usually take the form of a high-level national committee such as the proposed National Spatial Data Commission in India, although some countries have chosen to set up dedicated national centres for this purpose as is the case with respect to the National Centre for Geographic Information (CNIG) in Portugal.

Whatever the form of the mandate and nature of the machinery that is set up for coordinating the NSDI effort, it should not be forgotten that its principal task is to facilitate the evolution of the NSDI through the efforts of all its stakeholders. For this reason, it is important to try to avoid creating top heavy coordination structures as much as possible and to concentrate on developing initiatives that promote interagency collaboration and data sharing among the stakeholders.

Those involved in the coordination effort must also try to find the right balance between long- and short-term objectives. In particular, they must look for quick winners that produce visible results which demonstrate the potential benefits of the NSDI initiative and help to build up political support for the programme as a whole.

16.4 Metadata Services

The next most important step towards the implementation of an NSDI is the development of metadata services. This is because one of the biggest problems faced by users is the lack of information about information sources that might be relevant to their needs. Consequently, without appropriate metadata services which help them to find out this information, it is unlikely that an NSDI will be able to achieve its overarching objective of promoting greater use of geographic information.

There is also a very practical reason why the development of metadata services should be given a high priority in the implementation of an NSDI. This is because they can be developed relatively quickly and at a relatively low cost. In this respect, they can be regarded as a potential quick winner for those involved in NSDI implementation.

In practice, the development of metadata services is one of the most obvious NSDI success stories. This is particularly evident in the experience of the US National Geospatial Data Clearinghouse project which has exceeded all expectations. The decentralised model that has been implemented in this case has resulted in the establishment of more than 200 national, regional and local nodes. What is most surprising is that 70 of these nodes are not even located in the United States but are to be found in many different parts of the world, particularly in South America.

It is worth noting that there has been a lot of discussion about metadata documentation standards. At the outset of an NSDI initiative, it may be useful to distinguish between relatively simple user-orientated standards for discovery metadata and the more complex sets of professional technical standards developed by bodies such as ISO TC 211. It can be argued that some professionals tend to underestimate the importance of discovery metadata standards such as those developed by the global library community as the Dublin Core. It is also worth noting that the costs of implementing these standards are generally much lower than those involved in implementing even a minimal version of the ISO standards.

16.5 Capacity Building

The implementation of an NSDI initiative is also a process of organisational change management. Despite this the need for capacity building initiatives to be developed in parallel to the processes of NSDI implementation is often underestimated. This is particularly important in less developed countries where the implementation of NSDI initiatives is often dependent on a limited number of staff with the necessary geographic information management skills. For this reason, the experience of the Portuguese CNIG is particularly interesting as it was recognised that modernising government was one of the most important priorities for those involved if effective use was to be made of the new opportunities provided by the development of an NSDI. With this in mind, a great deal of effort has been devoted by the national GI centre to equipping public sector agencies and training staff at the central, regional and local levels of government.

Capacity building can be undertaken in various ways. In Portugal it has been closely integrated with the development and implementation of the NSDI by CNIG. Elsewhere, professional associations such as URISA (the US) and AURISA (Australia) have played an important role, as have national GI associations such as the British Association for Geographic Information. The AGI has also been instrumental in creating a Continuous Professional Development scheme to ensure that its members are continuously updating their skills. There is also a strong case for institutional- as well as individual-level capacity building as there is plenty of evidence that suggests that many local- and regional-level government agencies experience great difficulties in adapting to new responsibilities imposed on them by central government.

16.6 Data Integration

It may come as something of a surprise to find that matters relating to data integration come last on my list of issues. This is because the development and implementation of NSDIs involves much more than database creation. This is clearly evident from the preceding discussion. It should also be noted that the potential for data integration is heavily dependent on the specific institutional context of the country involved. Because of the distribution of responsibilities between the different levels of government in the United States, for example, a complex patchwork of local, state-wide and federal data has come into being rather than an integrated national database.

The creation of an integrated national digital database is also likely to be a very expensive task that takes place over a relatively long period of time. In the meantime, those involved in NSDI development must seek to create partnerships of stakeholders that promote interoperability. It will also be necessary to exploit alternative information sources such as remotely sensed data in addition to conventional survey technology. A great deal can be done in this way without incurring the delays that are inevitably associated with conventional data base creation.

16.7 Conclusion

The proposed National Geospatial Data Infrastructure is a major step forward for India. Its implementation will require the active involvement of all the geographic information stakeholders. It is important that those involved in this process build upon the experiences of other countries. In this process, particular attention must be given to the nature of the machinery for coordination purposes, the need to develop metadata services, the importance of capacity building and the need to promote data integration.

16.8 What Has Happened Since 2001?

The answer to this question is 'not a lot' despite the enthusiasm shown at the 2001 workshop. A National Spatial Data Infrastructure has come into being but has lacked the authority to get the main government departments to cooperate fully in such a project. This is evident in a collection of essays in 2009 entitled

'Indian NSDI: A Passionate Saga', which has been edited by Maj. Gen. (Dr.) R. Siva Kumar, the CEO of the NSDI. In this volume several contributors were critical of the progress that had been made in implementing a NSDI. For example, Mukund Rao argued:

> In this tenth year of NSDI, I really wish that the NSDI Secretariat takes decisive steps that will make NSDI move 'from a debating concept to a tool for governance'. I think the onus is on the Government to take corrective action (after having founded the Indian NSDI) by recognising the importance of NSDI across agencies/departments/societies. It is the responsibility of the Department of Science and Technology (the nodal department) to steer the NSDI in the right direction, so as to facilitate the making of the national SDI into a truly useful tool that supports governance, society and the nation.
>
> *(Rao, 2009, p. 26)*

More recently, Datta and Paul (2018, p. 22) have gone even further in their criticisms:

> The National Spatial Data Infrastructure was mooted in 2000 and launched in 2006 under DST (the Department of Science and Technology). The project envisioned a national infrastructure for the availability and access to organised spatial data and use of the infrastructure at all levels for sustained economic growth. Today 16 years later NSDI is a toothless body which has not achieved much, not even managed to get complete metadata in one place. Since departments were not mandated to share data, some did, many didn't.

Why is this the case? One possible answer is that India is a very large country, not a nation of 29 states and 7 union territories, but a state of many very different nations and territories. However, another and more important answer to this question can be found in the reflection on the machinery for coordination in my own commentary above. This stated that the machinery that is developed for coordination is one of the most important factors in the development and implementation of NSDIs and that NSDIs must be given clearly defined mandates by their respective governments if they are to be effective. This has clearly been not the case in India.

It is also worth reconsidering the four areas that I identified at the start of this chapter as being in need of special consideration with reference to the experiences of the India NSDI since 2001:

- The machinery for coordination is obviously the major source of most of the problem. Although the coordination exists in theory, its authority is not recognised by all the other stakeholders.
- Metadata services: Some progress, but also thwarted by lack of mandates to share data.

- Capacity building: This is less of a problem as there is lots of talent in India.

- Data integration: Not enough effort by those involved in NSDI development to create partnerships of stakeholders that promote interoperability.

Despite these critical comments, however, from what I have seen and heard about India, I would still agree with Professor Dasgupta's (2012, p. 37) comments that, 'Overall, the trajectory of India in geospatial space has been and continues to be very promising. However, considerable efforts are needed to realise these promises.'

References

Dasgupta, A., 2012, The Indian geospatial trajectory, *Geospatial World*, 2, 28–37.

Datta, A. and Paul, S., 2018, Houston, we have a problem, *Geospatial World*, 8, 16–25.

Masser, I., 1998, *Governments and geographic information*, London: Taylor and Francis.

Masser, I., 1999, All shapes and sizes: The first generation of national spatial data infrastructures, *International Journal of GIS*, 13, 67–84.

National Academy of Public Administration, 1998, *Geographic information in the 21st century: building a strategy for the nation*, Washington: National Academy of Public Administration.

Rao, M., 2009, Governance of NSDI – The way ahead, In R. Siva Kumar (Ed.) *Indian NSDI: A passionate saga*, Delhi: National Spatial Data Infrastructure. www.nsdiindia.gov.in (Accessed February 19, 2019).

Task Force on NSDI, 2001, *National Spatial Data Infrastructure: strategy and action plan*, New Delhi: Department of Science and Technology, Government of India.

17

Geographic Information and the Enlargement of the European Union

Four National Case Studies

Massimo Craglia and Ian Masser

17.1 Introduction

> For more than 40 years, the iron curtain divided the continent of
> Europe between a prosperous and free west and an impoverished and
> oppressed east. The European Economic Community, originally made
> up of six members, gradually expanded to take in almost all of the
> western part of the continent. More recently, it signalled its growing
> integration by changing its name to the European Union. When com-
> munism collapsed and the iron curtain came down in 1989, the EU
> pledged to embrace the countries of the east by admitting them to its
> club. This, it was hoped, would spread the peace, stability and prosper-
> ity enjoyed in the west to the east and 'reunify' the continent. More
> than a decade later, the Union looks likely to make good on its
> promise.
>
> *(Economist, 2001: 1)*

This extract from the Economist neatly summarizes the historic changes
that are taking place in the central and eastern European countries at the
present time. These changes are likely to have a profound impact on the
future development of the European Union (EU). By the end of this
decade, the number of members of this body may increase from 15 to as
many as 28 and its total population may grow by more than 25% to more
than 500 million. However, the procedures for accession are demanding.
Candidate member countries must satisfy an exhaustive set of criteria laid
down by the EU before they can be admitted to membership. Geographic
information (GI) has an important part to play in meeting many of these
criteria and requires strategic thinking on the part of the candidate
countries.

With these considerations in mind, this chapter explores some of the GI
policy issues associated with this unique set of circumstances. The first

section summarizes the main features of the procedures devised by the EU for evaluating progress toward membership and provides an overview of the countries that have applied for membership to the EU. The second section deals in more detail with GI policy and the emergence of a national spatial data infrastructure (NSDI) in Bulgaria, Hungary, Lithuania, and Slovenia. The final sections compare and evaluate the experiences in relation to the EU accession process and the development of national GI policies as a whole.

17.2 Context

The EU has expanded over the last 40 years. Originally, there were six countries that signed the Treaty of Paris in 1951 establishing the European Coal and Steel Community; in 1957, the Treaty of Rome was signed, which launched the European Economic Community and Euratom (the European Atomic Energy Community). There are currently 15 member countries. The gradual process of enlargement has not been without challenges but has also offered enormous opportunities to its members. The decision in 1997 by the European Council to initiate the process of enlargement to a further 13 countries (Bulgaria, Cyprus, the Czech Republic, Estonia, Hungary, Latvia, Lithuania, Malta, Poland, Romania, Slovakia, Slovenia, and Turkey) has no comparison to the previous process in terms of the number of countries involved, their area, population, and prosperity, as well as their different traditions and cultures.

The criteria for accession to the EU were designated at the European Council (the Council of Prime Ministers and Heads of State of the Member Countries) in Copenhagen in 1993. The criteria include:

- the stability of institutions guaranteeing democracy, the rule of law, human rights, and respect for and protection of minorities;
- the existence of a functioning market economy as well as the capacity to cope with competitive pressure and market forces in the union; and
- the ability to take on the obligations of membership, including adherence to the aims of political, economic, and monetary union.

The European Conference was established to provide a framework for the process of enlargement. The Conference is a multilateral discussion forum on issues of common interest such as foreign and security policy, regional cooperation and economic matters, and justice and home affairs (the three pillars set up by the Maastricht Treaty of the EU in 1992). The accession process itself involves the development of a pre-accession strategy, accession negotiations, and screening the extent to which candidate countries are adopting the Acquis communautaire, which is the body of laws and regulations enacted by the Union since its foundation.

The Acquis consists of 31 chapters describing policies ranging from agriculture and fishery policies to regional development on the one hand, and from financial and budgetary provisions to education and training on the other. The negotiation process takes the form of bilateral intergovernmental conferences between the EU Member States and each candidate country on a chapter-by-chapter basis. The meetings are held at the level of Minister or deputy for the Member States and Ambassadors or chief negotiators for the candidate countries. The Commission prepares yearly reports on the state of progress for each country toward adopting the Acquis. The results of the negotiations are incorporated in a draft treaty submitted to the European Council for approval and the European Parliament for assent. After signing, the accession treaty is submitted to the Member States and the candidate country for ratification. Once this is accomplished, the candidate country becomes a Member State.

The alignment of national legislation to the policies laid out by the Acquis is a major undertaking for the legislative system of the countries involved, the size of which cannot be overemphasized. More crucially, though, is implementing legislation once adopted and undertaking the necessary changes in administrative practices, cultures, and procedures for effective implementation, monitoring, and reviewing. This is a major challenge because the organization and the way things are done must change. Moreover, the changes will have a significant impact on the society and existing economic structures, so that engendering change and maintaining political support throughout the process is possibly the greatest challenge of all.

To assist in the process of adopting the Acquis, the EU has developed a framework of accession partnerships and national programs. Support for the process comes from three programs: PHARE (Pologne Hongrie Aide a la Restructuration Economique), ISPA (Instrument for Structural Policies for PreAccession), and SAPARD (Special Pre-Accession Assistance for Agriculture and Rural Development). PHARE has an annual budget of €1,560 million, and finances institution-building measures across all sectors not covered by the other two programs, including integrated regional development programs. PHARE is under the direct responsibility of the Directorate General for Enlargement of the European Commission (DG Enlargement), which also has overall coordination between the three programs. ISPA has an annual budget of €1040 million and is dedicated to major environmental and transport infrastructure. This comes under the responsibility of the Directorate General for Regional Development. SAPARD has an annual budget of €520 million and finances agricultural and rural development. It is under the responsibility of the Directorate General for Agriculture (DG Enlargement, 2000).

17.3 An Overview of the Accession Countries

The 13 accession countries include all of the former central and eastern European countries together with Slovenia (part of the former Yugoslavia), Turkey, and the Mediterranean islands of Cyprus and Malta. These countries are very diverse in terms of area, population, and relative wealth (Table 17.1). From Table 17.1, it can be seen that seven of the 13 countries (Cyprus, Estonia, Latvia, Lithuania, Malta, Slovakia, and Slovenia) are relatively small in terms of land area and population. The population of these seven countries varies from only 400,000 persons in the case of Malta to more than 5 million in Slovakia. The other six countries (Bulgaria, Czech Republic, Hungary, Poland, Romania, and Turkey) are generally much larger in terms of both the land area and population. Their populations range from 8 million persons in the case of Bulgaria to more than 65 million in Turkey. Turkey is bigger than all of the present members of the EU apart from Germany.

The 13 accession countries fall into two categories with respect to their relative wealth (Table 17.1). Six of the countries (Cyprus, Czech Republic,

TABLE 17.1

Basic 2000 data for the accession countries

	Area	Population	GDP	
	(1000 km²)	(Millions)	PPS/inh	% EU avg.
Bulgaria	111	8.3	4700	22
Cyprus	9	0.7	17,100	81
Czech Republic	79	10.3	12,500	59
Estonia	45	1.4	7800	36
Hungary	93	10.1	10,700	51
Latvia	65	2.4	5800	27
Lithuania	65	3.7	6200	29
Malta	0.3	0.4	n.a.	n.a.
Poland	313	38.7	7800	37
Romania	238	22.5	5700	27
Slovakia	49	5.4	10,300	49
Slovenia	20	2	15,000	71
Turkey	770	65.6	6200	29

Hungary, Slovakia, Slovenia, and Malta) have gross domestic product (GDP) levels that range from just under half the EU average (Slovakia) to more than 80% of the EU average (Cyprus). The GDP per capita in the latter is higher than that of two of the present EU members (Greece and Portugal). In contrast, the GDP per capita ranges from 37% to 22% of the EU present average in the other seven countries (Bulgaria, Estonia, Latvia, Lithuania, Poland, Romania, and Turkey). However, it should be noted that these categories cut across the previous group divisions based on size and population.

To illustrate the extent to which the accession countries are taking a strategic view of the importance of formulating and implementing GI policies and strategies, regardless of their size and level of economic development, four of the countries were selected for more detailed consideration. The countries described in this chapter are Bulgaria (a country relatively large but relatively poor with respect to the group as a whole), Hungary (which is relatively large and relatively rich), Lithuania (which is relatively small and poor), and Slovenia (which is relatively small and rich). The case studies are presented in a similar format to facilitate the comparison, including the structure of government, the state of public sector information legislation, the main providers of GI, and key elements of an NSDI (i.e., coordination, core data, and metadata services). The format is based on framework used by Masser (1998) in a comparative analysis of GI policies in Australia, Britain, the Netherlands, and the United States.

17.4 Case Studies of Four Nations

17.4.1 Bulgaria

Bulgaria is a parliamentary democracy with a unicameral National Assembly, or Narodno Sobranie. The country is divided into 28 administrative regions, headed by regional governors who are appointed by the Council of Ministers. Bulgaria's 278 municipalities constitute the basic units of the country's economic and political organization.

Important elements of the framework already in place include the Law of United Cadastre and Property Register of the Republic of Bulgaria and copyright legislation. Concerning the provision of digital data, there is no government policy that defines which organizations have rights for providing such data. According to the current State policy, part of the data is classified or restricted for use. This is regulated by the normative document The List of Facts, Records, and Objects Constituting the State Secret of the Republic of Bulgaria. The provisions of these more restrictive regulations are likely to change with the introduction of new legislation currently being considered by Parliament. This includes the Law for Protection of State and Military Secrets, which will create a common system for classified geographic data for the entire country, a law on Access to Public Information that will harmonize the existing rules with the European ones for using public sector information, and a new Law of Geodesy that will complete the legal framework with respect to the acquisition and use of GI.

The key providers of GI in Bulgaria are the Agency of Cadastre under the Council of Ministries, which organizes and maintains key administrative datasets, and the Main Office of Cadastre and Geodesy under the Ministry of Regional Development and Public Works, which organizes the collection and creation of the data at a large scale (1:5000 to 1:10,000) as well as the data for the 28 administrative regions. It is also in charge of dissemination of the data through the Central Cadastre office. Other key agencies are the Military Topographic Service of the Ministry of Defence (which provides topographic data at a 1:25,000 scale and lower), the Ministry of Agriculture and Forestry, and the National Statistical Office.

Coordination: The government sector is key in the development of the NSDI through its activities of data collection, maintenance, and dissemination. The state institutions are currently the only producers of spatial information; however, there is no central body for coordination of a national GI policy in Bulgaria. Some steps to coordinate responsibilities result from the new Law of United Cadastre and Property Register, which established the Cadastral Agency. This is responsible for the creation and maintenance of the National Integrated Collection of Geodetic, Cartographic, Cadastral, and Other Data. The agency collects data from ministries and other organizations, such as the Border Police (administrative data of the country borders), the 28 administrative regions, and the Ministry of Agriculture and Forestry (data on agriculture and forest lands).

All organizations collecting geographic data need to follow the regulations issued by the Ministry of Regional Development and Public Works. The only exception is the Military Topographic Service, which collects and organizes data using its own rules. The laws convey rights to the municipalities to assign tasks to companies and other organizations for collecting geodata in their own areas. Therefore, the municipalities are the owners of such data. Local data could also be produced by private companies, schools, and other organizations.

Core Data: Progress has been made in developing core datasets in digital form starting at the national level. In particular, a digital terrain model, geological data, and the forestry and agricultural cadastre are available for the entire country. The urban cadastre has only been completed for approximately 10% of the country.

Metadata: Metadata exist only to a limited extent. The European pre-standard developed by CEN TC 287 has been translated into the Bulgarian language and is gaining acceptance. New developments in this field, including the International Organization for Standardization (ISO) standards, will need to be taken on board; most importantly, the practice of widely documenting information resources in the public sector needs to be strengthened.

17.4.2 Hungary

Hungary is a parliamentary democracy with a unicameral National Assembly. Public administration is organized through 7 statistical regions, 19 counties plus the capital (which has a similar legal status to that of the counties), 218 districts, and 3144 communities. The Constitution of Hungary states that: 'In the Republic of Hungary everyone has the right ... to know and to disseminate data of public interest.' On the basis of this fundamental democratic civil right, the Protection of Personal Data and Accessibility of Public Data require all government agencies at a national or local level to facilitate access to information in their possession and 'regularly publish or make accessible data concerning their activities, data types held by them, and ... acts concerning their operation.' Regarding the financing, the Act has an important provision: 'For the conveyance of public data, the director of the data managing organization may establish a reimbursement of expenses – up to the level of the cost of the conveyance.' These laws therefore not only identify the right of citizens but also lay the foundations for the provision of metadata services and the pricing policy of public data.

The main provider of GI in Hungary is the Ministry for Agriculture and Regional Development. The Department of Lands and Mapping has the national responsibility for cadastral and topographic mapping, as well as servicing national land administration. Its institutional network includes 136 district and county land offices and the Institute of Geodesy, Cartography and Remote Sensing (FÖMI), a leading research and development institute in GI where the National Remote Sensing Centre (established in 1980) provides operational services for agriculture and the environment. In Hungary, there is a high-level mapping culture, a strong land registry, and a cadastral tradition of one and a half centuries long having its roots in the former Austrian-Hungarian Empire. The network of the Land Offices and FÖMI plays an important role in the implementation of land tenure, environmental, and agricultural policies. The Hungarian Institute for Town and Regional Planning, which also belongs to the Ministry of Agriculture and Regional Development, is responsible for data used for regional development, an increasingly important area of policy in all of the accession countries.

Coordination: Senior decision-makers in Hungary are acutely aware of the importance of digital information and the need to respond to the challenge of the information society. A discussion document, Hungary's National Informatics Strategy, was prepared in 1995 and 1996, leading a year later to the formulation of the Governmental Informatics Strategy. Following recommendation of the Governmental Committee on Telecommunication and Informatics, the National Geospatial Information Strategy was completed in 1998, and many of its major elements are now under implementation. A national policy on data access and sharing is being developed by the Inter-ministerial Committee on Informatics under the auspices of the Prime Minister's Office. This Office chairs key subcommittees including those developing and coordinating the National Strategy on Geographic Information, the Harmonisation and Geo-referencing of Addresses, and Data Dissemination, the latter also including the adoption of metadata standards and the development of a national clearinghouse.

Core Data: Several key datasets are already developed and available, including the geodetic reference systems and networks, digital elevation models, and remote sensing imagery, administrative boundaries, geographical names, and land cover data. Regarding the land and property databases, the text portion of the land registry is now in 100% digital format. Approximately 4% of the associated cadastral maps were available in digital format in 1998, and it is anticipated that 15% will be available by 2002. Base topographic maps at a 1:50,000 scale have full country coverage in digital format, while approximately 5% of the country is covered by mapping at a 1:10,000 scale in digital format, with significant activity being undertaken in digitizing existing large-scale maps.

Metadata: In Hungary, many public administration agencies operate Internet Web sites. The Inter-ministerial Committee for Information Technology established a Data Management Technical Committee with the task of facilitating exploitation of public sector information and developing tools supporting this policy. The primary task of this committee has been to establish a public administration data catalogue accessible for everyone through the Internet. In addition, two other metadata services are in place: one based at the Geological Institute of Hungary and the other with a server at the Institute of Geodesy, Cartography and Remote Sensing. Recently, these systems have been linked, thus strengthening the centrality of this service that has become a one-stop point providing access to a wide range of information including certain state registries such as cadastres of real estates, land properties, and enterprises.

17.4.3 Lithuania

Lithuania is a parliamentary democracy with a unicameral Parliament, or Seimas. Administratively, it is divided into 10 counties, which are further subdivided into 56 local government units or municipalities. The municipal councils are elected by the local population for a period of 2 years on the basis of universal, equal, and direct suffrage by secret ballot. Municipal elections took place on March 19, 2000. Parliamentary elections took place on October 8, 2000, on the basis of a new electoral law that abolished the second round in the uninominal constituencies.

The emerging national GI policy is part of a broader strategy to deliver an information-based society in Lithuania. The Government Program for 2001 to 2005 sees the establishment of an information society in Lithuania as a strategic undertaking with immediate priority. To this end, the Government has set up a separate budget line to finance information society projects and programs aimed at developing e-government services, including regulatory framework, physical infrastructure, and computer literacy among civil servants and citizens, and ensuring that, by 2005, all children finishing secondary school will be computer literate.

The National Service for Geodesy and Cartography (NSGC) is the main provider of GI in Lithuania. As the national mapping agency, the NSGC develops the national strategy for data acquisition and maintenance in the field of GI, coordinates activities in relation to standards, and oversees the protection of copyright. To carry out its operations, the NSGC established two enterprises dealing with production: the UAB Institute of Aerial Geodesy, and the GIS [Geographic Information System] Centre, the State Enterprise for Remote Research and Geoinformatics, which carries out activities in the fields of geodesy and cartography including georeferenced databases. Other key providers of GI are the National Geological Survey, which has a well-developed Internet-based information system, the National Forestry Institute, and the Department of Statistics. The Ministry of Environment and the Ministry of Agriculture have recently set up an Agency for the State Land Cadastre and Register.

Coordination: The coordination of activities can be divided between the generic information society and informatics activities led by the Department of Information and Informatics, and those more specifically addressing GI issues that are coordinated by the NSGC. At an operational level, an important initiative was the establishment of the Department of Information and Informatics within the Ministry of Administrative Reforms and Municipal Matters in 1998. The Department is charged with creating a national strategy on informatics and the information society, coordinating the provision of the infrastructure, and harmonizing legislation with the EU Acquis. With specific reference to a national GI infrastructure, the Department has developed specifications for geographic data to be included in the integrated geoinformation system approved by the Ministry of Administrative Reforms and Municipal Affairs in April 2000.

The specifications set the standards for the collection, coding, attribute structure, metadata, and data exchange of geographic data among agencies at both the national and local levels. This agreed-upon framework is extremely important for the development of a national GI infrastructure, and its specifications are updated regularly to take into account the proposals and comments made by all the agencies at the national and local levels and private enterprises involved in the creation of spatial databases. To prevent discrepancies in the classification and double coding of the geodata, the specifications are accessible to the public on the Internet.

The database of administrative units, settlements, streets, and addresses was 30% complete in October 2000. However, building and maintaining an integrated database of addresses for the entire country is one of the key tasks currently being undertaken, with the aim of having a single official source available to all registers, information systems, and users. This task involves the development of a series of datasets including the administrative boundaries, settlement outlines, street centerlines, building locations, down to the coordinates of individual apartments. The creation of these geographic layers is being done on the basis of large-scale digital maps for towns and cities where available, and the combination of vectorized orthophotos and ancillary raster data elsewhere. Linked of these datasets is an entire series of attribute tables ranging from the level of the individual dwelling unit to the settlement and territorial unit. These are currently being developed in close collaboration with the Land and Property Cadastre and the Registers of Buildings.

Core Data: Key datasets acting as the foundation of the national GI infrastructure are already in place or are being developed. They include topographic databases in 1:1 million and 1:200,000 scales, which are already available, and the 1:10,000 scale, which is under production. Digital orthophotos at a 1:10,000 scale cover the entire country and were developed in partnership to a large extent with the National Land Survey of Sweden.

Metadata: Metadata specifications are included in the integrated geoinformation system specifications as discussed above, but have yet to be fully developed.

17.4.4 Slovenia

Slovenia is a parliamentary democracy with a unicameral National Assembly. Its constitution was adopted in December 1991. The structure of local government is based on 192 municipalities.

Most of the legislative framework with respect to the handling of personal data and electronic commerce is now in place. This includes copyright legislation passed in 1995, the Data Protection Act approved in 1999, and legislation on electronic commerce and signatures passed in August 2000. The Slovenian Certification Authority was also established during 2000. With respect to pricing, there are some policy directives recommending that government information should be free of charge or charged only with respect to reproduction costs. However, as a whole, a coherent policy enshrined in law on data pricing is not in place, and data for private companies are charged based on price lists prepared independently by each data provider.

The main data provider in Slovenia is the Ministry of Environment and Physical Planning. This Ministry has a particularly important role, as most of the key providers of GI depend from it, including the Surveying and Mapping Authority (responsible for land cadastre, the basic geodetic system, and the cartographic and topographical database), the Office for Physical Planning (responsible for the development and control of state spatial plans), and the Environment Protection Agency. Among the other government ministries and agencies, the most important is the Statistical Office of the Republic of Slovenia.

Coordination: The Geoinformation Centre of the Ministry of the Environment and Physical Planning was established in 1991 with the following mission:

- To regulate and coordinate GI policy at the national level and cooperate with other national and international organizations with respect to GI-related standardization, legislation, policy, and legal and organizational aspects of data exchange and distribution.
- To develop user services including user requirement analysis, translation of requirements in terms of information processing, technical advice, linking information users and providers, and quality support (the preparation of quality manual, quality assurance, and quality audits).
- To develop metadata services, remote access to metadata catalogues and data provision through a distributed data warehouse system
- To raise awareness of the importance of an information technology infrastructure, including human resources management, research and development, provision of tools, training, and data integration.

Within the process of establishing a GI infrastructure, cooperation agreements have been signed between the Geoinformation Centre and six data providers at the government level as well as with a range of local communities to also develop a regional organization for GI.

Core Data: There are already a significant number of core datasets available in digital format for the entire country. They include the topographic databases in both raster and vector format, the administrative boundaries, and the databases of street addresses. Socioeconomic and statistical data are close to being completed and the attribute data of the land cadastre is fully digital while the geographic layer is approximately two-thirds complete. Additional efforts are being made to complete coverage of environmental data and the street network.

Metadata: The Ministry for Science and Technology has the responsibility for the development or adoption of standards. In the field of GI, the

Ministry adopted the CEN TC 287 prestandards in 1999. In that same year, the Minister for the Environment issued an order for all data providers within the Ministry of the Environment and Spatial Planning to update their metadata descriptions every 6 months based on CEN TC 287, and submit the metadata for entry in the Slovenian National Data Catalogue, which is the Slovenian National Spatial Data. The catalogue currently contains metadata on 407 information resources by 110 providers, classified into 43 thematic groups. The Directory contains information about the content, purpose, usage, quality, distribution, and all other information necessary to select and use available spatial data. Moreover, a specific tool has been developed for the entry of metadata by data producers (MPEdit). The Government Centre for Informatics and the Geoinformation Centre are working on a common project to develop a GI subportal within the Government electronic portals. The Geoinformation Centre is also collaborating with the Surveying and Mapping Authority to develop online access to the geodetic databases.

17.5 Discussion

17.5.1 Geographic Information Policy

The case studies illustrated here were chosen because these four countries exemplify the significant variations that exist within the accession countries in terms of levels of economic development. Bulgaria is the poorest country, with a GDP per capita of just over 20% of the EU average and over one quarter of the workforce employed in agriculture. The situation in Lithuania is not better, with a GDP per capita a little less than 30% of the EU average and 20% of the workforce in agriculture. By contrast, Hungary is already more fully industrialized, with only 7% of the workforce employed in agriculture and a GDP per capita running at 50% of the EU average. Slovenia, with a GDP at 70% of the EU average, is already ahead of Greece and a little behind Portugal, thus displaying similar levels of economic development as some of the existing EU Member States.

In spite of these differences, all four case studies illustrate the extent of awareness and political commitment of these countries with respect to the strategic role of GI. Clearly, the political importance of land restitution and registration following the demise of former communist regimes has helped to support a broader strategic commitment; however, the extent of this commitment is truly impressive and one that many other nations could learn from.

The main findings of the analysis are summarized in Table 17.2, which shows that Hungary and Slovenia have not only developed a clear framework for NSDI as part of a broader national information infrastructure, but have also gone furthest in implementing its key components that include coordination, core data, and metadata.

Hungary and Slovenia are followed by Lithuania, which has an explicit NSDI policy articulated by the government as part of its Information Society Strategies, which in itself is an indication of the strategic importance attached to GI policies. While the development of core data and metadata in particular still needs considerable progress, the most crucial battle (i.e., making the case for an NSDI, gathering the necessary political support, and crystallizing into legislation) appears to have already been won. Moreover, a specific budget line for the development of the Information Society has been set aside, and the commitment to get all secondary school children to be computer literate is an indication of the forward-looking strategy being pursued by Lithuania. In relative terms, Bulgaria has a less-developed framework than the other three case study countries,

TABLE 17.2

Summary of the key features of the four case study countries

	Bulgaria	Hungary	Lithuania	Slovenia
Selection criteria	Relatively large, but poor	Relatively large and rich	Relatively small and poor	Relatively small, but rich
Public sector information legislation	Legislation on access to public information under consideration	Protection of public data and accessibility to the public administration	GI part of National information strategy	Legislation mostly in place
Main data providers	Agency of the Cadastre Military Topographic Service	Ministry of Agriculture and Rural Development (including Institute of Geodesy, Cartography and Remote Sensing)	National Service for Geodesy and Cartography	Ministry of Environment and Physical Planning (including Surveying and Mapping Authority)
Coordination mechanism	Under consideration through cadastral agency	Inter-Ministerial Committee on Informatics, Subcommittee on National GI strategy	Department of Information and Informatics	Geographic Information Centre within MEPP
Core data	Limited data in digital format	Several key datasets developed in 1:50,000 topographic coverage	Several datasets in progress. Complete 1:10,000 orthophoto coverage	Significant number of datasets in digital format
Metadata	Limited	Public administration data catalogue on Internet	Not yet fully developed	Well-developed metadata services

but even in Bulgaria there are some indications of a dynamic process taking place that could rapidly alter the extent of NSDI development in this country.

With respect to other elements of the GI infrastructure, most countries already have small-scale topographic data available, as well as varying degrees of environmental and socioeconomic data. At present, the major emphasis lies in the development of land information and cadastral data in each of these countries (e.g., Bogaerts, 1997; Dale and Baldwin, 2000). This promises to become a key block of the infrastructure at the detailed level.

It is worth noting that metadata appears to be given a varying degree of priority, bearing in mind the effort needed to make organizations throughout the public sector appreciate the value of documenting data resources as part of the wider strategy to increase access to public sector information. Metadata services are well developed and have specific funding and policy support in Hungary and Slovenia, while they are still at a relatively early stage in Lithuania and Bulgaria.

17.5.2 GI and Accession

As argued at the beginning of this chapter, the accession of up to 13 countries to the EU represents the largest challenge facing Europe in the coming years and is not without critics both within the existing Member States, some of which fear increased in-migration or loss of regional aid, and within the accession countries themselves in view of the major structural, social, and economic reforms necessary to achieve this goal.

The findings of the analysis suggest that GI and related technologies such as GI systems have two key roles to play in the accession process. The first relates to the need to develop the infrastructures necessary to support the process of modernizing public administration. The term 'infrastructure' includes digital data (geographic, statistic, and administrative), computer systems, networks, procedures, people, and skills necessary to inform policy, target delivery, monitor progress, and evaluate impact. The existence of an efficient administration (e.g., the ability to perform ex-ante and ex-post evaluations) is a prerequisite to qualifying for regional aid, access to which is seen by many as one of the potential most direct benefits of accession; hence, the pressure from the EU to develop legal and administrative procedures such as the establishment of appropriate regional administrative units, the collection of relevant data, and the setting up of the necessary computerized monitoring systems.

The second role is even more direct as there has been a significant shift in policy at the EU level during the 1990s away from sectoral approaches and top-down regulatory mechanisms, which were manifestly unable to address the increasing complexity and interaction of environmental, economic, and social issues. What has emerged, particularly in light of increasing environmental concerns, is a more integrated approach to policy where the interactions and cumulative impacts of different policies and actions are assessed ex ante to increase their effectiveness. This shift to a more integrated approach is evident in all key policy areas. Directly flowing from the point above is the emergence of spatial planning at the regional scale as a powerful framework for analysis, coordination of intervention, and evaluation of the impact. The formulation of the European Spatial Development Perspective (Committee on Spatial Development, 1999) is the clearest embodiment of this approach, but its principles are also present in all other areas of policy. Regional planning in turn requires an increasing amount of spatial data for policy formulation, implementation, and evaluation.

Finally, major EU policy areas make direct requirements for the development of GI systems. Among them, the EU Common Agricultural Policy is particularly significant on two counts: first, because it is financially the most important EU policy, absorbing almost 50% of the EU yearly budget of 93 billion euros. Second, because agriculture is still a major economic sector in many of the accession countries, employing large sections of the population, the reform of which has major social implications with respect to rural development, and political implications with respect to land restitution, consolidation, and registration. One of the key requirements to access funds in this area is the development of a computerized Integrated Administration and Control System (IACS) to target intervention, administer the funds, and prevent fraud. The IACS has a strong geographic component, and its development clearly feeds on the GI-related policy developments discussed earlier.

Similar requirements are also increasingly common in environmental policy, including the management of river basins to protect water quality, nature conservation, and integrated coastal zone management, all of which specifically require Member States and relevant local agencies to set up a GI system for policy monitoring and evaluation. The strong emphasis on environmental policy in the EU is also leading to a major shift in emphasis toward a more decentralized approach to data management, leaving data at the level at which it can be more easily collected and updated (i.e., at the regional and local levels rather than at the EU level). Assuring access to such geographic and environmental data becomes in this scheme an absolute prerequisite. Hence, the initiative was announced by Directorate General for environment in April 2001 toward the development of an Infrastructure for Spatial Information in Europe (INSPIRE) embedded in community legislation. The path toward the development and implementation of this initiative is not going to be without challenges, but constitutes a significant milestone in European policy.

With these considerations in mind, it might not be surprising that all of the accession countries are taking such a strategic approach toward the development of GI policies and strategies. However, if EU policy were the main driver, we would see a similar strategic approach throughout the current members of the EU, which is not borne out by the findings of the review carried out by Craglia et al. (2000). Therefore, the process of accession and the development of national GI policies and infrastructures are not simply cause and effect, but rather parallel processes that feed into, and derive support from, each other.

17.6 Conclusion

In this chapter, we examined the experience of four of the 13 countries seeking EU accession with respect to the development of national policies on GI and NSDI. These case studies have been chosen as examples of the situation that is emerging across this diverse group of countries, all of which to a greater or lesser extent are making very significant strides both in the road to joining the EU, and in developing and implementing an NSDI.

It has been argued that the relationship between enlargement and the development of national GI policies is not one of cause and effect. Having said that, it is also clear that there are numerous areas of common ground. In particular, the accession process requires:

- the modernization of public administration;
- the development of land and property markets, supported by efficient cadastral systems;
- computerized support systems for policy monitoring and evaluation, such as the IACS, which includes inter alia agricultural parcel identification, and animal identification and registration to track the source and movement of animals and prevent the spread of disease, a very topical issue in Europe at present;
- specific GI system-based systems and geographic indicators for a range of environmental and agri-environmental policies, including nature protection, water quality protection, and integrated coastal zone management; and
- an increasing shift to spatial planning as the conceptual and analytical framework for policy integration across sectors.

GI systems and infrastructures are clearly crucial to all of the above either directly or indirectly. There is certainly a high degree of consensus that public administration benefits considerably from having modern information systems, not just on direct support of its activities but also by allowing a more open access to public sector information, which in turn enables more informed public participation and accountable administrations.

The difficulties faced by the accession countries in implementing GI strategies and infrastructures are largely the same as those of the EU Member States and include a lack of awareness across different levels of the public sector, lack of management support and technical skills, varying policies with respect to access to data and pricing, weak motivation and coordination across agencies. In addition, limited financial resources exacerbate these problems. However, for those who have argued that it is difficult to make a real business case for the development of national, or even global, spatial data infrastructures (Rhind, 2001), the experience of the accession countries is very instructive as it shows that a business case can indeed be made if such policies and infrastructures enable the transition to a fully working market economy and the accession to the largest single market in the world.

17.7 The Position in 2018

Since this chapter was written, all the countries, apart from Turkey, have joined the EU based on the criteria for the further enlargement of EU membership that had been agreed in 1993. Ten former central and eastern European countries (Czech Republic, Estonia, Hungary, Latvia, Lithuania, Poland, Slovakia, Slovenia) together with Cyprus and Malta became members in 2004, while Bulgaria and Romania joined in 2007 followed by Croatia in 2013. As a result of this evolutionary growth the number of members of the EU has risen from 15 to 28 countries in 2018. Currently, most of the former Yugoslav countries have joined the list of candidate countries or potential candidate countries that are currently attempting to satisfy the EU membership requirements.

The data on purchasing power standard (PPS) that are collected annually by Eurostat (http://ec.europa.eu/eurostat/web/products-datasets /-/tec00114) show that the four case study countries are all more prosperous than they were when they joined the EU. Bulgaria has done particularly well and increased its PPS to nearly half the European average in 2017 even though it is still the poorest of the four case study countries. In contrast, Lithuania's PPS in 2017 had increased to more than three quarters of the EU average. In the process it has not only overtaken Hungary, but also countries such as Greece and Portugal in the EU PPS ranking list. However, both of the two richer countries had experienced more modest rates of growth with Hungary moving to a position of just over two-thirds and Slovenia to over 85% of the EU average, respectively.

During this period, the groundbreaking INSPIRE Directive was approved by the European Parliament and the European Council of Ministers March 2007. Since that time, an ambitious program has been undertaken by the European Commission together with participants from the 28 national Member States to create the technical rules that are required so that the Member States can successfully implement the provisions of this Directive (see Chapter 19). By spring 2014 all of these measures were in place. Since 2010 the implementation of the Directive has been increasingly in the hands of the 28 national Member States.

Because of the nature of the monitoring arrangements that have been made for the monitoring INSPIRE implementation, it is possible to reexamine the current position of the four case study countries described earlier. From the outset it was clear that two sets of implementing rules would be required for monitoring purposes. The first would be based on a quantitative approach based on indicators derived from the list of spatial datasets and network services that is submitted annually to the Commission. Alongside the substantial body of statistical material that is created by these quantitative indicators, the INSPIRE Directive also recognized that it would be necessary for the Member States to provide qualitative information on their progress in the form of written reports every 3 years covering developments since the previous report. Three rounds of qualitative country reports have been completed so far: in 2010, 2013, and 2016. The latest reports for the four case study countries described earlier are available in microfiche form on the INSPIRE website (http://inspire.ec.europa.eu/) and give a good overview of the current position in the four states. This includes details of the statistical indicators together with comments from the Commission regarding progress.

17.6.1 Bulgaria

Coordination: Good progress has been made with respect to coordination in Bulgaria. It also notes that plans for speeding up the implementation process were agreed in June 2015. An expert working group below the intergovernmental council has been set up to oversee INSPIRE implementation. This supports its activity on a technical level.

Core data: Bulgaria has identified a total of 312 spatial datasets with relation to the themes listed in the INSPIRE annexes. There is a big increase of identified spatial datasets since 2014, under all data themes. However, the Commission notes that 'the identification still seems incomplete and Bulgaria could further improve by identifying and documenting spatial datasets required under the existing reporting and monitoring regulations of EU environmental law.'

Metadata: Overall, nearly 80% of the Bulgarian metadata conforms to the INSPIRE metadata specifications. 'The Commission also notes that the documentation of spatial datasets has further improved and shows a high level of maturity.' However, the documentation of digital services is lagging behind and should be addressed. To support data discovery for the end-users of the INSPIRE infrastructure, the Commission feels that Bulgaria should aim at achieving better technical conformity of the available metadata even though it has built the necessary capacity and competences to make data accessible through digital INSPIRE network services but the technical conformity of the available services with the INSPIRE network service specifications is poor.

17.6.2 Hungary

Hungary is lagging behind in INSPIRE implementation. According to its microfiche this is largely because

> With the exception of a limited set of spatial data sets, the existing Hungarian data policy does not allow for free data sharing between public administrations. This prevents cooperation between the different sectors in Hungary and creates an important obstacle for data-sharing.

As a result, not all the information needed for the evaluation and implementation of EU environmental law has been made available or is accessible.

Coordination: The national coordinating body that was set up to oversee INSPIRE implementation was disbanded in May 2010. As a result, 'currently no national SDI with harmonised rules exists in Hungary and each sector handles data according to their own regulations'. The Ministry of Agriculture and Rural Development plays the role of national contact point supported by the Institute of Geodesy, Cartography and Remote Sensing. The implementation of the INSPIRE Directive is also actively supported by the Hungarian Association for Geographic Information (HUNAGI).

Core data: Hungary has identified a total of 104 spatial datasets with relation to the themes listed in the INSPIRE annexes. Additional spatial datasets have been identified in 2015, mainly under Annex II data themes. A lot of relevant spatial datasets have already been identified for the different data themes. However, the identification still could further improve by identifying and documenting spatial datasets required under the existing reporting and monitoring regulations of EU environmental law.

Metadata: Overall, nearly half the metadata conforms to the INSPIRE metadata specifications and 35% of the available digital services also conform to the INSPIRE network service specifications. In addition, Hungary has 45% of its datasets accessible for viewing through a view service and 26% of its datasets accessible for download through a download service. However, the Commission feels that it is clear that Hungary has built the necessary capacity and competences to make data accessible through digital INSPIRE network services. However, 'the technical conformity of the available services with the INSPIRE network service specifications is still poor. Hungary should also boost their efforts to further improve the accessibility of their spatial data through digital INSPIRE services.'

17.6.3 Lithuania

The National Land Service under the Ministry of Agriculture is responsible for the development of infrastructure measures to ensure the functioning of the metadata, datasets, network services, sharing services for the themes referred to in the Directive and the access to the INSPIRE portal. According to the microfiche, 'the competence of spatial data providers has significantly increased since 2013, and more and more spatial information is being used to justify decisions.' This is demonstrated by the growing use of the LSI portal services and the changing nature of the queries made by the users.

Coordination: Nevertheless, some challenges were identified to improve coordination: address existing legal and organizational issues; organize coordination at the highest level; spatial datasets and related services corresponding to the themes in Annex III; and stronger integration of municipalities.

Core data: The number of identified spatial datasets decreased in 2015, mainly for Annex I and II. Many relevant spatial datasets have already been identified for the different data themes. However, the identification still seems incomplete and Lithuania could further improve by identifying and documenting spatial datasets required under the existing reporting and monitoring regulations of EU environmental law.

Metadata: Lithuania shows a high degree of maturity with respect to metadata. Lithuania has made nearly 90% of its datasets accessible for viewing through a view service; and nearly 60 of its datasets are accessible for download through a download service. More than half of the available digital services conform to the INSPIRE network service specifications. Lithuania shows that it has built the necessary capacity and competences to make data accessible through digital INSPIRE network services. However, accessibility of datasets through download services and overall conformity should be further improved.

17.6.4 Slovenia

Slovenia has made steady progress with respect to INSPIRE implementation over the last few years. The microfiche notes that 'Slovenia has indicated in the 3-yearly INSPIRE implementation report that the necessary data-sharing policies allowing access and use of spatial data by national administrations, other Member States' administrations and EU institutions without procedural obstacles are available but not fully implemented. Recently amendments were made to the Slovenian Public Information Access Act to implement the Directive on the re-use of public sector information.'

Coordination: Work on INSPIRE implementation **is** coordinated by the Ministry of the Environment and Spatial Planning surveying and mapping authority. A project for the establishment of INSPIRE compliant network services was carried out in 2015 and the services are in the final stages of testing.

Core data: Slovenia has identified a total of 56 spatial datasets with relation to the themes listed in the INSPIRE annexes. However, the Commission notes that 'the identification still seems incomplete and Slovenia could further improve by identifying and documenting spatial data sets required under the existing reporting and monitoring regulations of EU environmental law.'

Metadata: The public use of spatial data services and spatial data themselves has increased in 2014 whereby the INSPIRE geoportal currently provides only view network services. However, the Commission notes that:

> Slovenia shows that it has built the necessary capacity and competences to make data accessible through digital INSPIRE network services. However, ... a significant amount of the spatial data still has to be brought online. The technical conformity of the available services with the INSPIRE network service specifications is very high. Slovenia should boost their effort to further improve the accessibility of their spatial data through digital INSPIRE services.

The remarks on the microfiches for the four case study countries provide a useful overview of the current situation which to a large extent bear out the comments made more than 15 years ago in the initial review. The most surprising finding from the case studies is that Lithuania has overtaken Hungary, which is still trying to overcome its data sharing problems. Bulgaria is progressing steadily but is still having to catch up with the other three countries and Slovenia is also making good progress with its INSPIRE implementation strategies. Nevertheless, as the formal INSPIRE implementation process enters its final years, all four countries have quite a lot of work to do before the official INSPIRE road map ends in 2021.

References

Bogaerts, T., 1997, A Comparative Overview of the Evolution of Land Information Systems in Central Europe, *Computers Environment and Urban Systems*, 21, 109–131.

Committee on Spatial Development, 1999, *European Spatial Development Perspective: Towards Balanced and Sustainable Development of the Territory of the EU*, Brussels: Office for Official Publications of the European Communities.

Craglia, M., A. Annoni, and I. Masser, 2000, *Geographic Information Policies in Europe: National and Regional Perspectives*, Ispra: Joint Research Centre.

Dale, P., and R. Baldwin, 2000, Emerging Land Markets in Central and Eastern Europe, *Proceedings FIG Working Week*, Prague.

DG Enlargement, 2000, *European Union Enlargement: A Historic Opportunity*, Brussels: Office for Official Publications of the European Communities.

The Economist, 2001, *Survey: European Union Enlargement*, London: The Economist, May 17th.

Masser, I., 1998, *Governments and Geographic Information*, London: Taylor and Francis.
Rhind, D., 2001, Global and National Geographic Information Policies, Practice, and Education in a G-Metadata: Business World. In M. Konecny (Ed.), *Proceedings of the 4th AGILE Conference: GI in Europe: Integrative, Interoperable, Interactive*, Brno.

18

Operational SDIs

The Subnational Dimension in the European Context

Joachim Rix, Swetlana Fast, and Ian Masser

18.1 Introduction

The spatial data infrastructure (SDI) field goes back 20 years but it did not really take off until about 10 years ago. Since then it has been transformed by two momentous developments. The first of these is the accelerated diffusion of SDIs throughout the world during the last 10 years. As a result, most countries in Europe have now taken steps to implement at least one component of a national SDI. The INSPIRE initiative has played an important role in promoting this diffusion process in Europe but similar developments have taken place throughout the whole world.

The second momentous event is the shift in emphasis that has taken place in the second generation of SDIs from national (strategic) SDIs to subnational (operational) SDIs (Masser, 2009). Whereas a great deal of the discussion in earlier years revolved around talking about (national) SDIs much more time is currently being spent on discussing different ways of doing (subnational) SDIs, and success at the subnational level has become a crucial yardstick of overall success.

These two developments have been recognized in a number of recent European initiatives. These include a workshop on Advanced Regional SDIs that was held at the Joint Research Centre in Ispra in May 2008 (Craglia and Campagna, 2009) and the series of national and regional workshops organized throughout Europe as part of the eSDI-Net+ project (www.esdinetplus.eu/). This is a Thematic Network co-funded by the eContent*plus* Programme of the European Commission and coordinated by the Technical University of Darmstadt, Germany, which has promoted cross-border dialogue and stimulated the exchange of best practices on SDIs in Europe. The project started in September 2007 and ended in August 2010.

The findings of the eSDI-Net+ project are particularly interesting in that it brought together a substantial number of SDI players in a Thematic Network, which provided a platform for the communication and exchange of ideas and experiences between different stakeholders involved in the creation and use of SDIs throughout Europe. The network also promoted Europe-wide debates as well as subnational, national, and regional discussions within Europe. In the process it has made an important contribution toward the characterization of SDI implementation throughout Europe and collected information about more than 200 working, accessible, and intelligible solutions. To facilitate this task a unique SDI assessment methodology was developed by consortium.

With these considerations in mind this chapter describes some of the main findings of this project and discusses their implications for future SDI research. The findings of the eSDI-Net+ project are particularly interesting in that it brought together a substantial number of SDI players in a Thematic Network, which provided a platform for the communication and exchange of ideas and experiences between different stakeholders involved in the creation and use of SDIs throughout Europe. It is divided into three main sections, which consider the methodological issues, evaluate the experiences of the 12 SDIs that were selected for best practice awards, and discuss some of the main implications of this project for future SDI research, respectively. The final section discusses some developments since the project ended and considers some North American applications with reference to the Geospatial Maturity Assessment (GMA) indicators developed by the National States Geographic Information Council (NSGIC).

18.2 SDI Selection and Assessment Methodology

After an initial evaluation of the applications on a regional level, a number of promising SDIs were selected for detailed interviews to provide further information. Each interviewed SDI was evaluated by the national representatives of the eSDI-Net+ project, focusing on the key aspects such as

1. the technological, innovative level, and originality of the project;
2. implementation of and/or readiness for the INSPIRE principles;
3. the level of fostering cooperation between different users (proof of visibility and/or user feedback); and
4. possibility of extension or transfer to other countries and regions.

The five main stages of the evaluation process are summarized in Table 18.1. The first of these was devoted to the development of an initial methodology for describing subnational SDIs. This methodology helped in the selection of SDIs during the second stage to participate in a series of national and regional workshops that were organized by the different consortium partners throughout the whole of Europe. Following these workshops, an evaluation framework was developed and a table of indicators was created in the third and fourth stages, which was used by a jury consisting of representatives of the consortium members and the project's advisory board to select SDIs that they considered to be examples of best practices in Europe as a whole.

18.2.1 Evaluation Criteria and Indicators

A great deal of thought was given to the criteria for selecting SDIs and their evaluation in the eSDI-Net+ project. The organizers invited all types and sizes of stakeholders in charge of SDI developments from any region of Europe and at any level, from local through regional to national to apply to participate in the project in early 2008. The application should be submitted by organizations facilitating access to geographical content or providing geoinformation services to end-users.

Table 18.2 shows the 32 weighted indicators that were developed for this purpose with respect to five main criteria: SDI size in terms of quantity, SDI quality with respect to the extent to which it met user requirements, cooperation and subsidiarity, sustainability, and usability.

TABLE 18.1

SDI analysis, evaluation, and selection process

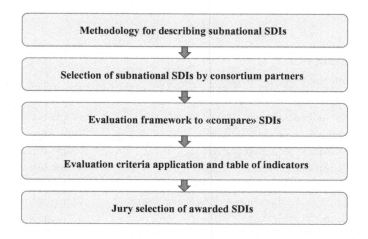

TABLE 18.2

Evaluation criteria and indicators

Five main criteria	32 weighted indicators
1. SDI 'size' (quantity)	6
2. SDI 'quality' (meeting user requirements)	7
3. Cooperation and subsidiarity	7
4. Sustainability	4
5. Users' usability	8

To be considered SDIs had to meet the following criteria:

1. They must have been operating for at least the last 1 year.
2. They should meet the overall profile outlined in the current invitation.
3. The SDI application must also be web-based.
4. The application must include an accessible web address.

With these considerations in mind, the partners in the eESDI-NET+ project were asked to identify SDIs in their areas and set up interviews with key officers in the SDIs using an agreed questionnaire. These considered the administrative context of each SDI, the extent of SDI usage, the user networks that had been created by the SDIs as well as their socioeconomic impacts, and their legal, organizational, and technical characteristics.

A questionnaire containing 106 questions was created to guide the selection process. The main topics that were included in the questionnaire are listed in Table 18.3.

TABLE 18.3

Information used to describe an SDI

Subnational SDI identity card	15 questions
SDI usage assessment	14 questions
Networking people assessment	10 questions
Socioeconomic impact analysis	9 questions
Organizational assessment	12 questions
Coping with legal aspects	6 questions
Technical functionalities-facilities-components	28 questions
Geoportal assessment	12 questions

18.2.2 National and Regional SDI Workshops

A series of workshops were organized throughout Europe during the second half of 2008 and the beginning of 2009 to facilitate the comparative analysis of the experiences of these SDIs. Some of these workshops were organized on national lines as was the case, for example in France and Italy, while others were organized on a regional (e.g., South Eastern Europe) or a common language basis (e.g., the UK and Ireland). Altogether, 12 national and regional SDI Best Practice workshops were organized throughout Europe (see Table 18.4 for details).

The workshops focused on common issues, usability, and socioeconomic impact of SDIs and addressed the integration between SDIs and e-government

TABLE 18.4

National and regional SDI workshops

European region(s) represented at the workshop	Workshop location	Date
France	Strasbourg, France	June 5–6, 2008
Hungary	Budapest, Hungary	August 29, 2008
Czech Republic, Slovakia	Brno, Czech Republic	September 10, 2008
Italy	Rome, Italy	September 25, 2008
Romania	Bucharest, Romania	December 11–12, 2008
Poland	Krakow, Poland	January 29, 2009
Portugal, Spain	Lisbon, Portugal	February 5, 2009
SE Europe: Bulgaria, Cyprus, Greece, Romania, Slovenia, Albania, Bosnia, Croatia, FYROM, Montenegro, Serbia, Turkey	Thessaloniki, Greece	February 4–6, 2009
United Kingdom, Ireland	Liverpool, UK	February 11, 2009
Germany, Switzerland	Darmstadt, Germany	February 12–13, 2009
Belgium	Brussels, Belgium	April 28, 2009
Scandinavia: Denmark, Finland, Iceland, Norway, Sweden	Stockholm, Sweden	April 27, 2009
France	Lille, France	June 29–30, 2009

policies. They brought together stakeholders, showed use cases, and addressed open questions. The outcomes of some of these workshops have been presented at national and international conferences (e.g., France in Salge et al. 2009 and the UK and Ireland in Waters and Masser, 2009).

18.3 SDI Best Practices

18.3.1 European SDI Best Practice Awards

Following the workshops, SDI officers were asked by the organizers to register their activities on a common eSDI-Net+ database. By October 2009, this database provided details of 135 SDIs from 24 different countries, which were subsequently evaluated by a jury consisting of three of the project partners and three members of the Advisory Board. The jury considered the submissions and finally selected the 12 SDIs who were subsequently invited to the Best Practice Awards Ceremony in Turin in November 2009 (Table 18.5). They also found that there were considerable differences between the selected SDIs and decided that all the selected SDIs were winners in terms of their own best practices and that it would be invidious to select overall winners from such a diverse group.

The extent of this diversity can be seen from a more detailed analysis of the presentations of the 12 SDIs. These were grouped into four broad categories:

- Technology, with particular reference to quantitative and qualitative aspects of data and service quality
- Organizational and institutional aspects including cooperation and subsidiarity as well as sustainability
- User involvement
- Thematic SDIs

18.3.2 Technological Aspects

Three of the presentations fell into this category. The first of these from the *Forth Valley GIS in Scotland* described the evolution of the present local authority public company from an informal collaborative agreement between three local authorities in 1993 to combine their GIS activities. This company has been driven by business needs to develop a wide range of applications in many different parts of Scotland as well as the components of an SDI for its three main shareholders. Its success in meeting these needs was recognized in a recent survey of local authority services in Scotland as a whole when it was described as the 'most frequently mentioned example of good practice'.

TABLE 18.5

Award winners

Award winners	Region and country	URL
Technological aspects		
Forth Valley GIS	Scotland, UK	www.forthvalleygis.co.uk
SNIG – Sistema Nacional de Informação Geográfica	Portugal	http://snig.igeo.pt
IDERIOJA: Infraestructura de Datos Espaciales del Gobierno de La Rioja	La Rioja, Spain	www.iderioja.org
Organizational and institutional aspects		
Centre Régional de l'Information Geographique (CRIGE-PACA)	Provence-Alpes-Côte d'Azur, France	www.crige-paca.org
GDI Nordrhein-Westfalen	North Rhine-Westphalia, Germany	www.geoportal.nrw.de
Infrastruttura per l'Informazione Territoriale (IIT) della Lombardia	Regione Lombardia, Italy	www.cartografia.regione.lombardia.it
User involvement		
IDEC Infraestructura de Dades Espacials de Catalunya	Catalunya, Spain	www.geoportal-idec.net
X BORDER GDI (Cross border Geo-data infrastructure XGDI)	Province of Limburg, Netherlands	www.x-border-gdi.org
Thematic SDIs		
National Land & Property Gazetteer and National Street Gaze	English Regions & Wales, UK	www.nplg.org.uk
SIG Pyrénées	Aquitaine, Midi-Pyrénées et Languedoc-Roussillon, France	www.sig-pyrenees.net
Plansystem.dk	Denmark	www.plansystem.dk
Norway Digital-ND	Norway	www.geonorge.no

The second presentation of *Portugal's Sistema Nacionale de Informacao Geografica* (SNIG) discussed the resurgence of one of the oldest SDIs. SNIG was set up by law in 1990 and played an important role during the nineties in modernizing local government in Portugal. In recent years, issues of affordability and sustainability together with education have been central to its latest phase of development.

The last presentation in this group considered the work of *IDERioja*, the SDI that has been developed for the autonomous region of Rioja in Spain. With a population of only 300,000, Rioja is a relatively small region. Its SDI has evolved over the last 10 years into a neat example of centralized GI management, which has won awards in Spain with respect to both good practice and eGovernment.

18.3.3 Organizational and Institutional Aspects

Three presentations were made of SDIs that were primarily selected as best practices with respect to their treatment of organizational and institutional aspects. The presentation of the *Centre Regional de Information Geographique for the Provence-Alpes-Cote d'Azur (CRIGE-PACA)* described the development of an SDI for the public sector in a large region extending over six departments in southeast France, where one job in every five is in the tourism industry. The strong thematic dimension to this SDI was evident from the 12 different applications that had been established and the staff saw one of their main objectives as coordinating communities of practice within the region.

The second presentation about the development of the *SDI for the state of Nordrhein-Westfalen* in Germany also covered a large area. Its population of more 18 million is more than that of many European Union member countries. An important feature of this SDI is the strong links that exist between the state organization and the municipalities in the region because the lower level authorities were responsible for the collection and main-tenance of cadastral information. The information that is held in this SDI is made widely available to private as well as public sector bodies and more than a million maps are downloaded from the SDI by users every month.

The final presentation in this section was by staff from *the Infrastruttura per l'Informazione Territoriale della Regione Lombardia* in Italy. This SDI was strongly driven by spatial planning considerations and its main emphasis was on the creation and maintenance of a regional topographic database which acts as a platform for other applications. Information held in this database was also made freely available to private sector users.

18.3.4 User Involvement

Two SDIs were selected with respect to their strong user involvement. The first of these presentations of the *Infraestructura de Dades Especals de Catalunya (IDEC)* in Spain described itself as 'a network of labelled web services'. The main objectives of this SDI were to facilitate the use of geographic information (GI) and to motivate all kinds of users. As a result of IDEC's activities more than half the municipalities in the region are actively making use of GI in their work and private sector users account for 40 percent of all usage. The second presentation in this group was made by staff from the *X-Border GDI* that is led by the province of Limburg in the Netherlands introduced another dimension into the discussions. As its name suggests, this SDI is a collaborative venture that involves four Dutch provinces, three Belgian provinces and 12 districts (Kreis) from Germany. Its activities are very much problem oriented and user driven, with particular reference to emergency management and spatial planning in a densely populated border region.

18.3.5 Thematic SDIs

This group raised important questions about the nature of SDIs. Some participants felt that they should have been disqualified on the grounds that they were not 'proper' SDIs at all but it was pointed out that 43 out of the original 135 submissions fell into this category and that many of them contained good examples of best practices. The latter is evident from the four shortlisted examples. The first presentation discussed the creation of the *National Land and Property Gazetteer and the National Street Gazetteer in England and Wales*. The initial stage of this project took 10 years to complete and required the active participation of nearly 500 local authorities to create databases to a common set of standards. This highly decentralized initiative provides a consistent platform for local authorities to develop a wide range of thematic applications.

There was also a strong application emphasis in the second presentation from the *French SIG Pyrenees* staff. This SDI recognized the different needs of five main groups of users from agriculture, forestry, climate, economy, and spatial planning, respectively, and created bespoke solutions for each of them using open-source software and content management system platforms such as Joomla! as well as conventional GIS software.

The main objective of the *Danish Spatial Planning System*, the third presentation in this group, was to eliminate duplication in the reporting of the 30,000 local plans that have been prepared by the 98 municipalities in Denmark. The basic philosophy of this system is summarized by the slogan 'data are available in one and only place'. Unfortunately, no one from the staff of the fourth group, *Digital Norway*, was able to attend the awards ceremony. This nation-wide program for cooperation with respect to the establishment, maintenance, and distribution of digital geographic data has attracted a great deal of attention in international circles in recent years. Its main objective is to enhance the availability and use of quality GI among a broad range of users, primarily in the public sector.

18.3.6 SDI Best Practice Database and Self-Assessment Framework

Details of more than 100 SDIs from various parts of Europe that were considered by the project's jury can be found in the SDI Best Practice database, and are available online at the eSDI-Net+ website under www. esdinetplus.eu/best_practice/database.html. Arrangements have also been made with the European Umbrella Organisation for Geographic Information (EUROGI) to maintain this database after the project ended in September 2010 and updates to existing SDIs and new SDI entries can also be added to it online.

In addition, an SDI self-assessment framework has been developed, which is derived from the general SDI assessment methodology. The SDI self-assessment framework is intended to help SDI's officials or SDI steering committees in characterizing and describing their SDI. It can also be seen as a checklist that can be used to better focus key issues involved in developing an SDI.

Various factors influencing the creation of an SDI and its implementation trajectory have been considered in the self-assessment framework. Some of them are structural, 'hard' factors, like favourable legislation, strength of local authorities, overall technological development of the country, and the economic situation. Others are 'soft' like attitudes of involved people and their willingness to cooperate.

The experiences with the SDI assessment show that 'each SDI is a special case' (Vico, 2009). To single out and to follow a successful implementation path in developing an SDI needs understanding of its own strengths and weaknesses. Self-understanding implies comparisons and measuring against others. The second aim of SDI self-assessment framework is strictly linked to this issue. The SDI self-assessment framework is intended to facilitate comparison among various SDI practices, and consequently to foster networking and sharing experiences among similar SDIs.

18.4 Discussion

The findings of the eSDI-NET+ project highlight both the number and the diversity of SDI and SDI-like activities that are currently underway throughout Europe. Consequently, one of the most significant outputs of the project is the SDI database containing the details of the SDIs that have been created during the project, which is a potentially valuable resource for SDI research.

However, it should be noted that the entries in the database must be treated with some caution as there are considerable variations between countries in the number of entries included. This reflects to a large extent the different perceptions of the national and regional organizers of the workshops but there is also an element of self-selection in some cases. Notwithstanding this, the findings of the project suggest that there are at least 200 SDIs in operation at the subnational level in Europe at the present time and that this number could rise to somewhere around 300 if all the possible candidates were included.

It can also be argued that some workshop organizers took a rather catholic view of what constitutes an SDI. This is particularly the case with respect to the inclusion of thematic SDIs, which accounted for nearly a third of the total number. Yet, as the experience of the four thematic SDIs described in the previous section shows, a great deal can be learnt from examining them as well as the experiences of the more conventional SDIs that have been created for local and regional administrative purposes.

The inclusion of thematic SDIs also brings a much stronger user perspective into the discussion as most thematic SDIs are driven to a considerable extent by specific sets of user requirements. When examining cases such as these, special attention must be given to the arrangements that have been made by the users to meet their requirements and the organizational structures that have emerged for this purpose.

The findings of the eSDI-NET+ project also draw attention to the importance of taking the dynamics of SDI development into account in future research. Many of the subnational SDIs considered in the project began life as relatively straightforward GIS applications which have evolved over time into SDIs. It is also worth noting that many of these developments have yet to use the term 'SDI' to describe their current activities. Findings such as these highlight the need for more longitudinal studies in SDI research.

Finally, the experience of the eSDI-NET+ project underlines the importance that must be attached to capacity building in SDI development and the creation of appropriate mechanisms to facilitate the exchange of ideas and experience between those involved. Participants in the national and regional workshops in particular felt very strongly that they had played an important role in this respect and that more activities of this kind were needed to further develop the field.

18.5 Some Reflections on Subnational SDIs

In the closing stages of the eSDI-NET+ project, it was agreed that the EUROGI should take over its work to ensure the long-term sustainability of the investment of the European Commission. Future activities included the maintenance of the website and the updating of the SDI database. It was also agreed that EUROGI should undertake further rounds of best practice awards at about two yearly intervals.

The second round of awards was announced during summer 2011 and the awards ceremony took place in Brussels in October (Masser, 2012). Some 46 submissions from 13 European countries were made for these awards using the SDI self-assessment framework devised during the original project. Most of these came from regional or municipal bodies although there were also seven submissions from thematic and administrative SDIs. As was the case in the previous round of best practice awards, there were considerable differences between the approaches that have been emerged in each of the SDIs, which reflect the local and national institutional circumstances in which they have developed.

Work on the topics considered in the eSDI-Net+ project has continued in recent years. For example, Franco Vico (2013) organized a workshop on best practices at the INSPIRE Conference in Florence in 2013. This brought together a number of SDI practitioners mainly from France, Italy, Portugal, and Spain to make presentations on three main questions: (1) how to promote data flow and sharing from local to regional SDIs; (2) examples to horizontal cooperation among regional SDIs, and (3) how to foster more effective use of SDIs by external users.

A particularly interesting presentation at the workshop described the network of French Infra-National SDIs that has been created under the umbrella of AFIGEO, the French national GI association (Salgé, 2013). The 13 active members of this group come mainly from the southern and western regions of France who work largely through a monthly teleconference and an annual face-to-face conference. They have created a catalogue of 56 subnational SDIs and also produced common specifications for metadata and interoperability as well as compatible rules for downloading datasets. Their discussions also recognize that the governance of the regional SDIs in France takes different forms. In some cases, these SDI activities take place under the auspices of an association or a memorandum of understanding. In others, a local government body or a Public Interest Group (GIP) or an ad hoc body may take the lead.

Studying subnational SDIs in large countries such as the United States is a mammoth task, which is made particularly difficult as a result of the devolution of responsibilities to state and local governments under the American government system. As a result, more than 80,000 agencies are involved in some way with GI in the public sector alone and thereby involved in some way in implementing the US NSDI (National Spatial Data Infrastructure). It is also important to recognize that many private sector companies in the utilities fields and that of land title insurance hold large quantities of data that would be held by public or semipublic agencies in countries such as Britain, the Netherlands, and Australia.

The activities of the NSGIC give some indication of the intermediate position that exists between the municipalities and counties on the one hand and the federal level on the other. The latter is the main focus for the work of the Federal Geographic Data Committee that was set up following the President Clinton's path-breaking Executive Order in 1994 (Executive Office of the President, 1994). The NSGIC is an organization of the states, the District of Columbia, and the territories that work to improve the use and sharing of geospatial data and GIS tools. Its purpose is 'to encourage effective and efficient government through the coordinated development of GI and technologies to ensure that information may be integrated at all levels of government'. Its members include state GIS coordinators and senior state GIS managers, representatives of federal agencies, local and county governments, the private sector, the academic sector, and other professional organizations. Among the NSGIC membership are experts, recognized nationally and internationally. Its major focus areas include:

- Support for the NSDI
- Establishing well-planned, comprehensive, national-scale data initiatives such as Imagery for the Nation (FGDC 2010)
- Creating standards and a national approach to address location data
- Bringing all 50 states, the District of Columbia, and the territories up to a common standard of GIS coordination

One of the tools developed by the NSGIC for monitoring the activities of its members is the NSGIC GMA. The NSGIC GMA provides them with a summary of geospatial initiatives, capabilities, and issues within and across state governments. The material collected in this way is intended to assist state governments with setting goals, identifying peer states for collaboration, identifying areas requiring attention, and connecting with opportunities and resources. Completing the GMA also offers state governments a chance to reflect on their geospatial strategy, operations, and progress.

The assessment has been carried out every 2 years since 2011. The current assessment results were based on questionnaires completed during the second half of 2017 (www.nsgic.org/NSGIC-GMA). The findings from the 2017 survey show that a strong indicator of a state's geospatial maturity is an established Geographic Information Officer (GIO) position. Eighty-three percent of states indicated a figure acting in this capacity and 43 percent of states identified an official, state-level GIO. Another strong indicator of geospatial maturity is the existence of a coordinating council focused on GIS and geospatial priorities, issues, and challenges. In 2017, 23 states claimed that they had an active state GIS Council recognized by either law and another five states had unofficial but active councils and three others reported that they had GIS user associations. Only five states reported that they had no active GIS council or body. These statistics suggest giving some indication of the complexity of subnational SDI governance in large countries such as America.

There has been a strong local government emphasis in much of this research in both Europe and the United States, which supports Nancy Tosta's (1998) 20-year-old dictum that

> Successful SDIs will be local in nature. This is as much a function of practical matters such as the challenges of coordinating large numbers of people over large areas, as it is recognising that all geography is local and issues, physical characteristics, and institutions vary significantly across nations and the world.

In overall terms, then, these examples point to the need for more case studies not only in Europe but also in other parts of the world that trace the factors that led to the evolution of particular SDIs over time and in some cases force them to reinvent themselves to respond to changing circumstances. Such studies could make a valuable contribution to the implementation of national and multinational SDI programs.

References

Craglia, M., and M. Campagna, 2009. *Advanced regional spatial data infrastructures in Europe*, Ispra: Joint Research Centre Institute for Environment and Sustainability.

Executive Office of the President, 1994. Coordinating geographic data acquisition and access: the national spatial data infrastructure, executive order 12906, *Federal Register*, 59, 17671–17674.

Federal Geographic Data Committee, 2010. *Imagery for the nation*, Reston, VA: Federal Geographic Data Committee.

Masser, I., 2009. Changing notions of spatial data infrastructures, In B. van Loenen, J. W. J. Bessemer and J. A. Zevenbergen (Eds.), *SDI Convergence: research, Emerging Trends, and Critical Assessment*, Delft: Netherlands Geodetic Commission, 219-228.

Masser, I., 2012. Regional and local SDIs in Europe: theory and practice, *GeoInformatics, 15, 3*, 28–31.

Salgé, F., 2013. The network of French Infra-National SDIs within AFIGEO, Presentation at the work shop on sub national dimensions of INSPIRE, *INSPIRE Conference*, Florence.

Salgé, F., E. Ladurelle-Tikry, L. Fourcin, and B. Dewynter, 2009. Review of sub-national SDIs in France: an outcome of the eSDI-Net+ project.

Tosta, N., 1998. NSDI was supposed to be a verb: a personal perspective on progress in the evolution of the US National Spatial Data Infrastructure, in B. M. Gittings (Ed.), *Integrating information infrastructures with geographic information technology*, London: Taylor and Francis, pp. 13–24.

Vico, F., 2009. Institutional and organizational aspects, *European SDI Best Practice Awards 2009*, Session 1, 26-27 November 2009.

Vico, F., 2013, Sub national EU SDIs: best practices gallery and criticism, Workshop on sub national dimensions of INSPIRE, *INSPIRE Conference*, Florence.

Waters, R and I. Masser, 2009. Sub national SDI best practice in the UK and Ireland - comparisons and contrasts, Proc AGI 2009 Conference, London: Association for Geographic Information.

19

Learning from INSPIRE

Ian Masser and Joep Crompvoets

19.1 Introduction

The development of spatial data infrastructures (SDIs) in Europe during the last 20 years has been dominated by the formulation and implementation of Directive 2007/2/EC of the European Parliament and of the Council of 14 March 2007, establishing an Infrastructure for Spatial Information in the European Community (INSPIRE) for improving environmental data management in the European Community by 2020 (Commission of the European Communities 2007). The basic objective of the INSPIRE Directive is to make harmonized high-quality geographic information readily available to support environmental policies along with policies or activities which may have an impact on the environment throughout Europe. It is a legally mandated programme managed by the European Commission's Environment Directorate General together with its Joint Research Centre and the European Environment Agency, which brings the 28 national Member States together to build a SDI based on 34 related data themes. The European Commission (EC) began working on the Directive in 2001 and it was formally approved by its Council of Ministers and the European Parliament in 2007. The Directive builds upon the infrastructures for spatial information already operated by the Member States and its detailed implementation is in their hands (Masser and Crompvoets 2015). It is anticipated that the basic elements of the Directive will be in place by the beginning of 2021, some 20 years after work first began on its formulation and 14 years after the Directive was approved by the Council of Ministers and the European Parliament.

The INSPIRE Directive is a multinational, multiagency, multidisciplinary, multiobjective information infrastructure initiative. It includes not only themes related to the ongoing work of the established mapping and surveying community but also a wide range of different activities related to the environment that make up the European Community's

Environmental Action Programmes (see, for example, Commission of the European Communities 2013) such as (public) organizations concerned with the disciplines of geology, hydrology, meteorology, oceanography, biogeography, and demography.

The implementation of the INSPIRE Directive involves collaboration between the national Member States with diverse cultures and professional backgrounds. These countries differ substantially in terms of wealth, resources, and access to technology. Such an initiative demands a collaborative and participatory approach which promotes capacity building. Looking back on INSPIRE from an insider's perspective at the Commission's Joint Research Centre, Craglia (2014, p. 32) has described it as

> an infrastructure built on those of 28 different countries in 24 languages by a truly democratic process. INSPIRE is a role model not only in relation to the developments of SDI but more generally to the formulation of public policy at the European level.

For these reasons, there is a great deal for those concerned with the future development and implementation of SDIs can learn from the INSPIRE experience. Given the large number of topics discussed at the yearly INSPIRE conferences and the workshops that have been convened to tackle specific problems, it would be difficult to do justice to the range of experiences that is associated with INSPIRE in such a short chapter. Consequently, this chapter focusses on four main topics that are associated with the overall framework that has been created for the INSPIRE implementation process. These are 1) the establishment of a legal framework for SDI development, 2) the procedures developed for creating the range of implementing rules (IR) that is required for the implementation process, 3) the mechanisms developed for monitoring progress in each of the member states, and 4) the methods used to evaluate the overall performance of the initiative.

At the outset it must be emphasized that the development and implementation of INSPIRE is based on five fundamental principles (Masser and Crompvoets 2015, Chapter 3). These apply to all SDIs:

1. Data should be collected only once and kept where it can be maintained most effectively.

2. It should be possible to combine seamless spatial information from different sources across Europe and share it with many users and applications.

3. It should be possible for information collected at one level/scale to be shared with all levels/scales; detailed for thorough investigations and general for strategic purposes.

4. Geographic information needed for good governance at all levels should be readily and transparently available.

5. Easy to find what geographic information is available, how it can be used to meet a particular need, and under which conditions it can be acquired and used.

19.2 The Creation of a Legal Framework for Implementing INSPIRE in the EU Member States

One of the axioms of SDI development is its insistence upon the need for a firm legal basis for the whole exercise. This identifies the overall objectives of the SDI and sets out the framework within it will operate together with the responsibilities of the main participants. It should also be noted that the existence of a legal framework does automatically mean that it will be used in practice. The discussion of the Indian experience in Chapter 16 highlights the difficulties that may arise in some situations.

The development and implementation of the INSPIRE Directive took advantage of the existing procedures for handling such matters in the European Union. The INSPIRE Directive came into operation on May 15, 2007 and the member states were given two years from this date to complete the tasks of transposing its provisions into national legislation. The basic objectives of the EU transposition process were as follows:

> The EU cannot achieve its policy goals if EU law is not effectively applied by the Member States. The respective responsibilities for the Commission and the Member States are clearly defined in the Treaties. According to Art. 17 of the Treaty on European Union (TEU), the Commission is the guardian of the Treaties and has a duty to ensure the application of EU law under the control of the Court of Justice of the European Union (the Court). Hence, the Commission has the responsibility for monitoring Member States' efforts and ensuring compliance with EU law, including resorting to formal legal procedures.
>
> *(European Environment Agency 2014, p. 18)*

Given the complexity of the legal procedures that are involved in such a process, it is not surprising to find that there is only limited information about them in practice. The main source of information is the EUR-Lex website (http://eur-lex.europa.eu/legal-content/GA/NIM/?uri=ce lex:32007L0002), which gives details of the legislation submitted by each Member State. This contains all the documents submitted by the national Member States in their national languages.

In practice, the process of transposition was by no means simple, especially in states with federal systems of government. For example, 28 separate measures were needed to meet the requirements of the Commission in Germany, as well as 21 in Lithuania and 12 in Belgium. Even the EU's smallest state, Malta required three separate measures. Despite this, nine Member States managed to deal with transposition in a single measure. There are also some differences between the types of measure itself. Most countries went through the statutory procedure and passed enabling legislation through their respective national or state level decision-making bodies, but some countries, notably the UK and Malta, bypassed these bodies and relied on statutory instruments to deal with this task.

The information available about what happened during the transposition phase of the INSPIRE Directive can be found in two short sections in the Mid-Term Review report (European Environment Agency 2014) and the Staff Working Document (Commission of the European Communities 2016b), which accompanied the Commission's response to this report (Commission of the European Communities 2016a).

The main conclusions from these sources are as follows:

> Only one Member State (Denmark) completed this task on time. 16 Member States finalised their transposition within a year, and 3 Member States of the remaining 10 needed almost two years before communicating their laws. Croatia became the 28th Member State of the EU on 1 July 2013 and notified transposition of the INSPIRE Directive on time (i.e. in May 2013). As main reasons for the delays, Member States informed of political (e.g. change of government following elections), legal (e.g. constitutional requirements to transpose in parliamentary processes at national and regional level) and administrative (e.g. extended consultation procedures or delays in the administrative processes) delays.
>
> *(Commission of the European Communities 2016b, p. 16)*

19.3 The Development of Implementing Rules

It was recognized from the outset that the initial implementation of the INSPIRE Directive was a complex process that would take at least a decade. The first stages were particularly important in this process as they cover the development of IR to guide the detailed implementation of the Directive and their formal approval by the European Commission. From the outset, the Commission recognized that the drafting of IR would require the participation of a large number of stakeholders from the member states. To assist the drafting teams and to make the process as inclusive as possible, it established a network of Spatial Data Interest

Communities (SDICs) throughout Europe to operate alongside the Legally Mandated Organizations (LMOs), which are formally charged with one or more elements of INSPIRE implementation such as national mapping, cadastral, statistical, and environmental agencies.

With this in mind, the Commission set up drafting teams in the years following 2007 to prepare IR for each of the main topics listed in the Directive. Each team was made up of members of the SDICs and LMOs together with specialists from the Commission's Joint Research Centre and other EC agencies to address one of the main elements set out in the INSPIRE Directive.

The first set of IR for Metadata was approved by the INSPIRE Committee in May 2008 (see Table 19.1) after extensive consultations. The Committee gave a unanimous positive opinion to the Regulation which was formally approved by the European Parliament and the Council in December 2008. This bound all member states in the EU to implement its provisions. The Metadata Regulation sets out the definitions of the terms used and discusses the key elements involved. These include the classification of datasets and services and the use of keywords such as geographic location and temporal referencing, as well as details of the responsible organization and points of contact.

During the following years, drafting teams went through the IR process outlined above with interoperability of spatial datasets and services, network services, data sharing, and monitoring and reporting. The most interesting of these are the IR that have been developed for the 34 spatial data themes listed in Annexes I–III of the INSPIRE Directive (Table 19.2) The most important difference between these categories is with respect to the timetable set out in the Directive. It was planned that the data specifications for the nine Annex I data themes would be completed by the end of 2009, while those for the 25 Annex II and III themes which mainly involve environmental data were set a date in 2013. Prior to the development of data specifications for the nine Annex I reference data themes, generic conceptual models and encoding guidelines were established to provide a baseline for the work of the nine thematic working groups set up for each of the themes. The draft guidelines for these themes were released for consultation in December 2008 and published in revised form in September and October 2009.

The most complex of these IRs concerned the proposed legislation for the harmonization and interoperability of the 25 data themes in INSPIRE Annexes II and III, which were formally approved in October 2013 (CEC 2013). These Regulations cover 267 pages. This is more than all the other INSPIRE IR put together. The basic concepts involved are outlined in a short opening section. This is followed by two sections outlining a number of further amendments to the original Regulations. The remainder of the document deals with the requirements of each of the remaining Annex II and III data themes. As was the case with respect to the Annex I data themes, some of these are quite short, while others deal with more complex topics such as Geology which alone occupies more than 40 pages. In addition to these Regulations, the Commission produced nearly 5000 pages of technical guidelines which elaborate on the details in the text.

The process whereby this massive and highly complex IR came into being is regarded by the Commission as 'the biggest participatory data interoperability and harmonisation development ever done' (Nunes de Lima 2013). More than 200 technical experts participated under the supervision of the Joint Research Centre in 19 thematic working groups. Over 8000 comments were received from all over the world as a consequence of the consultation process and resolved by these groups. Some 160 organizations participated in the testing process, which involved 240 real spatial datasets from a wide range of different sources and countries. These figures highlight the key role that the data specifications play in the INSPIRE implementation process as a whole and demonstrate the degree of stakeholder involvement in the development of the

TABLE 19.1

Main decisions by the INSPIRE Committee 2008–2014

1. Commission regulation (EC) No 1205/2008 of 3 December 2008 implementing Directive 2007/2/EC of the European Parliament and of the Council as regards metadata, *Official Journal of the European Union, L326*, 12–30.

2. Commission decision of 5 June 2009 implementing Directive 2007/2/EC of the European Parliament and of the Council as regards monitoring and reporting, *Official Journal of the European Union, L148*, 18–26.

3. Commission regulation (EC) No 976/2009 of 19 October 2009 implementing Directive 2007/2/EC of the European Parliament and of the Council as regards the network services, *Official Journal of the European Union, L274*, 9–18.

4. Commission Regulation (EU) No 268/2010 of 29 March 2010 implementing Directive 2007/2/EC of the European Parliament and of the Council as regards the access to spatial data sets and services of the Member States by Community institutions and bodies under harmonised conditions, *Official Journal of the European Union, L83*, 8–9.

5. Commission Regulation (EU) No 1089/2010 of 23 November 2010 implementing Directive 2007/2/EC of the European Parliament and of the Council as regards interoperability of spatial data sets and services, *Official Journal of the European Union, L323*, 11–102.

6. Commission regulation (EC) No 102/2011 of 4 February 2011 amending regulation (EU) No 1089/2010 of 23 November 2010 implementing Directive 2007/2/EC of the European Parliament and of the Council as regards interoperability of spatial data sets and services, *Official Journal of the European Union, L31*, 13–34.

7. Commission regulation (EC) No 1253/2013 of 21 October 2013 amending regulation (EU) No 1089/2010 of 23 November 2010 implementing Directive 2007/2/EC as regards interoperability of spatial data sets and services (Annex II and III), *Official Journal of the European Union, L331*, 1–267.

8. Commission regulation (EC) No 1312/2014 of 10 December 2014 amending regulation (EU) No 1089/2010 implementing Directive 2007/2/EC as regards interoperability of spatial data services, *Official Journal of the European Union, L354*, 8–16.

TABLE 19.2

Data themes defined in the INSPIRE Directive

Annex I

1. Coordinate reference systems
2. Geographical grid systems
3. Geographical names
4. Administrative units
5. Addresses
6. Cadastral parcels
7. Transport networks
8. Hydrography
9. Protected sites

Annex II

1. Elevation
2. Land cover
3. Orthoimagery
4. Geology

Annex III

1. Statistical units
2. Buildings
3. Soil
4. Land use
5. Human health and safety
6. Utility and Government services
7. Environmental monitoring facilities
8. Production and industrial facilities
9. Agricultural and aquaculture facilities
10. Population distribution – demography
11. Area management/restriction/regulation zones and reporting units
12. Natural risk zones
13. Atmospheric conditions
14. Meteorological geographical features
15. Oceanographic geographical features
16. Sea regions
17. Bio-geographical regions
18. Habitats and biotopes

(Continued)

TABLE 19.2 (Cont.)

19. Species distribution

20. Energy resources

21. Mineral resources

Source: Commission of the European Communities, (CEC), 2007. Commission regulation (EC) No 976/2009 of 19 October 2009 implementing Directive 2007/2/EC of the European Parliament and of the Council as regards the network services, Annexes I–II, *Official Journal of the European Union, L274, 9-18*.

IR. Together with the members of the other IR teams, this has led to the creation of a European wide community of professionals with shared values and a common commitment to INSPIRE implementation.

19.4 Monitoring and Reporting INSPIRE Implementation

Two types of IR were needed for this purpose. The first was based on a quantitative approach centred on indicators derived from the list of spatial datasets and network services. This approach is set out in the definitive Commission Decision (Commission of the European Communities 2009) regarding INSPIRE monitoring and reporting of June 5, 2009 (Articles 3–10). These indicators must include already conformant datasets and services as well as those that still had to be brought into conformity and reflect the Member State's plans for the implementation of INSPIRE. They cover topics such as the existence of metadata for spatial datasets and services, the geographic coverage of spatial datasets, the accessibility of metadata for spatial datasets and services through discovery services, and the use of network services.

Alongside the substantial body of statistical material that is created each year by these quantitative indicators, it was also recognized that it would be necessary for the Member States to provide qualitative information on their progress in the form of written reports every three years covering developments since the previous report. The provisions for reporting are set out in Article 21(2) of the Directive itself. 'No later than 15 May 2010 Member States shall send to the Commission a report including summary descriptions of five topics:

1. How public sector providers and users of spatial datasets and services and intermediary bodies are coordinated, and the relationship with the third parties and of the organisation of quality assurance

2. The contribution made by public authorities or third parties to the functioning and coordination of the SDI.
3. Information on the use of the infrastructure for spatial information
4. Data-sharing agreements between public authorities
5. The cost and benefits of implementing this Directive.'

These topics cover a much wider range of issues than those covered by the quantitative monitoring indicators. Three rounds of qualitative country reports have been completed so far: in 2010, 2013, and 2016. The first two rounds of reports have been translated into English. The materials that they contain have also been the subject of several comparative studies (see, for example, Masser 2011; Crompvoets et al., 2018; Masser and Crompvoets 2018).

The first two rounds of country reports in 2010 and 2013 and the annual submission of the statistical indicators were dealt with routinely by the Commission (see, for example, Borzacchiello et al., 2011; Eiselt, 2011). The efforts of the European Environment Agency that took over the handling of these reports during the third round of submissions, and the ongoing activities of the INSPIRE Maintenance and Implementation Group (MIG) that was set up in 2013, introduced a much more proactive dimension into these monitoring and reporting tasks (INSPIRE, 2016). In the process, the whole monitoring and reporting process has been streamlined (Cetl et al., 2017). This reflects the feeling that the current system delivered too much textual information and that its outputs did not produce comparable results across the member states. As a result, the information contained in the 2016 reports and the system of annual statistical indicators were combined into a system of key performance indicators, dashboards, and country fiches. In addition, bilateral meetings were organized with the national member states to discuss specific issues and they were asked to develop action plans setting out their strategies for the rest of the initial implementation period. These new procedures seem to be already working reasonably successfully in the annual 2018 round of monitoring (see, for example, Rubio Iglesias, 2018).

The 2017 status report also shows that the implementation of INSPIRE Directive requires Member States to concentrate on four main steps in relation to the management of spatial datasets which fall under the Directive:

1. Identify spatial datasets: By mid-2016, Member States had identified more than 90,000 spatial data sets with relation to the themes listed in the INSPIRE annexes. A lot of progress can be seen from 2013 onward (Cetl et al., 2017, p. 14).
2. Document these datasets (metadata): Documentation on data and services in EU is constantly improving. Overall, 87% of the metadata

(datasets and services) conform to the INSPIRE metadata specifications (p. 16)

3. Provide services for identified spatial datasets (discovery, view, download): The number of digital spatial data services across EU is evolving slowly. More than 40,000 view services and more than 30,000 download services are available (p. 19). However, many of identified spatial datasets are still not accessible through the services and there is the space for further improvement. The overall technical conformity of the existing services is more than 50%, which is low and should be also further improved.

4. Make spatial datasets interoperable by aligning them with the common data models: Almost 14,000 datasets in EU reported to be conformant to the INSPIRE interoperability specifications (p. 24). It shows that Member States already started preparations to meet 2017 and 2020 data interoperability deadlines. However, significant efforts need to be made by all Member States in order to meet these upcoming obligations.

These steps are essentially sequential in nature. Consequently, a delay in meeting the deadlines for the completion of the work required for each of these steps has a knock on effect on all subsequent steps.

19.5 The INSPIRE Mid-Term Review

Article 23 of the INSPIRE Directive states that the Commission shall 'present a report on the implementation of this Directive by 15 May 2014.' This report evaluates the extent to which the Directive has achieved its original objectives and may lead to remedial policy action to adapt current approaches to align them better to changing circumstances. This policy evaluation is also effectively a mid-term review given that the implementation process began in 2007 and will continue until at least 2020 (European Environment Agency 2014). This self-evaluation stimulated a formal response from the European Parliament.

As part of the preparations for the review, a public consultation was launched in December 2013 to obtain the views of all stakeholders on the extent to which the actions already under way to implement the INSPIRE Directive are still on course to meet the objectives pursued. Nearly 700 responses were received to the questionnaire by the end of February 2014. Two-thirds of these were from a wide range of public sector organizations and the remainder from private sector organizations, academic institutions, individual citizens, and INSPIRE national coordination organizations (European Environment Agency 2014). The responses show that more than 90%

of the respondents felt that the original objectives of INSPIRE are still pertinent as to the continuing relevance of INSPIRE's objectives and two-thirds of them also felt that the actions foreseen by INSPIRE are still appropriate. There was also strong support for the view that INSPIRE has improved the availability of spatial data and services and more than four out of five respondents also felt that INSPIRE has contributed to a more open policy for public sector data. However, some concern was expressed about the delay by the Member States in putting in place measures neces-sary to remove obstacles to the sharing of data at the point of use among public administrations. Only about half of the data producers indicated that such policy measure had been put in place in their organization.

The overall findings of the review suggested that implementation had reached its half way stage with generally positive outcomes. Three of the five original objectives had undergone a positive evolution. Increased availability of metadata had led to improved documentation, and consid-erable progress had also been made with establishing internet-based net-work services. Interoperability was improving, even though most of the measures required had yet to be implemented. However, organizational, legal, and cultural barriers still restricted data sharing and the arrange-ments that have already been made for coordination need strengthening at the EU, national, and local levels.

The evidence presented in the report clearly showed the uneven imple-mentation of the INSPIRE Directive across the EU and the markedly different progress between Member States. The report concluded that some form of additional support is likely to be needed to help Member States that are lagging.

The formal response to this mid-term evaluation took the form of a 13-page report on the implementation of the INSPIRE Directive that was approved by the European Council and the European Parliament in 2016 (Commission of the European Communities 2016a). The report was accompanied by a substantial Staff Working Document (Commission of the European Communities 2016b) containing the detailed reasoning behind its conclusions

The Commission's main report describes the findings with respect to the Commission's own Regulatory Fitness and Performance Programme (Com-mission of the European Communities 2012) to assess whether the Direc-tive remained fit for purpose at the halfway mark of its implementation.

These findings were evaluated with respect to the regulatory fitness of the INSPIRE implementation so far with respect to the following criteria: effectiveness, efficiency, relevance, coherence, and EC-added value. The section on effectiveness shows that the number of spatial datasets that is available across all the Member States had increased to more than 56,000 as against nearly 1,400 in 2007 but also that progress was very uneven.

The report also argued that efficiency is best measured by a quantitative evaluation of the costs and benefits arising from implementation, but given the current stage most of the currently available studies are based on estimates or predictions, and quantified benefits data were scarce. However, likely benefits arising from INSPIRE implementation include more efficient access to information, an improved evidence base for policy development, better cooperation between public authorities and sectors, and building up technological skills.

The Commission also felt that the objectives of the INSPIRE Directive 'have become increasingly relevant over time, and are included in Commission priorities relating to the 2015 EU Digital Single Market strategy'(p. 10) and that they

> identified the need to increase cross-sector interoperability in the public sector (with the revision of the European Interoperability Framework) where INSPIRE is of major relevance. Promoting eGovernment services and the need to apply the 'digital by default' and 'use once' principles are all enshrined in the INSPIRE Directive.
>
> *(p. 10)*

It was felt that it was not possible to assess the EU-added value arising from INSPIRE implementation given the timing of the report but nevertheless it notes that

> The potential improvements in EU and cross-border spatial data management offered by the INSPIRE Directive remain significant, not just in the environmental field. Whether it is sharing data on air quality or flood risk management, environmental solutions often need cross-border collaboration.
>
> *(p. 11)*

However, 'collaboration between the Commission and Member States has generally been seen as positive but can be strengthened further by, for example, developing implementing tools and components together rather than each Member State "reinventing the wheel"' (p. 11).

19.6 Discussion

During the last 20 years, the development and implementation of the INSPIRE Directive has substantially raised awareness of the importance of sharing geographic information throughout Europe. It has also created the necessary community to further develop SDIs. Given the success of these developments, it is particularly worth noting some of the general lessons that can be learnt from the experience of those involved in the INSPIRE formulation and implementation process described above. These are largely concerned with the procedures that will be involved in developing any SDI. However, the findings of the analysis also highlight the complexity of operations involved in managing the implementation of large-scale multinational II initiatives such as the INSPIRE Directive. It also underlines the need for strong and lasting commitments to projects of this kind. By the end of the formulation of INSPIRE and its initial implementation reached in early 2021, more than 20 years will have elapsed since the first discussions were organized in 2001 and 14 years will have been spent by a large number of people from 28 different countries on the detailed implementation process. During this period, those managing the implementation process will have had to deal with massive changes not only in the technologies but also in the environment as a whole. For example, the European Union consisted of only 15 member states in 2001, whereas it had nearly doubled to 28 member states in 2018.

With these considerations in mind, it is worth bearing in mind that these questions arise in any information infrastructure. Montiero et al. (2014) have pointed out 'that IIs are also typically stretched across space and time: they are shaped and used across many different locales and endure over long periods (decades rather than years).' It also goes without saying that the successful implementation of the INSPIRE Directive is largely due to the fact that is it is underwritten by law at most stages. This can be seen in the formal approval of the Directive itself in 2007 by the Council of Ministers and the European Parliament. The first of the implementation measures, the transposition of the legislation into the laws of the 28 national member states, reinforces the importance of them recognizing the legal rules behind the process. At the European level, the requirement for a full review of progress toward implementation every 7 years gives the European Council and the European Parliament the opportunity to evaluate the objectives of the whole exercise and to take action if necessary. Similarly, the fact that the IR themselves are legally binding on the member states once they had been approved by the INSPIRE Committee makes it clear that member states are obliged to implement them. This combination of measures has proved to be a formidable set of instruments in practice, which supports successful implementation.

Nevertheless, it does not guarantee that all obstacles can be overcome in practice. For example, the implementation of the INSPIRE Directive has been a massive learning process for all the participants and there is still a considerable amount of capacity building to be undertaken in some European countries to enable them to fully participate in the implementation process particularly at the local and municipal levels. There are also a number of obstacles that are still inhibiting implementation. For example, there are still quite a lot of formal and informal barriers that inhibit data sharing in some countries. These reflect entrenched organizational cultures that may take years to resolve.

Given these difficulties, it is a tribute to those involved that so much progress has been made in connection with the implementation of the

INSPIRE Directive. As a result of their efforts, a strong community of professionals involved in the implementation process has come into being. However, the successful implementation of the Directive also raises question about what will happen after the current initiative ends in early 2021 and a continuing commitment will be required to maintain and update the data resources that have been created by INSPIRE to keep up the head of steam that has been built up during the last few years.

References

Borzacchiello, M. T., M. Craglia, and R. S. Smith, 2011. Monitoring INSPIRE implementation in Europe: analysis of current methodologies and outcomes, Proc INSPIRE Conference, Edinburgh.

Cetl, V., V. Nunes de Lima, R. Tomas and M. Lutz With J. D 'Eugenio, A. Nagy, and J. Robbrecht, 2017. *Summary report on status of implementation of the INSPIRE directive in EU*, JRC Technical Report, EUR 28930 EN. Luxembourg: Publications Office of the European Union.

Commission of the European Communities (CEC), 2007. Directive 2007/2/EC of the European Parliament and of the Council of 14 March 2007 establishing an Infrastructure for Spatial Information in the European Community (INSPIRE), *Official Journal of the European Union*, L108, 1–14.

Commission of the European Communities (CEC), 2009. Commission regulation (EC) No 976/2009 of 19 October 2009 implementing Directive 2007/2/EC of the European Parliament and of the Council as regards the network services, *Official Journal of the European Union*, L274, 9–18.

Commission of the European Communities (CEC), 2012. *Communication from the Commission to the European Parliament, the Council European Economic and Social Committee and the Committee of the regions: EU Regulatory Fitness*, COM (2012) 746 final.

Commission of the European Communities (CEC), 2013. *Decision No1386/2013/EU of the European Parliament and the Council of 20 November 2013 on a general union environment action programme to 2020: living well, within the limits of our planet.* Brussels: Commission of the European Communities.

Commission of the European Communities (CEC), 2016a. *Report from the Commission to the Council and the European Parliament on the implementation of Directive 2007/2/EC of March 2007 establishing an Infrastructure for Spatial Information in the European Community (INSPIRE) pursuant to article 23*, COM (2016) 478 final/2.

Commission of the European Communities (CEC), 2016b. *Commission staff working document: evaluation accompanying the document Report from the Commission to the Council and the European Parliament on the implementation of Directive 2007/2/EC of March 2007 establishing an Infrastructure for Spatial Information in the European Community (INSPIRE) pursuant to article 23*, SWD (2016) 273 final/2.

Craglia, M., 2014. INSPIRE: Towards a participatory digital earth, *Geospatial World*, 4, 7, 32–36.

Crompvoets, J., G. Vancauwenberghe, S. Ho, I. Masser, and W. Timo de Vries, 2018. Governance of national spatial data infrastructures in Europe, *International Journal of SDI Research*, 13, 253–285.

Eiselt, B., 2011. Is Europe getting Inspired, *GEO Informatics*, 14, 5, 22–28.

EUR-Lex, National transposition measures communicated by the Member States concerning Directive 2007/2/EC of the European Parliament and of the Council of 14 March 2007 establishing an Infrastructure for Spatial Information in the European Community (INSPIRE) http://eur-lex.europa.eu/legal-content/GA/NIM/?uri=celex:32007L0002 (last accessed February 16 2019)

European Environment Agency, 2014. *Mid-Term Evaluation report on INSPIRE implementation: joint EEA-JRC report*, EEA Technical Report 12/2014, Luxembourg: Publications Office of the European Union.

INSPIRE, 2016. Towards a Maintenance and Implementation work programme for the period from 2016–2020, Maintenance and Implementation Working Party 2016–2020. www.inspire.ec.europa.eu (last accessed February 16 2019)

Masser, I., 2011. Developments of national spatial data infrastructures in Europe, *GIS Professional*, 40, 16–19.

Masser, I., and J. Crompvoets, 2015. *Building European spatial data infrastructures*, Third edition. Redlands, CA: Esri Press.

Masser, I., and J. Crompvoets, 2018. Qualitative monitoring of information infrastructures: a case study of INSPIRE, *Environment and Planning B*, 45, 2, 330–344.

Montiero, E., N. Pollock, and R. Williams, 2014, Innovation in information infrastructures: introduction to a special issue. *Journal of the Association for Information Systems*, 15, i-x.

Nunes de Lima, V., 2013. Building up data interoperability in Europe. Proc INSPIRE Conference, Florence.

Rubio Iglesias, J.M., 2018. The 2018 INSPIRE monitoring round: discussing the outcomes, Proc INSPIRE Conference, Antwerp.

Index